Potato: Nutrition, Food Security and Agriculture (Volume I)

Potato: Nutrition, Food Security and Agriculture (Volume I)

Lloyd Gaston

Larsen & Keller
www.larsen-keller.com

Potato: Nutrition, Food Security and Agriculture (Volume I)
Lloyd Gaston
ISBN: 979-8-88836-020-0 (Hardback)

Larsen & Keller
Published by Larsen and Keller Education,
5 Penn Plaza,
19th Floor,
New York, NY 10001, USA

Cataloging-in-Publication Data

Potato : nutrition, food security and agriculture (Volume I) / Lloyd Gaston.
p. cm.
Includes bibliographical references and index.
ISBN 979-8-88836-020-0
1. Potatoes. 2. Crop science. 3. Potatoes--Breeding. 4. Potatoes--Diseases and pests. 5. Potatoes--Varieties. 6. Potatoes--Harvesting. I. Gaston, Lloyd.
SB211.P8 P68 2023
635.21--dc23

For more information regarding Larsen and Keller Education and its products, please visit the publisher's website www.larsen-keller.com

Table of Contents

Preface

Potato is a versatile, carbohydrate-rich food which is a good source of nutrition since it contains dry matter, edible energy and edible protein. It is also a good source of vitamins such as Vitamin B1, B3 and B6, and minerals such as potassium, phosphorus and magnesium. Additionally, it is a moderate source of iron, and its high vitamin C content promotes the absorption of iron in the body. Potatoes are cultivated in many countries around the world and have the potential to significantly improve future food security since they can be grown in a wide range of agroclimatic conditions. This book attempts to understand the multiple advantages of potato as a crop and how it is important for ensuring food security. The topics included herein on the nutritional aspects of potato are of utmost significance and bound to provide incredible insights to readers. The extensive content of this book provides the readers with a thorough understanding of the subject.

This book is the end result of constructive efforts and intensive research done by experts in this field. The aim of this book is to enlighten the readers with recent information in this area of research. The information provided in this profound book would serve as a valuable reference to students and researchers in this field.

At the end, I would like to thank all the authors for devoting their precious time and providing their valuable contribution to this book. I would also like to express my gratitude to my fellow colleagues who encouraged me throughout the process.

Lloyd Gaston

Part I
Global and Dietary Relevance of Potato

1

Sustainable Agri-Food Systems: Role of Potato Research and Innovation

André Devaux, Jean-Pierre Goffart, Athanasios Petsakos, Peter Kromann, Marcel Gatto, Julius Okello, Victor Suarez, and Guy Hareau

Abstract In the coming decades, feeding the expanded global population nutri tiously and sustainably will require substantial improvements to the global food system worldwide. The main challenge will be to produce more food with the same or fewer resources. Food security has four dimensions: food availability, food access, food use and quality, and food stability. Among several other food sources, the potato crop is one that can help match all these requirements worldwide due to its highly diverse distribution pattern, and its current cultivation and demand, par ticularly in developing countries with high levels of poverty, hunger, and malnutrition. After an overview of the current situation of global hunger, food security, and agricultural growth, followed by a review of the importance of the potato in the

A. Devaux (✉)
International Potato Center, Quito, Ecuador
e-mail: a.devaux@cgiar.org

J.-P. Goffart
Walloon Agricultural Research Center, Gembloux, Belgium
e-mail: j.goffart@cra.wallonie.be

A. Petsakos
Formerly CIP, Seville, Spain

P. Kromann
International Potato Center, Nairobi, Kenya
e-mail: p.kromann@cgiar.org

M. Gatto
International Potato Center, Hanoi, Vietnam
e-mail: m.gatto@cgiar.org

J. Okello
International Potato Center, Kampala, Uganda
e-mail: j.okello@cgiar.org

V. Suarez · G. Hareau
International Potato Center, Lima, Peru
e-mail: v.suarez@cgiar.org; g.hareau@cgiar.org

current global food system and its role played as a food security crop, this chapter analyzes and discusses how potato research and innovation can contribute to sustainable agri-food systems with reference to food security indicators. It concludes with a discussion about the challenges for sustainable potato cropping considering the needs to increase productivity in developing countries while promoting better resource management and optimization.

1.1 Introduction: The Current Situation of Global Hunger, Food Security, and Agricultural Growth

A growing earth population and the increasing demand for food is placing unprecedented pressure on agriculture and natural resources. Today's food systems do not provide sufficient nutritious food in an environmentally sustainable way to the world's population (Wu et al. 2018). Around 821 million are undernourished while 1.2 billion are overweight or obese. At the same time, food production, processing, and waste are putting unsustainable pressure on environmental resources. By 2050, a global population of 9.7 billion people will demand 70% more food than is consumed today (FAO et al. 2018). Feeding this expanded population nutritiously and sustainably will require substantial improvements to the global food system—one that provides livelihoods for farmers as well as nutritious products to consumers while minimizing today's environmental footprint (Foley et al. 2011). A critical challenge is to produce more food with the same or fewer resources.

According to the Global Hunger Index (GHI), substantial progress has been made in terms of hunger reduction for the developing world (Von Grebmer et al. 2017). The GHI ranks countries on a 100-point scale with 0 being the best score (no hunger) and 100 being the worst. Whereas the 2000 GHI score for the developing world was 29.9, the 2017 GHI score is 21.8, showing a reduction of 27%. Yet, there are great disparities in hunger at the regional, national, and subnational levels, and progress has been uneven.

Sub-Saharan Africa (SSA) and South Asia (SA) have the highest 2017 GHI scores, at 29.4 and 30.9, respectively. These scores are still on the upper end of the serious category (20.0–34.9), and closer to the alarming category (35.0–49.9) than to the moderate one (10.0–19.9). These data show that persistent and widespread hunger and malnutrition remain a huge challenge in these two regions. In other parts of the developing world within the low range, are also countries with serious or alarming GHI scores, including Tajikistan in Central Asia (CA); Guatemala and Haiti in Latin America and the Caribbean (LAC); and Iraq and Yemen in the Near East and North Africa (NENA) regions. Black et al. (2013) estimate that undernutrition causes almost half of all child deaths globally.

The current rate of progress in food supply will not be enough to eradicate hunger by 2030, and not even by 2050. Despite years of progress, food security is still a serious threat. Conflicts, migration, and climate change are hitting the poorest people the hardest and effectively maintaining parts of the world in continuous crisis.

The 2017 GHI report emphasizes that hunger and inequality are inextricably linked. Most closely tied to hunger, perhaps, is poverty, the clearest manifestation of societal inequality. Both are rooted in uneven power relations that often are perpetuated and exacerbated by laws, policies, attitudes, and practices.

According to FAO (2002) "*Food security exists when all people, at all times, have physical, social and economic access to sufficient, safe and nutritious food to meet their dietary needs and food preferences for an active and healthy life.*" Food security has four key dimensions: (a) food availability, (b) food access, (c) food quality and use, and (d) food stability.

Food availability refers to the supply of food at the national or regional level which ultimately determines the price of food. Improved availability of food is necessary to reduce food insecurity and hunger but is insufficient to completely end malnutrition, particularly because access to other services such as potable water, sanitation, and health services is also required.

Food access refers to the ability to produce one's own food or buy it, which implies having the purchasing power to do so. Given that a large portion of the poor worldwide are farmers, there remains considerable attention to promoting agriculture to enhance food access. The emphasis on an agricultural pathway to increase food access is twofold, since increased agricultural production provides income to purchase food as well as direct access to food for consumption obtained from own production.

Food use and quality refers to the level of nutrition obtained through food consumption from a nutritional, sanitary, sensory, and sociocultural point of view.

Food stability incorporates the idea of having food access at all times thus incorporating issues such as price stability and securing incomes for vulnerable populations (FAO 2006a).

This widely accepted FAO definition reinforces the multidimensional nature of food security that requires multisector approaches. Such approaches should combine the promotion of broad-based agricultural growth and rural development with programs that directly target the food insecurity as well as social protection programs focused on nutrition including a gender approach (Salazar et al. 2016). Agricultural growth results in rural development and prosperity through a series of multiplier effects, that is, through backward and forward linkages, due to increased incomes. These effects typically stimulate enhanced investment in both farm and non-farm sectors (Hazell and Haggblade 1989; Pandey 2015). Growth in rural farm sector increases demand for goods and services produced by the non-farm sector, further increasing purchasing power and effective demand, thus deepening growth in non-farm sector. Further, Haggblade et al. (2007) argue that the increased income earned in rural non-farm sector can kick off a series of reverse linkages in which such income is invested in agriculture to further strengthen its growth and improve livelihoods of farm households.

During the 2014 World Economic Forum, in a debate on "Rethinking Global Food Security," Shenggen Fan, Director of the IFPRI, argued that tackling hunger and malnutrition is not only a moral issue but also one that makes economic sense. The world loses 2–3% Gross Domestic Product (GDP) per year because of hunger, while investing US$1 in tackling hunger yields a return of US$30. Ajay Vir Jakhar, Chairman of Bharat Krishak Samaj (Farmers' Forum) in India, added that farmers do not think in terms of food security at the global level, but in their own households. While policy makers tend to think in terms of global and national issues and solutions, localized solutions and help from the public and private sectors are also needed to support the bulk of farmers who are farming small plots of land and which have a critical role as engines of food productivity growth and social development. By declaring 2014 the International Year of Family Farming, the United Nations acknowledged the importance of family farming in reducing poverty and improving global food security. Localized, technical, and commercial solutions with the support of both public and private sectors are needed in combination with global food security policies.

Therefore, enhancing food security requires policies that improve households' ability to obtain food through production and better income. Growth in agricultural productivity is key to reducing rural poverty since most of the poor depend on agriculture and related activities for their livelihoods. Because the potato is one of the global crops with a most diverse distribution pattern (Haverkort et al. 2014) and is grown in areas with high levels of poverty, hunger, and malnutrition, it can be particularly effective crop for enabling smallholder families to attain food security and climb out of poverty. Hence, innovations based on potato science can be a significant vehicle for targeting the poor and hungry as part of a broader set of research and development activities.

This chapter first presents the importance of the potato in the current global food system and its value as a food security crop. It then discusses the role of agriculture and the potato for their contribution to food security in its different dimensions: analyzing opportunities and challenges on how potato research and innovation can enhance productivity and how potato agri-food systems can contribute to food security at a global scale using natural resources in a sustainable way. A list of key research and technology options that can contribute to sustainable agri-food systems intensification approaches is suggested. The chapter concludes with a discussion about the challenges for sustainable potato cropping combining the needs to increase productivity in developing countries while promoting better input management and optimization. These conclusions emphasize also the need to integrate better agriculture sustainable intensification and food security indicators.

1.2 The Potato in the Global Food System

Potato is currently grown on an estimated 19 million hectares of farmland globally, and the potato production worldwide stands at 378 million tons (Table 1.1). The highest concentrations are found in the temperate zone of the northern hemisphere where the crop is grown in summer during the frost-free period. In these regions, potato is mainly grown as a cash crop and is therefore an important source of income. In tropical regions, the crop is significant in the highlands of the Andes, the

Table 1.1 Potato production indicators

Region	2014–2016			Average annual growth rate								
	Production (000 tons)	Area (000 ha)	Yield (t/ha)	Production (%)	Area (%)	Yield (%)	Production (%)	Area (%)	Yield (%)	Production (%)	Area (%)	Yield (%)
				1961–1963 vs. 1988–1990			1988–1990 vs. 2014–2016			1961–1963 vs. 2014–2016		
Africa	25,270	1756	14.4	4.9	3.7	1.1	4.5	3.2	1.1	4.7	3.6	1.1
Asia[a]	190,617	9975	19.1	3.7	2.4	1.3	4.3	2.8	1.4	4.0	2.6	1.3
Europe	119,551	5547	21.6	−1.0	−1.8	0.8	−1.3	−2.4	1.3	−1.1	−2.2	1.1
North America	24,430	763	32.0	1.2	−0.2	1.4	0.8	0.6	0.1	1.0	0.2	0.8
Latin America and Caribbean	18,334	1023	17.9	2.2	0.0	2.2	1.5	0.1	1.4	1.8	0.0	1.8
World	378,202	19,063	19.8	0.1	−0.8	0.9	1.3	0.2	1.1	0.7	−0.3	1.0

Source: FAOSTAT (2017), accessed Oct 2018
n.a. = not available, a = U.S. Department of Agriculture, 2016
[a]Asia + Oceania

African highlands, and the Rift valley, and the volcanic mountains of West Africa and Southeast Asia, where production is both for food and cash (Muthoni et al. 2010). In the subtropics, the crop is grown as a winter crop during the heat-free period such as in the Mediterranean region, North India, and southern China. It is only in the tropical lowlands that potato is not a main staple, largely because the temperatures in these areas are too high for tuber development and growth in traditional potato cultivars (Haverkort et al. 2014). Figure 1.1 illustrates the current pattern of the potato distribution worldwide (You et al. 2014; FAO 2016).

Potato is now the world's third most important food crop in terms of human consumption, after wheat and rice (FAOSTAT 2013) despite the large proportion of potato produce used for seed and as animal feed (Fig. 1.2). Consumption of fresh potatoes accounts for approximately two-thirds of the harvest, and around 1.3 billion people eat potatoes as a staple food (more than 50 kg per person per year) including regions of India and China.

1.3 Potato Production and Demand Trends by Region

Across global landscapes, the versatility of the potato crop coupled with notable increases in production in many countries over the last two decades is unparalleled, although this increase has been mainly driven by area expansion and secondarily by yield improvements. Global statistics also indicate that potato production is shifting towards developing countries especially with strong increase in production in Asia and Africa, especially in East Africa (Fig. 1.3). In fact, the developing world's potato production exceeded that of the developed world for the first time in 2005

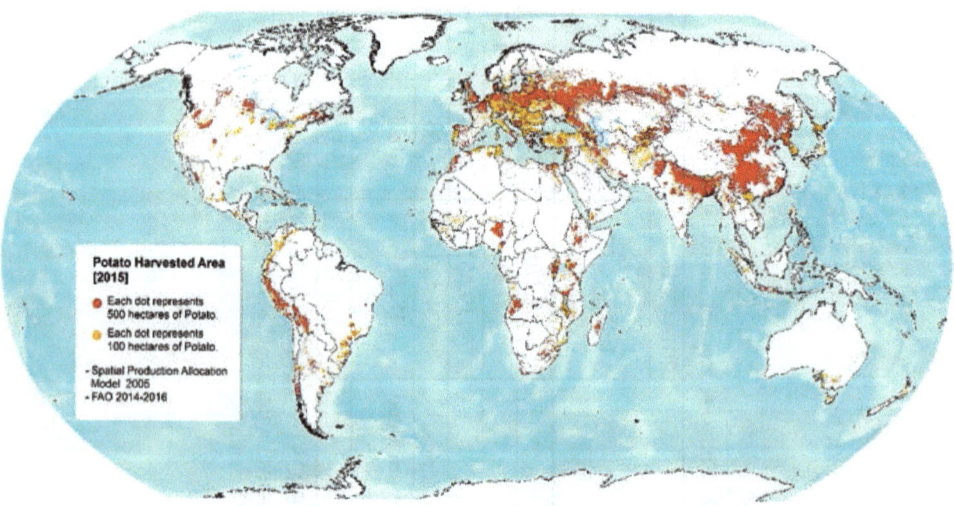

Fig. 1.1 Potato distribution worldwide, harvested area (You et al. 2014; FAOSTAT 2016)

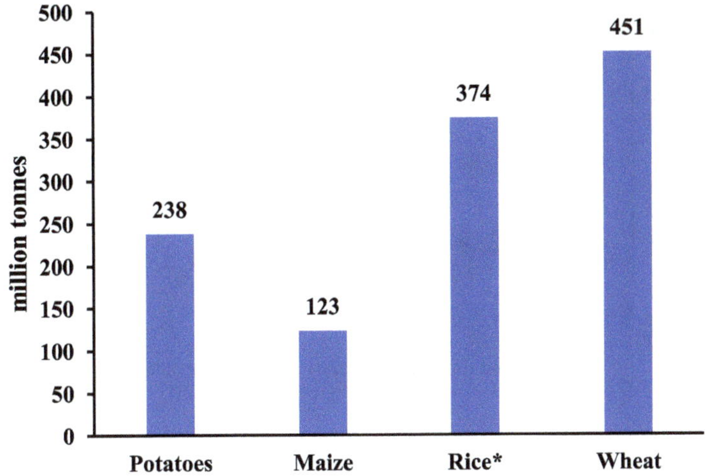

Fig. 1.2 World food consumption per year for the four main nutritious crops. (Source: FAOSTAT 2013)

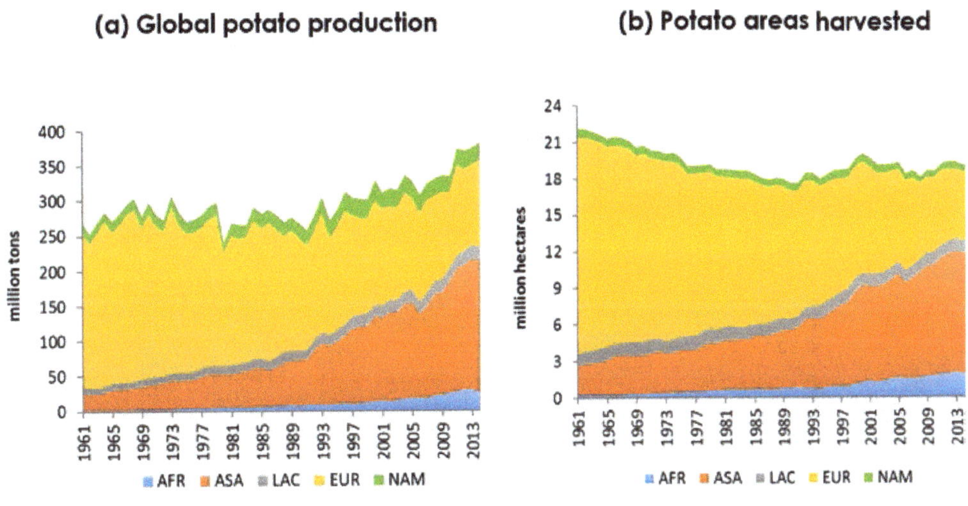

Fig. 1.3 Potato supply (**a**) and (**b**) 1961–2013. *AFR* Africa, *ASA* Asia, *LAC* Latin America and Caribbean, *EUR* Europe, *NAM* North America. (Source: FAOSTAT 2017, accessed Nov 2017). (**a**) Global potato production, (**b**) potato areas harvested

(FAO 2014). It reaffirms the increasing importance of potatoes as a source of food, employment, and income in Asia, Africa, and Latin America.

As shown in Fig. 1.4a, Africa has registered large increases of harvested area over the last 20 years, but despite the impressive growth, total production and harvested areas are still much smaller compared to Europe and Asia (Fig. 1.3a). In Africa, the increase in potato production has largely been through increase of area under production, which more than doubled since 1994 and now exceeds that of the Latin

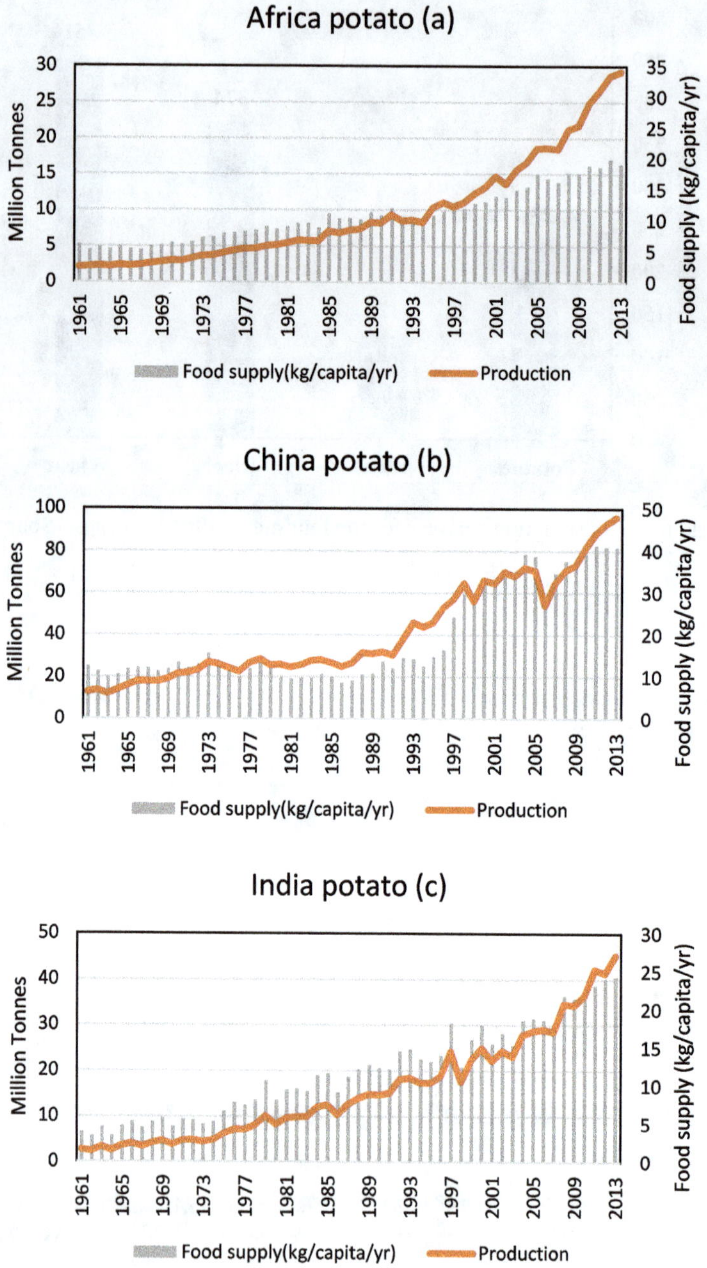

Fig. 1.4 Relative development of potato production and food supply (kg capita^{-1} year^{-1}) in Africa (**a**) China (**b**), and India (**c**). FAOSTAT (2013)

America and Caribbean (LAC) region. In the tropical highlands of East Africa, farmers grow potato both for food and cash (Muthoni et al. 2010). The increase of potato production in East African countries over the last years has been impressive, suggesting a higher contribution of the crop to local food systems. In Tanzania, for instance, potato supply has almost tripled between 2000 and 2014 (FAOSTAT 2017), while in Rwanda potato is included in the national priority list of crops due to its role in national food security (approximately 125 kg per capita consummed per year; FAO 2009). As world population levels are predicted to show the greatest rise in Africa in the coming decades, increased contribution of potato to local food systems in this region is of considerable importance (Birch et al. 2012).

In Asia, China and India have experienced nearly a half century of steady growth in potato production (Fig. 1.4b, c). Both countries also have ambitious growth targets for future years. For some decades now, the Chinese state has been working to increase national potato consumption, also launching a campaign since 2014 to promote both the cultivation and the consumption of this tuber (The Wall Street Journal 2015). China became the world's largest potato producer in 1993 and currently accounts for almost one quarter of global potato production and about 28% of total cultivated areas (FAO 2015a, b). Potato in China is mainly used for food, both as a vegetable and in processed forms, while a smaller part is also consumed as animal feed (Scott and Suarez 2012a). Potato in India is mainly grown in the Indo-Gangetic plain, either as monoculture or in rotation with maize, wheat, and/or rice and it is regarded as both an important staple and a cash crop. Following the growth in production volumes, potato yields in India have also increased significantly, at an average of 2% per year, because of successful breeding programs, quality seed systems, and storage infrastructure that have reduced post-harvest losses (Scott and Suarez 2011).

Looking at other Asian countries, potato is the principal vegetable in Bangladesh and the second most important crop behind rice. Its cultivation is widely distributed across the country where it is grown mainly as a cash crop (Scott and Suarez 2012b). Potato production in Bangladesh has greatly expanded during the last decades, especially after 2000 when output surged from about 1.5 million tons to more than 8 million tons in 2013 (FAO 2015a, b). This impressive growth, besides the rising domestic demand because of population growth and the "westernization" of dietary preferences in urban areas (Pingali 2006), can also be attributed to the introduction of several improved high-yielding varieties and the development of cold storage facilities which facilitated near year-round availability of potato. At the same time, producers also gained significant price advantages (Reardon et al. 2012). In Nepal, potato is the second most important staple food crop after rice. The potato has also become a significant source of rural income in Pakistan where production is concentrated in Punjab, with spring and autumn crops accounting for 85% of the harvest. Expansion of irrigated Pakistani land has resulted in substantial increases in potato production (up 254% from 1990 to 2009) and area under cultivation.

Regarding potato production in LAC over the past 60 years, the annual average potato domestic supply has increased from 7.2 million tons in the 1961–1963 period to 19.6 million tons in 2011–2013, which represents an average annual growth rate of 2%. By way of comparison, growth rates for potato production in ASA and AFR averaged over 4% for a similar period, i.e. more than double those of LAC (Scott 2011). Most of the production is oriented towards human consumption (74%, maintaining this trend throughout the period) and it highlights a relatively low processing level of 1% (FAOSTAT 2017).

The role that potato plays in the diets in LAC vary—from basic staple, producer/consumers in the Andean highlands to complementary vegetable for urban households in most of South America, to a relatively expensive complimentary vegetable in much of Central America and the Caribbean, and to a popular fast food in the form of French fries in urban markets throughout the region (Scott 2011). Per capita consumption of potatoes in Latin America increased slightly from 22 kg/person on average between 1961 and 1963 to 25 kg/person between 2011 and 2013. But these regional trends do not reflect the important differences in trends at the subregional and country levels. Peru, is one of the countries where potato consumption has grown significantly, reaching in 2015 a figure of 85 kg/person. This is due to various public–private policies, rural infrastructure, expansion of supermarket trade focused on potatoes and a strong relationship with the gastronomy sector promoting Andean food including the native potato and its products. Brazil and Mexico have increased their consumption, although their absolute values, 18.5 and 14.8 kg/ person respectively still remain low compared to other countries in LAC. The cases of Argentina and Colombia are showing a downward trend.

The United States is the fifth largest potato producer in the world with more than 420,000 ha harvested in 2013 and a total output of almost 20 million tons (FAO 2015a, b). Although in the United States potato is no longer the traditional staple of the past, it is nevertheless gaining increased appreciation by nutritionists because of its nutrient density and its contribution to a more balanced diet (Bohl and Johnson 2010). There is also a large demand by the processing industry for producing commodities like frozen French fries and chips for both the local and foreign markets. Potato yields in the United States have more than doubled over the last 50 years, rising from 22 tons ha^{-1} in 1961 to 49 tons ha^{-1} in 2016. This increase in yields has been suggested to be primarily driven by improvements in management rather than genetic improvements, since most breeding programs have traditionally focused on quality traits such as dry matter content and storage longevity to meet the demands of the processing industry and the consumer (Douches et al. 1996).

In Europe, Germany, France, Netherlands, the United Kingdom, and Belgium are together the strongest potato producers in the European Union (EU), due to potato yields higher than 40 tons ha^{-1} in this area of northwestern Europe (Fig. 1.5) and to the strong links of production with the dynamic European potato processing industry. Potato is also prevalent in Eastern European countries, particularly in Russia, Ukraine, and Poland where per capita consumption has traditionally exceeded 100 kg annually. Although Eastern Europe constitutes the region with the highest use of potato as animal feed globally, feed use of potato has been steadily declining over the last 20 years and being replaced by cereals, most notably in Poland. This decline in feed use, together with the shift of diets towards low-calorie food and a trend to spend less time on cooking observed in Western European countries, has led to a significant decrease in demand for fresh potatoes, and therefore potato production in the continent is falling (European Commission 2007). Despite the aforementioned decline, some European countries like France, Denmark, and Belgium have increased production over the last decade, due to growth of the processing industry (French fries, crisps) and starch production (Eurostat 2017).

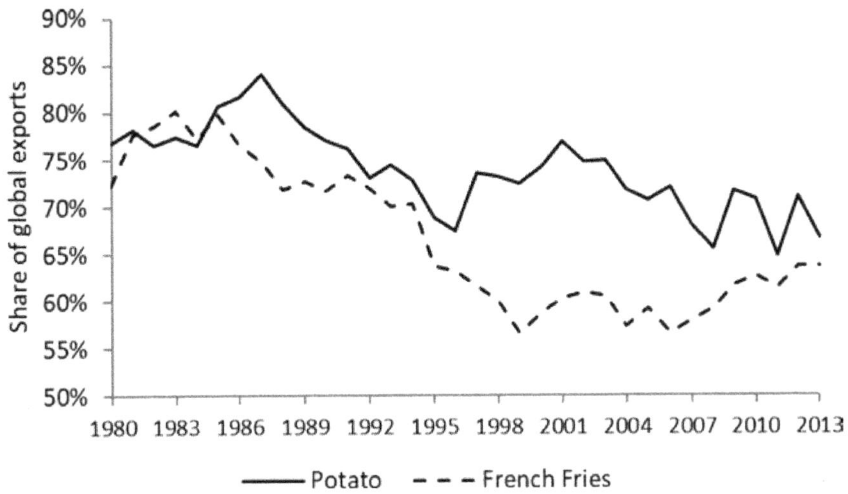

Fig. 1.5 European share of global exports for potato (fresh and seed) and French fries 1980–2013. (Source: FAOSTAT 2017)

Table 1.2 Potato utilization, consumption and trade indicators

	Utilization				Consumption	Trade	
Region	Food (%)	Feed (%)	Seed (%)	Other (%)	Per capita (kg year^{-1}) 2011–2013	Import Quantity (000 tons) 2011–2013	Export Quantity (000 tons)
Africa	69	4	8	19	19.2	967	830
Asia[a]	67	12	5	11	29.1	6,042	2,683
Europe	52	19	15	11	83.4	17,890	19,477
North America	84	1	6	9	55.3	3,615	5,354
Latin America and Caribbean	73	2	8	16	23.8	2,040	439
World	64	12	9	12	34.4	30,554	28,784

Source: FAOSTAT (2017), accessed Oct 2018
n.a. = not available, a = U.S. Department of Agriculture, 2016
[a]Asia + Oceania

Moreover, the competitiveness of the potato industry has established Europe as the world's biggest net potato exporter, amounting for more than 60% of all exports of fresh potato and a similar percentage of global exports of French fries (Fig. 1.5). These statistics concern mainly intra-EU trade, and also export of seed potato to non-EU countries, primarily to Mediterranean African countries like Egypt and Algeria (FAOSTAT 2017).

Tables 1.1 and 1.2 give a synthesis of the potato indicators by region confirming the expansion of potato in Asia, which is now the major potato producer continent. In Africa, the potato growth rate has also been strong with Egypt, Malawi, South Africa, Algeria, and Morocco producing more than two-thirds of all the potatoes in the region. In many countries of Latin America and the Caribbean, potato areas have

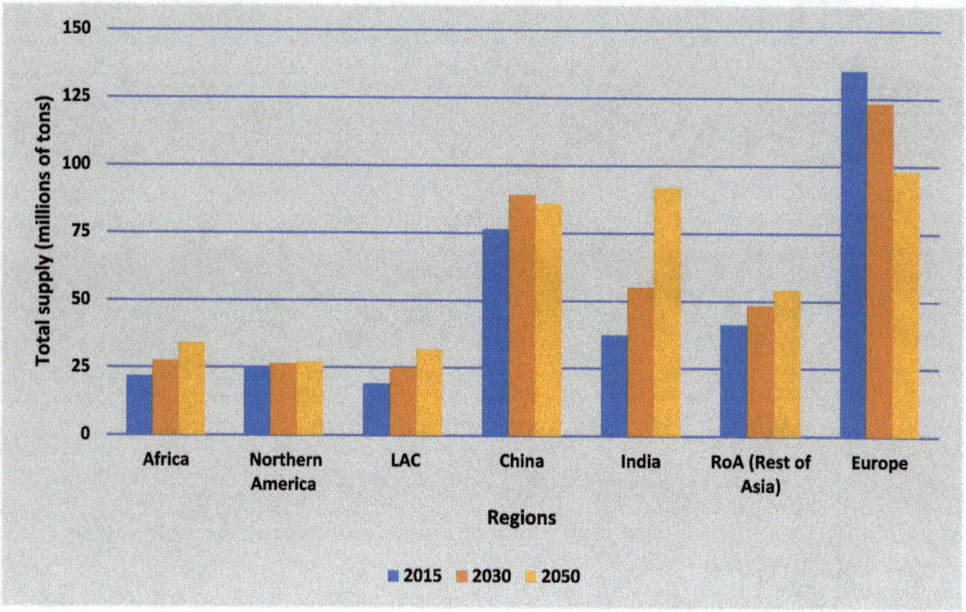

Fig. 1.6 The future of potato production. Adapted from Rosegrant et al. (2017)

declined although output has risen due to improvements in productivity. Potato production growth rate in Europe and North America has declined due to a significant decrease in demand for fresh potatoes that was partially compensated by a large demand from the processing industry as described before. In North America, yields increased rapidly between 1961 and 1990 and then somehow stagnated, suggesting yields are near their potential in the region and there may exist genetic limits. In Europe, on the contrary, the relative larger yields growth rates occurred after 1990.

Future trends by region indicate a major production increase in Asia and Africa as compared to other regions (Fig. 1.6). Considering some underlying assumptions such as population growth, climate change, and economic growth pathways, the UN projects a population decline in China and growth of per capita GDP which will affect diet composition. Therefore, the future supply of potato in China will not continue to grow faster than in the past. According to Rosegrant et al. (2017), it is in India where potato supply will almost triple because of the very high population growth, especially under certain socioeconomic scenarios.

1.4 The Potato Remains a Food Security Crop in the Developing World's "Nutrition Transition"

In many developing countries, and especially in urban areas, the globalization, the emergence of fast food outlets and supermarkets and the rising levels of income are driving a "nutrition transition" led by major shifts in the availability, affordability, and acceptability of different types of food, especially toward more energy-dense foods and prepared food products. It is translated into major and rapid shifts in dietary patterns. As an example of this nutrition transition process, demand for potato is increasing in many developing countries in Africa and Asia (Fig. 1.3). In South Africa, potato consumption has been growing in urban areas as part of the staple food consumption of the middle class, although maize remains the primary staple in rural areas. In China, higher income and increased urbanization have led to increased demand for processed potatoes. In this context, potato plays a role in diet diversification in many countries where family agriculture and smallholders continue to supply local markets with fresh and affordable agricultural produce. Potato is still an important staple in rural food systems where it is produced, and emerging food systems which are urbanized but where consumers still rely on staples such as potatoes. In Industrial food systems, highly urbanized in Northern countries in Europe and North America with the development of the processing industry there is a lower dependence on traditional staples (Gillespie and Van den Bold 2017).

As mentioned by Haverkort and Struik (2015) potato used to be a "local for local" crop and it still is in many countries because of the bulkiness and the limited storability of the seed and ware tubers. Compared to other staples, and except for processed potato products, fresh potato is a thinly traded commodity in global markets and is absent in major international commodity exchanges. It is therefore subject to less price volatility at global scale. Thus, potato can be relied upon to smooth the disruptions in global food supply and demand that have an impact on other commodity prices, such as witnessed during the 2007–2008 and subsequent food price spikes (FAO 2009).

Potato also improves food security because it is a source of employment and income, both of which have direct links to household food access and nutrition (Pinstrup-Andersen 2014; Kanter et al. 2015). The comparatively short maturity period, nutritious characteristics, employment, and income opportunity that characterize potato make it a resilient crop that can secure vulnerable livelihoods under the effects of climate change and changing market environments. Moreover, potato yields more food more rapidly on less land than any other major crop.

Where other staple crops are available to meet energy requirements, potato should not replace them but rather supplement the diet with its vitamins and mineral content and high-quality protein. Potato can be promoted as a healthy and versatile component of a nutritious and balanced diet including other vegetables and whole grain foods. Likewise, it contributes to combat micronutrient deficiency, also referred to as hidden hunger, which is a major global public health problem, affecting an estimated two billion people globally (Bailey et al. 2015). Potato contains interesting amount of health promoting components such as vitamin C, phenolic compounds, and iron and has protein content comparable to that of cereal grains (Burlingame et al. 2009). When eaten with its skin, a single medium-sized potato of 150 g provides nearly half the daily adult requirement (100 mg) of vitamin C. It is also a good source of vitamins B_1, B_3, and B_6; minerals such as iron, potassium, phosphorus, and magnesium; and contains folate, pantothenic acid, and riboflavin. Vitamin B deficiency has both short- and long-term impacts, including poor cognitive and pregnancy outcomes and poor child development and life outcomes. Vitamin C, on the other hand, is important for body metabolism and iron absorption (FAO 2009).

As an example, potato plays an important role in the food basket of most Peruvians; it continues to make up a relatively high proportion of daily calorie availability, reflecting its importance as traditional source of energy. To enhance the nutritional contribution of the potato in rural Andean highlands, the International Potato Center (CIP) coordinated activities to promote innovation in potato-based systems with the objective of contributing to food and nutrition security of rural highland populations enhancing native potato production and promoting dietary diversity. In rural areas of Peruvian highlands, there is a high prevalence of chronic malnutrition amongst children (42% in children under 2 years according to INEI Peru 2017). Native potato varieties are commonly grown in the Andes and are a significant part of the local diets. Some of them have higher contents of micronutrients (Zn and Fe) and are rich in antioxidants compared to commercial improved varieties. Creed-Kanashiro et al. (2015) explored the relationships between agricultural production characteristics and nutritional status of young children of families in rural areas of Peru whose livelihood is based on potato production systems combining native and commercial varieties. The results showed a positive relationship between percentage of recommended dietary intake (RDI) for both Fe and Zn intakes by children and production of native potatoes for home consumption, raising small animals for consumption and sale (e.g., guinea pigs and poultry) and the area of production of commercial potatoes that allowed the families to improve their incomes and diversifying the family diet.

Finally, it should be mentioned that potatoes have been related to increased risks of obesity mainly because of their high glycemic index. Recent reviews of clinical intervention and observational studies centered on the potato concluded that these studies did not provide convincing evidence to suggest an association between intake of potato and risks of obesity, Type II diabetes (T2D), or cardiovascular disease (CVD) (Borch et al. 2016). However, as part of the trend towards urbanization and associated lifestyles, raising incomes, and greater consumption of "convenience foods", demand for fried potatoes is increasing. Overconsumption of these high-energy products, along with reduced physical activity, can lead to overweight and obesity. Therefore, the role of fried potato products in the diet must be taken into consideration in efforts to prevent overweight and diet related noncommunicable diseases, including heart disease and diabetes. For these reasons, even though the potato is a nutritious staple crop, it has often been associated to a meal component with no specific attributes in many societies, even in developing countries. This image is reinforced by some negative myths such that potato is fattening, requires intense use of fertilizers and pesticides for its cultivation and can contribute to soil erosion. Negative images can be mitigated with better information about the nutritional value of the potato and its importance in the diet, while promoting sustainable and environmentally friendly production systems, thus addressing the question of natural resource management.

1.5 Policies and Strategies for the Development of the Potato as a Food Security Crop

The role potato can play as a food security crop at national scale has been addressed in some developing countries with different policies, either sectoral and crop-specific or at the macro level. In its quest to improve food security for a rising population, the Government of China is developing a national plan to increase production and consumption of potato and promoting the crop as a staple instead of a vegetable. This status can give access to important complementary policies and resources at national and regional level and to subsidies from the central government. It also recognizes the double role of potato in current China. Potato is still a major staple for poor rural areas where local governments continue to provide subsidized inputs (e.g., clean seeds of selected varieties), while at the same time being at the forefront of an increasing private sector-led processing industry, accompanying rising incomes in urban populations and diversification of diets (Scott and Suarez 2012a, c).

In Peru, the major center of origin of the potato, a large effort began in early 2000s to develop a competitive and inclusive native potato value chain for domestic markets. Initially led by the International Potato Center, the initiative gathered several private and not-for-profit actors to add value to the native potato grown by small farmers while developing a niche market. Several new products were developed in the process, for example selected native potato varieties for fresh consumption sold as gourmet potatoes in innovative packages in large supermarket chains, snacks such as colored native potato chips, and culinary innovations in the gastronomy sector featuring native potato as a central component of sophisticated dishes. The innovations in the value chain continued and a second round of new products emerged, including for example frozen native potato fries, native potato-based liquors, and even cosmetics made from potato. Although no specific sectoral policies were behind this initiative, the development of the native potato value chain took advantage of Government policies at the macro level promoting private sector and market-led developments and the fast growth of Peru's economy and of the purchasing power of the population since the beginning of the twenty-first century. While the Government of Peru focused on public investments to promote export-oriented agricultural growth, the experience with the native potato value chain has proven successful to link small potato farmers to domestic markets and to develop a more inclusive growth strategy of the highly diverse agricultural sector of the country (World Bank 2017).

A final example of policies that have been adopted to promote the potato sector in developing countries is through seed laws and regulations. Seeds are an important input of production of the potato crop and can affect yields since they are vehicle for important diseases. Seed degeneration due to viruses is one of the most common constraints affecting potato productivity, and therefore large research and extension efforts are made to improve availability and access to quality seeds for small farmers. One particular aspect of potato seed is that it is vegetatively propagated. In most developing countries, however, seed systems were first established following developed country standards for cereals and grains, promoting the use of certified seed under a formal seed system. This has led to very low use (less than 10%) of certified potato seed in most developing countries. To increase access to quality seed by small farmers, the Food and Agriculture Organization (FAO) of the United Nations promoted the definition of a new seed category, the Quality Declared Seed (QDS), that relaxes some of the standards required for Certified Seeds and recognizes the importance of seed producers in providing seed of enough quality through the informal seed system (FAO 2006b). Ethiopia adopted the QDS definition in a new seed law passed in 2016 without distinction of crops. Peru (in 2018) and Ecuador (in 2013) have modified the seed regulation for potato to accept the use of QDS. However, differences still exist on how countries are beginning to adopt this category. While in Peru, the new regulation defines the category as QDS similar to the FAO definition, in Ecuador the regulation defines the new category as "common seed" and introduces aspects of the FAO definition for QDS.

Other countries are updating the regulations regarding seed quality assurance systems for potato to increase availability and access of quality seed by farmers. A broad range of changes are proposed, from relaxing some of the standards required for certified seed to allowing the use of private inspection services to increase the number of seed producers that can be inspected each season (e.g., Kenya). One of the motivations of these changes is the increasing recognition of the role of the potato crop for national food security. Some concerns have been raised, however, on the potential consequences of relaxing seed quality standards on the incidence of seed-borne diseases (e.g., *Ralstonia solanacearum*).

1.6 Food Security Challenges and Perspectives for Potato Research and Development

1.6.1 Potato in a Global Food Security Context

The potato, because of its adaptability, its yielding capacity and its nutrition contribution, and as an important component of diversified cropping systems, has a long history of helping relieve food insecurities, and contributing to improve household incomes in times of crisis and today's population expansion. Among important issues and challenges at global level, the European Association for Potato Research (EAPR) Conference in 2017 identified three broad concerns: (1) food security and food safety for a growing population considering consumer's needs; (2) sustainable and environmentally friendly production addressing the question of natural resource management taking advantage of new technologies available such as breeding techniques, biocontrol and big data management; (3) innovation in practice turning scientific results into products and processes to improve the performance of agri-food systems (Andrivon 2017).

Potato's support to food availability can be achieved through improved productivity, either by increasing yields or expanding production areas, combined with technologies that reduce post-harvest losses. Newly developed potato technologies and production concepts can potentially help create solutions in areas where there is a huge need to increase food production. Figure 1.7 shows the global distribution of yields and the low yield levels in most of the developing world, where observed actual yield is usually much lower than the attainable yield. The actual yield is the expression of a potato cultivar in a specific agro-ecological environment and depends on availability of inputs, the economically optimal use of available inputs given the farmers' conditions, and externalities such as local meteorological variations and climate change as well as the technology level and the quality of crop management. Actual yields range from below 5 tons of fresh tubers ha^{-1}(median yield in Uganda: Gildemacher et al. 2009) to well above 100 tons fresh tubers ha^{-1} (in Columbian Basin, USA: Kunkel and Campbell 1987).

The yield gap, expressed as the difference between actual yield in farmers' fields and the attainable yield—using best agricultural practices—leaves a great potential for improvement considering that, in developing countries, the full expression of the crop's yielding capacity has not yet been achieved. Much improvement is needed in agronomic practices, quality seed production, and varieties tolerant/resistant to abiotic and biotic threats (Birch et al. 2012). The high nutritional value of potato mentioned above reinforces the potential of potato to respond to food security challenges.

To reach impact on food availability, access and better use in diets, proper selection of target areas for potato research and identification of the most important constraints to potato production are crucial for defining priority interventions. As a step towards achieving this and to prioritize options for potato research, CIP led an expert survey to assess priorities for potato research across the developing world (Kleinwechter et al. 2014), and an ex-ante evaluation of the economic relevance of

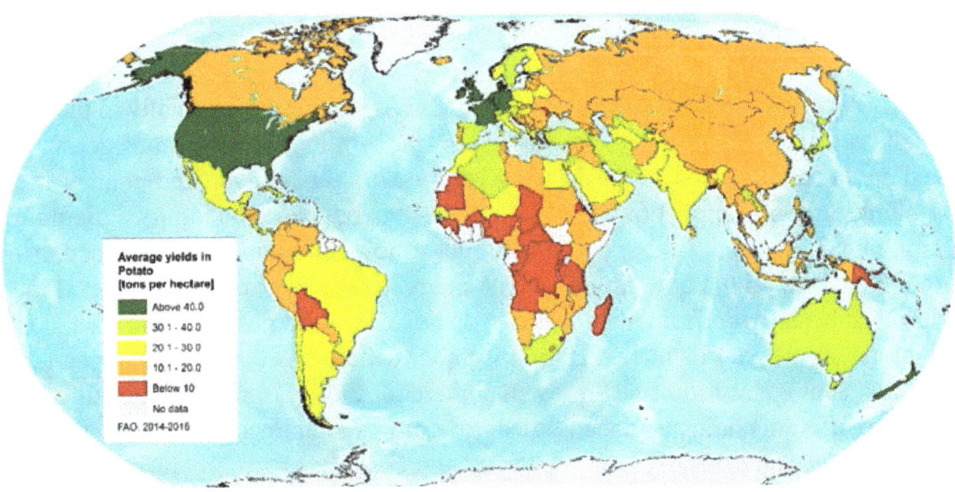

Fig. 1.7 Global distribution of potato yields (tons ha^{-1}). FAOSTAT, 2014–2016

these options (Hareau et al. 2014). Most of these research options for potato have been defined to be relevant in numerous countries of Asia, Africa, and LAC. They include research in potato late blight, drought tolerance/water use efficiency, seed systems and development of farmer organizations and farmers' links to markets (e.g., Harahagazwe et al. 2018 for research priorities in Africa). However, the current and upcoming contextual changes, especially considering the climate change requires to revisit some of these issues with a new perspective. Many simulation exercises based on IPCC (Inter-governmental Panel on Climate Change) scenarios and biology models are underway and suggest that future potato cropping systems could differ from those we know today with the implication that new cultivars will be required to respond these new conditions (Andrivon 2017; Quiroz et al. 2018).

1.6.2 Research and Innovation for Sustainable Potato Cropping

As suggested by Haverkort and Struik (2015), the future prospects of food security challenges for agricultural production can be expressed by the formula: $P = G \times E \times M \times S$ where Performance (P) is determined by Genotype or varieties (G), the Environment or agro-ecological conditions where the crop is grown (E)—which as mentioned above is evolving and will further change in the future—its Management and adaptation to local socioeconomic conditions (M) and the Societal requirements (S) driven by society's demands for food and the need to make agriculture more environmentally and consumer friendly with a focus on food safety. The societal or consumer requirements will vary according to the context. In high-income countries, consumers are looking principally for healthy and easy to prepare foods while in the developing world, the consumers' needs are driven by the food and nutrition challenges and demand concerns principally food availability in both quantity and quality. The actual performance of potato can be expressed as yield of fresh or dry matter per unit area or yield of the finished product per unit area recovered from the raw material after processing.

This Performance analysis ($P = G \times E \times M \times S$), suggested originally in the context of high-income countries, can be used for the developing world considering the need in most countries to principally respond to the hunger and malnutrition problems through a more efficient agri-food system but based on family agriculture. Family farms are the backbone of agriculture in low- and middle-income countries in Africa, Asia, and Latin America. For many years, the trend in the developed world has been towards intensification to achieve more outputs per unit of land but the sustainability of this intensification is under debate especially considering agriculture's ecological footprint. In low-income countries, sustainable intensification (SI) is a different challenge because it starts from a much lower level of inputs than in developed countries. This is especially the case in Africa where potential for increasing production through area expansion is diminishing, partly due to high population growth (Headey and Jayne 2014). For instance, Wu et al. (2018) and Jayne et al. (2014) argue that even though Africa has a high cropping intensity gap[1], closing this gap sustainably must focus on input intensification rather than area expansion. The relevant question is how to promote technology options that allow for increased output quantity and quality (especially from the nutrition point of view), while considering agriculture's environmental impact, preserving land and other resources in both developed and developing countries. In this context, sustainable intensification of potato cropping goes beyond production aspects and considers strong socioeconomic, demographic, and environmental trade-offs to optimize performance. Institutional incentives to support innovation involving diverse stakeholders, with emphasis on research partnerships, are also required to respond to hard-to-find compromises that will vary between different cropping systems (Hall et al. 2001). Multidisciplinary approaches contribute to recognize and solve practical problems at the level of the crop, the cropping system and the agri-food system to achieve sustainable food security in its four dimensions.

The issues mentioned above indicate that sustainable potato production and efficient use of resources will require future adjustments and redesigns of the cropping and processing systems. Two main options can be considered to increase food security, which remains a main goal for future potato research: (1) produce more with less through better input management and optimization; (2) to produce just as much but waste less both before and after harvest through better value chain management, better storage, processing and marketing operations, and responding to increased involvement and awareness of consumers (Andrivon 2017).

As an attempt to analyze how to combine and score different research and technology options according to their effect on sustainable agri-food system indicators in developing countries and their relation to the four dimensions of food security, Table 1.3 attached suggests a list of priority research and technology options in the spheres of G and M. The assessment is based on the literature and the authors' expert opinions. We scored key options using some critical indicators of sustainable agri-food system intensification related to productivity, agriculture income, human well-being and environmental sustainability according to Smith et al. (2017). We scored the research and development options using a simple scale of high, medium, and low effects according to their relation to sustainable intensification indicators and their main contribution to one of the four dimensions of food security.

The performance of any of these technologies or development approaches will, as expressed by the Performance equation ($P = G \times E \times M \times S$), depend on external factors both socio economic and agro-ecological. Their potential contribution to food security will be strongly influenced by the environment (E) and the societal (S) requirements in the local context where they are implemented—e.g., enabling environment, policies for financial and nonfinancial services are also key components for achieving efficient food systems.

---

[1] Wu et al. (2018, p. 2) define cropping intensity gap as "the amount of incremental cropping intensity that is possibly available if all croplands in a given region are fully intensively used."

Table 1.3 Qualitative evaluation of key research options to enhance food security dimensions through the performance of potato production systems (based on the authors appreciations and the literature related to indicators of contribution to sustainable agri-food systems intensification)

	Food security dimensions						
	Availability				Access	Use/quality	Stability
	Indicators of contribution to sustainable agri-food systems intensification						
Key research options to enhance the performance of potato production systems	Water use efficiency	Land use efficiency	Nitrogen and phosphorus use efficiency	Pesticide use efficiency	Farmer incomes	Increase in calorie and nutrient production efficiency	Reduction of environmental footprint (soil, water, air)
Breeding and variety development (Genotype)							
High yield		***	Neutral/negative. High productive crops might need additional N	Neutral/negative. High productive crops might need additional pesticides	***	**	
Disease resistance (e.g., late blight, viruses)		***		***	**	**	***
Tolerance to drought/ heat/salinity	***	***	*		**	**	*
Biofortification (e.g., Fe and Zn)					*	***	
Earliness	***	***	***	**	***	***	***
Potato seed production (Management)							
High quality seed production and distribution		**	*	*	***	*	Neutral/negative high-quality seed use might motivate increased chemical input

Farmer-based seed production		*	*	*	**	*	**
Decision support tools and farming practices (Management)							
Decision support and diagnostic tools for pest and disease control		*		***	**		***
SMART agriculture approaches (sensor use for precision in NPK and water use)	***	**	***		**	*	***
Intensification of cereal-based systems with introduction of early potato varieties	*** No irrigation needed, higher productivity per unit of water	** Uses current fallow land already under agriculture	*	**	**	**	**
Sustainable resource management							
Ecosystem management and biodiversity use	*		*	**	* Policy dependent	*** Use of potato biodiversity can increase nutrient production efficiency	***
Soil and water management	***	**	**		**	***	**
Efficient and inclusive value-chains (Management and Society)							
Value-chain innovation					***	**	***
Post-harvest management (Management and Society)							
Post-harvest losses assessment and reduction					**	***	***

*** = High effect; ** = Medium effect; * = Low effect

In the sections below, some key research and technology options identified in Table 1.3 are briefly described.

1.6.3 Potato Breeding, a Driving Force Towards More Efficient Potato Production

For genotype development (G), priority should be given to achieve a combination of traits to enhance stress tolerance and nutritional aspects to better respond to contextual changes, especially climate and local needs. The recent development of participatory breeding helps to best define the crucial trait combinations required and to facilitate acceptance of new genotypes by growers (Schulte-Geldermann et al. 2012). With the recent findings on the potato genome sequence (PGSC 2011) and the possibilities occurring with new breeding technologies (NBTs), potato breeding appears as the number one opportunity to improve potato production for global food security (Birch et al. 2012).

In many developing and in-transition countries governments have substantially invested in breeding improved varieties. A total of about 840 improved varieties[2] with various combination of traits have been released, most of which in Asian countries like China and India (Table 1.4). CIP has considerably contributed to global crop improvement through supporting NARS and providing access to advanced breeding material. About 43% of total releases are CIP-related (i.e., NARS-bred varieties distributed/facilitated by CIP, NARS selection from CIP crosses, NARS crosses from CIP progenitors). At the regional level, CIP-related varieties are most prominent in Africa where 70% of total releases are CIP-related. This points to the importance of CIP in the regions and the support many national breeding programs require.

Table 1.4 Total and CIP-related number of releases by region

Region	Year	Total	CIP-related[a]	
Asia[b]	2015	518	180	35%
Africa[c]	2010	178	124	70%
Latin America[d]	2007	141	60	43%
Total		**837**	**364**	**43%**

[a]NARS-bred varieties distributed/facilitated by CIP, NARS selection from CIP crosses, NARS crosses from CIP progenitors; calculations for Asia based on Gatto et al. (2018), for Africa on Labarta (2015), for Latin America on Thiele et al. (2008)
[b]Bhutan, Bangladesh, China, India, Indonesia, Nepal, Pakistan, Philippines, Sri Lanka, Vietnam
[c]Burundi, D.R. Congo, Ethiopia, Kenya, Madagascar, Malawi, Rwanda, Tanzania, Uganda
[d]Bolivia, Colombia, Ecuador, Peru, Venezuela

[2]The current true total number of releases is likely higher given that the most recent data available dates back more than a decade.

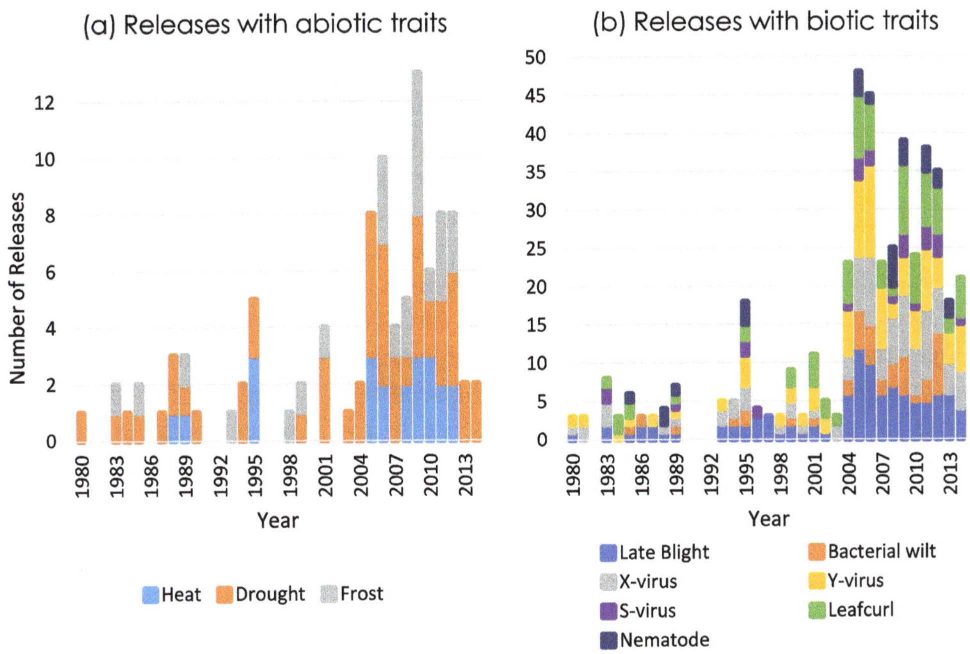

Fig. 1.8 Total releases with abiotic and biotic traits in Asia between 1980 and 2014. Notes: release is for *high* resistant category only. *Medium* and *low* resistant, and *susceptible* categories are not shown. (Source: Gatto et al. 2018)

The development of early and high-yielding varieties with resistance to *P. infestans* has been a longstanding potato breeding objective. Genotypes with resistance to viruses (PVY, PLRV, PVS, PVX), nematodes (mainly *Globodera* and *Meloidogyne* species), bacterial wilt, and a broader spectrum of cultivars tolerant to abiotic stresses like heat, drought, and saline conditions, and focus on beneficial root traits, can increase productivity and expand potato production to new areas.

The importance to develop and release varieties with high tolerance to abiotic stresses and high resistances to biotic stresses has increased, as Fig. 1.8 depicts for Asia. Especially starting in the early 2000, major traits have been bred into released varieties likely as a result of adjusting breeding objectives aiming increasingly at mitigating the adverse effects associated with climate change and variability.

New resilient varieties will potentially expand potato production to new areas and produce more nutritious food under current and future stress factors. Genetic biofortification through conventional and new breeding techniques can help to overcome micronutrient malnutrition and support the consumption of better-quality tubers. This crop improvement approach aims to positively influence human health, as a complement to diet supplementation and food fortification. In recent years, CIP has initiated the development of Fe and Zn biofortified potatoes, under the umbrella of the HarvestPlus Program (http://www.harvestplus.org/), a global interdisciplinary alliance for developing biofortified varieties of staple crops. Food security programs working to deploy biofortified crops will strongly benefit from nutritional education efforts and awareness programs considering gender roles in the beneficiary communities.

1.6.4 Seed Quality and Availability, the Key to Harvest Success

As sustainable potato production depends on a constantly renewed supply of disease-free planting material, improving quality seed production and seed distribution is another strong avenue of research opportunity related to crop management (M) in potato development. The conceptual framework underpinning the concept of seed security contemplates different types of seed insecurity: poor seed quality, lack of availability, limited access to high quality seed, lack of access to preferred and adapted varieties, inefficient seed systems (FAO 2016; CGIAR/RTB 2016). As a vegetative propagated crop, the growth, development, yield and quality of the potato is strongly influenced by the quality of the seed tubers planted affected by their physical, physiological and health status. More than 90% of seed potatoes in developing economies is produced in the farmer-based category and is considered to be of poor quality (Thomas-Sharma et al. 2015). Potato seed production systems should support the access to high quality seed potato tubers of improved varieties by combining rapid multiplication technologies (e.g., aeroponics or sand hydroponics) with decentralized seed multiplication, e.g., promotion of quality declared seed systems (FAO 2006b; Fajardo et al. 2010). It should be complemented with on-farm seed maintenance (e.g., positive selection, small seed-plot technique and improved storage) in an integrated approach (Gildemacher et al. 2011; Schulte-Geldermann et al. 2012; Thomas-Sharma et al. 2015; Obura et al. 2016). Improving technologies for farmer-based seed production and distribution of high quality planting material of existing and new varieties have the potential to reach high numbers of beneficiaries with strong impacts on poverty reduction and food availability.

1.6.5 Potato Crop Management and Farming Practices to Increase Productivity and Sustainability

The highly adaptable potato can fit to many types of environments (E) from sea level to high mountain conditions where small-scale farmers predominate. Beside temperature regime and solar radiation, there are many factors that affect its productivity as for example, soil characteristics, water use efficiency, nutrient availability and hazards such as night frost or heat waves that may drastically impact resource use, and thereby sustainability. The adaptation of the crop depends on the genotype but also on the crop management practices (M) that need to evolve according to the specific agro-ecological conditions, the socioeconomic context and the local production systems. Crop management is context specific and should consider local knowledge that can be improved promoting new tools and approaches.

Smart agriculture is a novel avenue for resource use optimization based on new monitoring and decision support tools. Remote sensing and global information system (GIS) tools coupled with decision support systems (DSS) and precision agriculture technologies may contribute to increased productivity while interaction among biophysical and social disciplines for sustainable food production intensification can at the same time contribute to resource use optimization. Fertilizer (N, P and K) recommendation systems are now using field scale models as well as tractor, drone and satellite embedded spectral sensors to monitor crop nutrient status to supplement fertilization according to inter and within field variability (Goffart et al. 2008, 2017). These more sophisticated technologies are still mainly used in the high-income countries, but massive and varied data management could foster new models to be developed and contribute to decision support systems under developing country conditions. CIP is adapting such system to the Andean condition developing a strategy to manage late blight that combines host resistance and a decision support system to optimize the use of fungicides. The vulnerability of the crop to many pests and diseases, which the current global climate change can worsen, remains one of the most severe threats for a wider potato diffusion and its sustainable cropping. To improve crop health, portable molecular diagnostic tools and decision support systems for early warning and control of pests and diseases (e.g., for efficient fungicide use to control late blight) will contribute to better crop production monitoring and input use efficiency. Research to develop biocontrol is very active and considered to grow substantially in the coming decade, but there are still few confirmed successes from the field, and specific management tools (Decision Support Tools) are still missing (Velivelli et al. 2014). CIP and EAPR are coordinating actions in Europe and Latin America to promote biocontrol and compare the efficiency of biocontrol agents using defined protocols (Devaux et al. 2017).

Specific intensification practices can be developed under specific cropping systems such as in the cereal-based systems in India through "Double-Transplanting (DT)" of rice and planting early maturing potato between the two rice crops as a valid alternative to the traditional potato-boro rice and kharif (monsoon) rice-boro rice. This cropping pattern contributes to enhance system productivity without sacrificing area or productivity of either of the two crops, thus creating new opportunities for potato cultivation for small-scale producers (Arya et al. 2015).

To enhance ecological sustainability, the objective is to implement management practices that increase the level of provision of ecosystem services such as natural soil fertility and biological control. Natural regulation of pests and diseases is an important element in potato agro-ecosystems. In organic farming, it has been demonstrated that the development of natural antagonist associations of the Colorado beetle, such as auxiliary insects and useful pathogens, can significantly improve the control of such a pest for the potato crop (Crowder et al. 2010). Another example of ecosystem management is the delivery of nitrogen through natural fixation and mineralization, which can be enhanced by cropping practices such as cover crops, legume-based intercropping systems and application of organic soil amendments before the potato crop. Biodiversity based agricultural approaches that rely on the design and management of on-farm agrobiodiversity to generate ecosystem services is another avenue to reduce potato's ecological footprint and increase farmer's resilience to cope with frost risks as it is the case in the Andes. The management of ecosystems at field and landscape level can provide a series of production benefits to reduce the need for off-farm inputs. The analysis of beneficial microbial communities and their impact on potato plant phenotypes expressions still needs to be developed as discussed at a EAPR-CIP workshop on biostimulant and biocontrol agents (Devaux et al. 2017).

1.6.6 Integrating Food Security and Value Chain Development

Although potato remains a staple food in rural areas in developing countries, it is also increasingly becoming a cash food for farmers in Asia, Africa and LAC (DeFauw Sherri et al. 2012). The majority of potato producers are smallholders who depend strongly on agriculture, including the potato crop, for income, food security and employment. Potato production reaches consumers via multilevel marketing systems, not directly from the farmer's field. Thus, the challenge to achieve food and nutrition security as well as prosperity for these smallholders will be obtained or lost by the way agricultural value chains are coordinated. Value chain development and organized markets through farmer associations, storage facilities, and better links with traders and consumers are then required to allow potato producers to access better value markets to get higher and steady incomes from their production. In the recent years, research activities to improve the efficiency of the value chain and coordination among its actors have evolved to achieve more inclusiveness in the value chain development approaches (Devaux et al. 2018). Several factors are contributing to this evolution: changes in consumer demands, new or emerging markets with strict standards, including food safety, processing technologies, and better access to market information. To respond to these changes and the need to make agriculture more environmentally and consumer friendly responding to the Society's requirements (S), research should be characterized by an interaction between natural and social sciences and should be market-driven considering the needs and challenges of all value chain stakeholders. In a compendium about perspectives on the status of innovation for Value Chain Development, Devaux et al. (2016) analyze the opportunities emerging from new markets for agricultural produce and identify challenges to smallholder participation in these markets, approaches for increasing access to markets through strengthening value chain stakeholders' relationships, enhancing innovation and improving an enabling environment. Linked to the value chain efficiency, the assessment of food losses across the value chain and the quality of marketed potatoes also require further research efforts to optimize food availability and consumer access to quality potato products.

1.6.7 Post-harvest Management: Reducing Food Losses

As indicated above, another way to face the food security challenge is to produce just as much, but waste less through better post-harvest management. Post-harvest management in potato, including storage, processing and value chain efficiency, is a much larger problem than cereals and deserves special attention. Reduction of food losses appears as a key opportunity. The basics of storage management have not changed, but the implementation and application of the basics are evolving worldwide, according to diversity in location, climate and market criteria, that will influence storage management structures and management decisions (Olsen 2014). In developing countries, recent studies have analyzed food loss across the potato value chain, as for example in Ecuador and Peru, by collecting qualitative and quantitative data to provide a comprehensive identification and characterization of losses. The results show that the most important losses occur in the production node, ranging from 90 to 95% of the total losses in the chain. On average farmers suffer this highest loss across the value chain ranging between 8 and 20% of their production at or before harvest before moving on to the next node of the chain. The main causes of losses are poor crop and harvest management, infested tubers by pest and diseases, high percentage of small tubers and weather conditions: frost and heavy rains (Delgado et al. 2017).

1.7 Concluding Remarks: Towards Future Potato Research for Global Food and Nutrition Security

The analysis of leveraging potato agri-food systems for global food security issues and challenges in this chapter emphasizes the need for making agricultural research programs and food system interventions more responsive to food security dimensions. The multidimensional nature of food security requires multisector innovation in approaches that allow to use the knowledge available and transform scientific results into products and processes to improve the performance of agri-food systems, considering the challenge to produce more food with the same or fewer resources.

In both developed and developing countries, innovations resulting from potato research should be incremental through a step by step improvement of an existing structure promoting technologies adapted to the local context. This is particularly true for smallholder family agriculture in developing countries where there is a great need to increase potato production in a sustainable way. While this approach has the advantage of not destabilizing an existing system, it may also suffer of a systemic lock-in or a lack of enabling environment that keeps agriculture and food systems on less efficient pathways as developed by Baret (2017). An example of lock-in is the use of pesticides and their promotion by agro-chemical companies and technical support services that influence farmers' decision making, restraining the use of more environmental friendly options such as decision support tools for efficient pest control with a more rational pesticide use. The valuable use of varieties tolerant or resistant to pests and diseases can also be limited by processing companies that promote varieties for their processing characteristics regardless of their environmental footprint. In developing countries, low infrastructure quality, weak institutions and policies create also huge limitations to the adoption of new and more sustainable technologies. To reach food security goals, a stronger emphasis must be put towards promoting evidence-based policies for communicating information and influence decision makers. It is also important to favor affordable and better-adapted technologies that can respond to the needs of small-scale farmers and, significantly limit negative impacts towards the environment. The research and technology options proposed in this chapter will require policy support, financial and nonfinancial services to have a chance to be adopted and used by local farmers. They will also necessitate a better access to discovery and creative ideas through better services from the potato research community at national and international levels.

Local calibration/validation and demonstration are two essential phases towards local end-user uptake, either involving farmers or extension services representatives. Public or private investments are also required to support such actions to enable farmers to have access to new technologies, and to be trained in their use. With the upcoming of new communication technologies such as smartphones, expansion of mobile broadband and access to local online platforms integrating large amounts of local data and links to Decision Support Systems, we have yet to fully exploit the potential of information technologies especially in developing countries. Local farmers, especially the younger ones, are expected to be able to have increasing access to such new adapted tools, i.e. for Late Blight management in the Andes. But this will only be possible globally if technological innovation is accompanied by capacity building and institutional innovation (associativity, access to credit, communication network) in rural areas.

There is a great dichotomy between research activities in developed versus developing countries that highlights the need for more exchange, knowledge sharing, and collaboration. Since 2014, the interaction between EAPR-linked research organizations and CIP has been looking at mechanisms to enhance partnership between European partners and CIP involving research partners in the Southern hemisphere to promote collaborative research activities, links between research networks such as Euro and Latin Blight (Acuña et al. 2017) as well as facilitating short and long-term training with universities in Europe (Durroux-Malpartida 2014).

To reach the strongest impact on food security, potato research and development efforts need to move towards food systems engineering, rather than focus explicitly on technology/solution development. In this paper, we are analyzing the different components that contribute to the performance of the potato using the key relation $P = G \times E \times M \times S$, enabling a list of key research and technology options to guide agriculture research and technology development toward sustainable intensification approaches responding to farmers' needs both for food security and better income. The argument is that agricultural programs need to integrate better agriculture sustainable intensification and food security indicators considering also other dimensions such as quality, diversity of products, health impacts and climate change effects. Multidisciplinary approaches and a better understanding of the evolving food systems are required to recognize and solve practical problems of the whole potato value chain to achieve sustainable food security. Policies, investments and services that support agricultural productivity, sustainability and expand risk management capacity are also required to give potato farmers the best chance to meet future needs, while increasing their adaptability and resilience to foster food security.

Acknowledgements The authors are grateful and sincerely appreciate the valuable contributions of Mr. Henry Juarez and Dr. Jorge Andrade-Piedra. Their inputs, opinions, and suggestions have greatly contributed to improve this chapter.

References

Acuña I, Restrepo S, Lucca F, Andrade-Piedra J (2017) Recent developments: late blight in Latin America. In: Schepers HTAM (ed) Proceedings of the sixteenth EuroBlight Workshop, Aarhus, Denmark, 14–17 May 2017. PAGV special report no. 18

Andrivon D (2017) Potato facing global challenges: how, how much, how well? Potato Res 60:389. https://doi.org/10.1007/s11540-018-9386-z

Arya S, Ahmed M, Bardhan Roy SK, Kadian MS, Quiroz R (2015) Sustainable intensification of potato in rice-based system for increased productivity and income of resource poor farmers in West Bengal, India. Int J Trop Agric. ISSN 0254-8755 33(2):203–208

Bailey RL, West KP Jr, Black RE (2015) The epidemiology of global micronutrient deficiencies. Ann Nutr Metab 66(Suppl 2):22–33

Baret PV (2017) Acceptance of innovation and pathways to transition towards more sustainable food systems. Potato Res 60:383. https://doi.org/10.1007/s11540-018-9384-1

Birch PRJ, Bryan GJ, Fenton B, Gilroy EM, Hein I, Jones JT, Prashar A, Taylor MA, Torrance L, Toth IK (2012) Crops that feed the world 8: potato: are the trends of increased global production sustainable? Food Secur 4:477–508. https://doi.org/10.1007/s12571-012-0220-1

Black RE, Victora CG, Walker SP, Bhutta ZA, Christian P, de Onis M, Ezzati M, Grantham-McGregor S, Katz J, Martorell R, Uauy R (2013) Maternal and child undernutrition and overweight in low-income and middle-income countries. Lancet 832(9890):427–451

Bohl WH, Johnson SB (2010) Commercial potato production in North America. The Potato Association of America Handbook. http://potatoassociation.org/wp-content/uploads/2014/04/A_ProductionHandbook_Final_000.pdf

Borch D, Juul-Hindsgaul N, Veller M, Astrup A, Jaskolowski J, Raben A (2016) Potatoes and risk of obesity, type 2 diabetes, and cardiovascular disease in apparently healthy adults: a systematic review of clinical intervention and observational studies. Am J Clin Nutr 104(2):489–498

Burlingame B, Mouillé B, Charrondiére UR (2009) Review: nutrients, bioactive non-nutrients and anti-nutrients in potatoes. J Food Compos Anal 22:494–502

CGIAR Research Program on Roots, Tubers and Bananas (2016) Multi-stakeholder framework for intervening in RTB seed systems: user's guide. Lima (Peru), 13 p. RTB Working Paper. ISSN 2309-6586, no. 2016-1

Creed-Kanashiro H, Hareau G, Devaux A, Maldonado L, Ordinola M, Fonseca C, Suarez V, Astete L, Marin M, Penny M (2015) Agriculture-nutrition linkages: analyzing nutritional outcomes of interventions in potato-based production systems of Peru. In: 2nd international conference on global food security, October 11–14, 2015. Cornell University, Ithaca

Crowder DW et al (2010) Organic agriculture promotes evenness and natural pest control. Nature 466:109–112

DeFauw Sherri L, He Z, Larkin RP, Mansour SA (2012) Sustainable potato production and global food security. In: He Z, Larkin R, Honeycutt W (eds) Sustainable potato production: global case studies. Springer, Amsterdam, 531 pp

Delgado L, Schuster M, Torero M (2017) The reality of food losses: a new measurement methodology. IFPRI discussion paper 1686. International Food Policy Research Institute (IFPRI), Washington, DC. http://ebrary.ifpri.org/cdm/ref/collection/p15738coll2/id/131530

Devaux A, Torero M, Donovan J, Horton DE (eds) (2016) Innovation for inclusive value-chain development: successes and challenges. International Food Policy Research Institute (IFPRI), Washington, DC. https://doi.org/10.2499/9780896292130

Devaux A, Goffart JP, Kromann P, Toth I, Braguard C, Declerck S (2017) Report on CIP-EAPR Workshop 2017 on biocontrol and biostimulant agents for the potato crop. Potato Res 60:291. https://doi.org/10.1007/s11540-018-9385-0

Devaux A, Torero M, Donovan J, Horton D (2018) Agricultural innovation and inclusive value-chain development: a review. J Agribus Dev Emerg Econ 8(1):99–123. https://doi.org/10.1108/JADEE-06-2017-0065

Douches DS, Maas D, Jastrzebski K, Chase RW (1996) Assessment of potato breeding progress in the USA over the last century. Crop Sci 36(6):1544–1552

Durroux-Malpartida V (2014) EAPR and CIP strengthen collaboration to tap potatoes' potential. Potato Res 57:367. https://doi.org/10.1007/s11540-014-9276-y

European Commission (2007). The potato sector in the European Union. Commission staff working document. SEC(2007) 533

Eurostat (2017) The EU potato sector—statistics on production, prices and trade. http://ec.europa.eu/eurostat/statistics-explained/index.php/The_EU_potato_sector_-_statistics_on_production,_prices_and_trade#Publications

Fajardo J, Lutaladio N, Larinde M, Rosell C, Barker I, Roca W et al (2010) Quality declared planting material: protocols and standards for vegetatively propagated crops: expert consultation, Lima, 27–29 November 2007. Food and Agriculture Organization of the United Nations, Rome

FAO (2002) The state of food insecurity in the world 2001, Rome, pp 4–7

FAO (2006a) Food security. Policy brief. FAO, Rome

FAO (2006b) Quality declared seed system. FAO plant protection and protection paper # 185, Rome

FAO (2009) International year of the potato 2008: new light on a hidden treasure. http://www.fao.org/potato-2008/en/events/book.html

FAO (2014) FAO statistical databases FAOSTAT. http://faostat3.fao.org/

FAO (2016) FAOSTAT Database. http://www.fao.org/faostat/en/#data/QC

FAO (2015a) Household seed security concepts and indicators. Discussion paper, 10 pp

FAO (2015b) FAOSTAT database. http://www.fao.org/faostat/en/#data/FBS

FAO (2016) Seed security assessment: a practitioner's guide, Rome

FAO, IFAD, UNICEF, WFP and WHO (2018) The state of food security and nutrition in the world 2018. Building climate resilience for food security and nutrition. FAO, Rome

FAOSTAT (2013) Food balance sheet. http://www.fao.org/faostat/en/#data/FBS

FAOSTAT (2017) Food and agriculture data. http://www.fao.org/faostat/en/#data/QCinfo

Foley JA et al (2011) Solution for a cultivated planet. Nature 478:337–342

Gatto M, Hareau G, Pradel W, Suarez V, Qin J (2018) Release and adoption of improved potato varieties in Southeast and South Asia. International Potato Center (CIP), Lima, Peru, 42 p. Social sciences working paper no. 2018-2. ISBN 978-92-9060-501-0

Gildemacher PR, Kaguongo W, Ortiz O, Tesfaye A, Woldegiorgis G, Wagoire WW, Kakuhenzire R, Kinyae P, Nyongesa M, Struik PC, Leeuwis C (2009) Improving potato production in Kenya, Uganda and Ethiopia: a system diagnosis. Potato Res 52(2):173–205

Gildemacher PR, Schulte-Geldermann E, Borus D, Demo P, Kinyae P, Mundia P et al (2011) Seed potato quality improvement through positive selection by smallholder farmers in Kenya. Potato Res 54:253–266

Gillespie S, Van den Bold M (2017) Agriculture, food systems, and nutrition: meeting the challenge. Global Chall 1:1600002. https://doi.org/10.1002/gch2.201600002

Goffart JP, Olivier M, Frankinet M (2008) Potato crop nitrogen status assessment to improve N fertilization management and efficiency: past–present–future. Potato Res 51(3):355–383. https://doi.org/10.1007/s11540-008-9118-x

Goffart JP, Gobin A, Delloye C, Curnel Y (2017) Crop spectral reflectance to support decision making on crop nutrition. Paper presented to the International Fertiliser Society at a conference in Cambridge, United Kingdom, on 7th December 2017. Proceedings 812, pp. 29. www.fertiliser-society.org © 2017 International Fertiliser Society—ISBN 978-0-85310-449-0

Haggblade S, Hazell PBR, Reardon T (2007) Research perspectives and prospectives on the rural nonfarm economy. In: Haggblade S, Hazell PBR, Reardon T (eds) Transforming the rural nonfarm sector: opportunities and threats in the developing world. The Johns Hopkins University Press, Baltimore

Hall A, Bockett G, Taylor S, Sivamohan MVK, Clark N (2001) Why research partnerships really matter: innovation theory, institutional arrangements and implications for developing new technology for the poor. World Dev 29(5):783–797

Harahagazwe D, Condori B, Barreda C, Bararyenya A, Byarugaba AA, Kude DA, Lung'aho C, Martinho C, Mbiri D, Nasona B, Ochieng B, Onditi J, Randrianaivoarivony JM, Tankou CM, Worku A, Schulte-Geldermann E, Mares V (CIP), de Mendiburu F, Quiroz R (CIP) (2018) How big is the potato (*Solanum tuberosum* L.) yield gap in Sub-Saharan Africa and why? A participatory approach. Open Agric 3(2):180-189. ISSN 2391-9531 .

Hareau G, Kleinwechter U, Pradel W, Suarez V, Okello J, Vikraman S (2014) Strategic assessment of research priorities for Potato. Lima (Peru). CGIAR Research Program on Roots, Tubers and Bananas (RTB). RTB working paper 2014-8. www.rtb.cgiar.org

Haverkort AJ, Struik PC (2015) Yield levels of potato crops: recent achievements and future prospects. Field Corp Res 182:76–85

Haverkort A, de Ruijter FJ, van Evert FK, Conijn JG, Rutgers B (2014) Worldwide sustainability hotspots in potato cultivation. 1. Identification and mapping. Potato Res 56:343–353

Hazell P, Haggblade S (1989) Agricultural technology and farm–nonfarm growth linkages. Agric Econ 3:345–364

Headey DD, Jayne TS (2014) Adaptation to land constraints: is Africa different? Food Policy 48:18–33

INEI Peru (2017) IV Censo Nacional Agropecuario (CENAGRO). Lima. INEI Perú (2017). Encuesta demográfica y de salud familiar 2016. Lima, Peru

Jayne TS, Chamberlin J, Headey DD (2014) Land pressures, the evolution of farming systems, and development strategies in Africa: a synthesis. Food Policy *48*:1–17

Kanter R, Walls HL, Tak M, Roberts F, Waage J (2015) A conceptual framework for understanding the impacts of agriculture and food system policies on nutrition and health. Food Secur 7(4):767–777

Kleinwechter U, Hareau G, Suarez V (2014) Prioritization of options for potato research for development—results from a global expert survey. Lima (Peru). CGIAR Research Program on Roots, Tubers and Bananas (RTB). RTB working paper 2014-7. www.rtb.cgiar.org

Kunkel R, Campbell GS (1987) Maximum potential potato yield in the Columbia Basin, USA: model and measured values. Am Potato J 64(7):355

Labarta R (2015) The effectiveness of potato and sweetpotato improvement programmes from the perspectives of varietal output and adoption in Sub-Saharan Africa. In: Walker T, Alwang J (eds) Crop improvement, adoption, and impact of improved varieties in food crops in Sub-Saharan Africa. CABI, Wallingford

Muthoni J, Mbiyu MW, Nyamongo DO (2010) A review of potato seed systems and germplasm conservation in Kenya. J Agric Food Inform 11(2):157–167

Obura B, et al (2016) Cost benefit analysis of seed potato replacement strategies among smallholder farmers in Kenya. Tropentag, September 18–21, 2016, Vienna, Austria. http://www.tropentag.de/2016/abstracts/links/Obura_eAhVYqsS.pdf. Accessed 7 Sept 2018

Olsen N (2014) Potato storage management: a global perspective. Potato Res 57:331–333. https://doi.org/10.1007/s11540-015-9283-7

Pandey A (2015) Rural farm and non-farm linkages in Uttar Pradesh. J Land Rural Stud *3*(2):203–218

PGSC (The Potato Genome Sequencing Consortium) (2011) Genome sequence and analysis of the tuber crop potato. Nature 475:189–195

Pingali P (2006) Westernization of Asian diets and the transformation of food systems: implications for research and policy. Food Policy 32(3):281–298

Pinstrup-Andersen P (2014) Food systems and human nutrition: relationships and policy interventions. In: Thompson B, Amoroso L (eds) Improving diets and nutrition: food-based approaches (Chapter 2). CABI, Wallingford, pp 8–20

Quiroz R, Ramírez DA, Kroschel J, Andrade-Piedra J, Barreda C, Condori B et al (2018) Impact of climate change on the potato crop and biodiversity in its center of origin. Open Agric 3:273–283

Reardon T, Chen K, Minten B, Adriano L (2012) The quiet revolution in staple food value chains: enter the dragon, the elephant and the tiger. Asian Development Bank (ADB) and International Food Policy Research Institute (IFPRI), Mandaluyong City

Rosegrant MW, Sulser TB, Mason-D'Croz D, Cenacchi N, Nin-Pratt A, Dunston S, Zhu T, Ringler C, Wiebe K, Robinson S, Willenbockel D, Xie H, Kwon H-Y, Thomas TS, Wimmer F, Schaldach R, Nelson GC, Willaarts B (2017) Quantitative foresight modeling to inform the CGIAR research portfolio. Project report for USAID. International Food Policy Research Institute (IFPRI), Washington, DC

Salazar L, Aramburu J, González M, Winters P (2016) Food security and productivity: impacts of technology adoption in small subsistence farmers in Bolivia. Food Policy 65:32–52

Schulte-Geldermann E, Gildemacher PR, Struik PC (2012) Improving seed health and seed performance by positive selection in three Kenyan potato varieties. Am J Potato Res 89:429–437. https://link.springer.com/content/pdf/10.1007%2Fs12230-012-9264-1.pdf. Accessed 13 Sept 2018

Scott G (2011) Growth rates for potatoes in Latin America in comparative perspective: 1961–07. Am J Potato Res 88:143–152

Scott GJ, Suarez V (2011) Growth rates for potato in India and their implications for industry. Potato J 38(2):100–112

Scott GJ, Suarez V (2012a) From Mao to McDonald's: emerging markets for potatoes and potato products in China 1961-2007. Am J Potato Res 89(3):216–231

Scott GJ, Suarez V (2012b) The rise of Asia as the center of global potato production and some implications for industry. Potato J 39(1):1–22

Scott GJ, Suarez V (2012c) Limits to growth or growth to the limits? Trends and projections for potatoes in China and their implications for industry. Potato Res 55(2):135–156

Smith A, Snapp S, Chikowoa R, Thorne P, Bekunda B, Gloverd J (2017) Measuring sustainable intensification in smallholder agroecosystems: a review. Glob Food Sec 12:127–138

The Wall Street Journal (2015) Pushing the potato: China wants people to eat more 'Earth Beans'. https://www.wsj.com/amp/articles/BL-CJB-25584?responsive=y

Thiele G, Hareau G, Suarez V, Chujoy E, Bonierbale M, Maldonado L (2008) Varietal change in potatoes in developing countries and the contribution of the International Potato Center: 1972–2007. International Potato Center (CIP), Lima, Peru. Working paper 2008-6, 46 p

Thomas-Sharma S, Abdurahman A, Ali S, Andrade-Piedra JL, Bao S, Charkowski AO et al (2015) Seed degeneration in potato: the need for an integrated seed health strategy to mitigate the problem in developing countries. Plant Pathol 65:3–16. https://www.researchgate.net/publication/280913576_Seed_degeneration_in_potato_The_need_for_an_integrated_seed_health_strategy_to_mitigate_the_problem_in_developing_countries. Accessed 13 Sept 2018

Velivelli SLS, Sessitsch A, Prestwich BD (2014) The role of microbial inoculants in integrated crop management systems. Potato Res 57(3–4):291–309

Von Grebmer K et al (2017) Global hunger index: getting to zero hunger. Welthungerhilfe/International Food Policy Research Institute/Concern Worldwide, Bonn/Washington, DC/Dublin. https://doi.org/10.2499/9780896292710

World Bank (2017) Gaining momentum in Peruvian agriculture: opportunities to increase productivity and enhance competitiveness. World Bank, Washington, DC

Wu W, Yu Q, You L, Chen K, Tang H, Liu J (2018) Global cropping intensity gaps: increasing food production without cropland expansion. Land Use Policy *76*:515–525

You L, Wood-Sichra U, Fritz S, Guo Z, See L, Koo J (2014) Spatial Production Allocation Model (SPAM) 2005 v3.2. http://mapspam.info

2

Human Health and the Importance of the Potato

Gabriela Burgos, Thomas Zum Felde, Christelle Andre, and Stan Kubow

Abstract Potato has contributed to human diet for thousands of years, first in the Andes of South America and then in the rest of the world. Its contribution to the human diet is affected by cooking, potato intake levels, and the bioavailability of potato nutrients. Generally, the key nutrients found in potatoes including minerals, proteins, and dietary fiber are well retained after cooking. Vitamins C and B_6 are significantly reduced after cooking while carotenoids and anthocyanins show high recoveries after cooking due to an improved release of these antioxidants.

In many developed countries potatoes are consumed as a vegetable with intakes that vary from 50 to 150 g per day for adults. On the other hand, in some rural areas of Africa and in the highlands of Latin American countries, potato is considered a staple crop and is consumed in large quantities with intakes that vary from 300 to 800 g per day for adults. These marked differences in the potato intake affect significantly the contribution of potato nutrients to the human dietary requirements.

In recent years, information about nutrient bioaccessibility and bioavailability from potatoes has become available indicating higher bioaccessibility of minerals and vitamins in potato as compared with other staple crops such as beans or wheat. Bioavailability refers to the fraction of an ingested nutrient that is available for utilization in normal physiological functions and/or for body storage while bioaccessibility refers to the amount that is potentially absorbable from the gut lumen.

In addition, potatoes have shown promising health-promoting properties in human cell culture, experimental animal and human clinical studies, including anticancer, hypocholesterolemic, anti-inflammatory, anti-obesity, and antidiabetic

G. Burgos (✉) · T. Zum Felde
International Potato Center, Lima, Peru
e-mail: g.burgos@cgiar.org; t.zumfelde@cgiar.org

C. Andre
The New Zealand Institute for Plant and Food Research Limited/Luxembourg Institute of Science and Technology, Auckland, New Zealand
e-mail: christelle.andre@plantandfood.co.nz

S. Kubow
McGill University, Quebec, Canada
e-mail: stan.kubow@mcgill.ca

properties with phenolics, anthocyanins, fiber, resistant starch, carotenoids as well as glycoalkaloids contributing to the health benefits of potatoes.

2.1 Introduction

Diverse studies have demonstrated that potato is an important source of carbohydrates, resistant starch, quality proteins, vitamins C and B_6 as well as potassium (Camire et al. 2009). Potato is also a source of antioxidants that can contribute to prevent both degenerative and age-related diseases with lutein and zeaxanthin being present in high levels in yellow-fleshed potatoes (Burgos et al. 2009) and anthocyanins being present in purple and red-fleshed potato landraces (Burgos et al. 2013b) commonly grown and eaten in the Andean highlands of Peru, Bolivia, Ecuador, and Colombia. Potatoes also contain glycoalkaloids, which in high concentrations can be toxic to humans but in low concentrations can have beneficial effects such as inhibition of the growth of cancer cells (Friedman 2015). The nutritional composition of potatoes is summarized in Fig. 2.1. The concentration of energy, starch, protein, lipids, dietary fiber, potassium, phosphorus, magnesium, iron, zinc, vitamin C, vitamin B_6, chlorogenic acid, and glycoalkaloids has a range of variation independent from the flesh color. Yellow-fleshed potatoes have a carotenoid concentration higher than white-fleshed potatoes while purple potatoes have a higher anthocyanin concentration than red- or white-fleshed potatoes.

Like other plant foods, the nutritional composition of potatoes is affected by different pre-harvest (environment, cultural practices, maturity at harvest, biotic and abiotic stresses, etc.) and post-harvest (processing, storage, transport, etc.) conditions.

Potato has contributed to the human diet for thousands of years, first in the Andean region and then in the rest of the world. Its contribution is affected by cooking, the amount of potato intake, and the bioavailability of the nutrients. Generally, the key phytonutrients found in potatoes including minerals, proteins, and dietary fibers are well retained after cooking. Vitamins C and B_6 are significantly reduced after cooking while carotenoids and anthocyanins show high recovery after cooking due to an improved release of these antioxidants from the food matrix after cooking (Tian et al. 2016). In this chapter, the range of nutrient concentrations is expressed on a fresh weight (FW) basis and ranges refer to both raw and cooked potatoes. However, for calculating their contribution to the diet, only values of cooked potatoes are considered.

The worldwide mean potato intake is equivalent to 93 g per day (FAO 2013). However, this value has a large range of variation. In many developed countries potatoes are consumed as a vegetable and served as a part of a larger meal with intakes that vary from 50 to 150 g per day for adults. On the other hand, in some rural areas of Africa and in the highlands of Latin American countries, potato is considered a staple crop and consumed alone in large quantities as a complete meal with intakes that vary from 300 to 800 g per day for adults (De Haan et al. 2019).

Energy:	96 to 123 Kcal	Magnesium:	16 to 40 mg
Starch:	16 to 20 g	Iron:	0.29 to 0.69 mg
Protein:	1.76 to 2.95 g	Zinc:	0.29 to 0.48 mg
Lipids:	0.1 to 0.5 g	Vitamin C:	7.8 to 20.6 mg
Dietary fiber:	1.8 g to 2.1 g	Vitamin B6:	0.299 mg
Potassium:	150 to 1386 mg	Chlorogenic acid:	19 to 399 mg
Phosphorus:	42 to 120 mg	Glycoalkaloids:	0.7 to 18.7 mg

Fig. 2.1 Nutritional composition of potatoes per 100 g FW

Implications of the contribution of potatoes to the human diet as related to the magnitude of potato intake are also described in this chapter.

Bioavailability refers to the fraction of an ingested nutrient that is available for utilization in normal physiological functions and/or for its contribution towards body stores (La Frano et al. 2014). Many factors affect the bioavailability of a compound; these may be divided into exogenous factors such as the complexity of the food matrix, the chemical form of the compound of interest, structure and amount of co-ingested compounds as well as endogenous factors including mucosal mass, intestinal transit time, rate of gastric emptying, intestinal and hepatic metabolism, and the extent of conjugation and protein-binding in blood and tissues (Holst and

Williamson 2008). A prerequisite for bioavailability of any compound is its bioaccessibility in the gut, defined as the amount that is potentially absorbable from the lumen (Fernández-Garcia et al. 2009). Bioavailability can be limited by a low bioaccessibility, which can be affected by the nature of the food matrix, location within the plant, food processing, gastric and luminal digestion, in addition to the physicochemical properties of the compound itself. In this chapter, the bioaccessibility and bioavailability of phytonutrients in potato will be reported and discussed.

The contribution of potato to human health will be described in terms of the evidence concerning the anticancer, hypocholesterolemic, anti-inflammatory, anti-obesity, and anti-diabetic role of potatoes.

2.2 Contribution to Diet

2.2.1 Energy

The energy provided by 100 g of boiled tubers of potatoes varies from 96.33 to 123.17 kcal (De Haan et al. 2019), which is similar to the energy provided by 100 g of cooked rice (130 kcal) but lower than the energy provided by 100 g of wheat (361 kcal), 100 g of cooked cassava (160 kcal) and soybeans (173 kcal) (King and Slavin 2013). Potato has a low energy density with 100 g of boiled potatoes contributing between 4 and 6% of the requirement of energy of an adult of between 50 and 90 kg of weight (considering 1.90 as basal metabolic rate factor, FAO/OMS/UNU 2004). However, preparing and serving potatoes with ingredients with a high fat content raises greatly the caloric value of the dish. One hundred grams of potato chips and French fries provide 529 and 564 kcal, respectively.

In areas where potato is considered as a staple food, the amount of potato intake is high and consequently the contribution of potatoes towards meeting dietary requirements is much higher. In Huancavelica, a location in the Peruvian central highlands, women have an average daily consumption of 840 and 645 g during abundance and scarcity period of potato, respectively. In those regions, potatoes are mainly eaten as boiled and provide between 28 and 38% of the recommended total energy requirements for women (De Haan et al. 2019).

2.2.2 Carbohydrates

2.2.2.1 Starch

Starch is the predominating carbohydrate in potato ranging from 16.5 to 20.0 g/100 g FW (Liu et al. 2007). Biochemically, potato starch is composed of amylose and amylopectin with the latter molecule typically making up 70–80% of the available starch in the tuber and the remaining portion being composed of amylose (Zeeman et al. 2010). Starch can also be classified by levels of digestibility within the human intestinal tract, i.e. rapidly digested (RDS), slowly digested (SDS), or resistant (RS) starch (Englyst et al. 1992). RDS and SDS represent the portion of starch digested within the first 20 and 21–120 min post-ingestion, respectively. The remaining resistant starch (RS) is undigested and fermented when it reaches the large intestine with the production of short-chain fatty acids (Raigond et al. 2014). Because of the resistance of the amylose structure to digestion, more of the RS component is expected to be composed of amylose rather than amylopectin. The rapid breakdown of amylopectin to digestion is the reason that it is more prevalent in RDS and SDS fractions (Bach et al. 2013). Potential health benefits attributed to SDS include satiety, improved physical performance, glucose tolerance enhancement and blood lipid level reduction in healthy individuals and in those with hyperlipidemia (Miao et al. 2015). Possible health benefits of RS include prevention of colon cancer, hypoglycemic effects, substrate provision for growth of gut probiotic microorganisms, reduction of gall stone formation, hypocholesterolemic effects, inhibition of fat accumulation, and increased absorption of minerals (Sajilata et al. 2006).

Monro et al. (2009) determined the RDS, SDS, and RS concentration of freshly cooked potatoes from nine potato varieties and found concentrations ranging from 9 to 15 g/100 g FW, from 0 to 1.72 g/100 g FW, and from 0.58 to 1.05 g/100 g FW, respectively. These authors also found that cooking and then cooling potatoes significantly increased SDS and RS (up to 7.7 g and 1.96 g/100 g FW, respectively), while RDS was significantly reduced to 7.3 g/100 g FW. This latter phenomenon is referred to as starch retrogradation, which is based upon rearrangement of the molecules of amylose and amylopectin causing decreased starch digestion (Leeman et al. 2005).

Glycemic index (GI) is a measure of the extent of the change in blood glucose content (glycemic response) following consumption of digestible carbohydrate, relative to a standard such as glucose (Venn and Green 2007). A higher GI value represents a more rapid entry of a larger quantity of glucose from a test food into the bloodstream. Based on in vivo postprandial GI, high RDS content in foods has been significantly correlated with a high glycemic index response (Champ 2004), whereas low RDS levels are associated with low to medium GI values (Lynch et al. 2007). A wide variability in GI values of potatoes has been noted ranging from high to medium to low values based on cultivar differences (Ek et al. 2012). Such variations could partly be related to differences in the amylopectin to amylose ratio as amylose-rich starches are digested more slowly due to their difficulty to gelatinize and swell as opposed to starches with a high amylopectin content (Brennan 2005). Bach et al. (2013) defined low RDS and high SDS as the optimal profile for potatoes that leads to low GI values, and identified two genotypes with this profile. Tuber cooking followed by cooling (forming retrograded starch) is also another way for the consumer to obtain lower postprandial glucose levels, and thereby benefit from reduced GI following potato intake (Fernandes et al. 2005). Lowering the dietary GI load has been associated with body weight loss, improved blood pressure, and decreased risk of cardiovascular diseases, whereas habitual intake of high GI foods has been linked to type 2 diabetes and other chronic heart issues (McGill et al. 2013). As the GI does not take into account the typical portion size, the GI value and the quantity of carbohydrates are combined to generate the glycemic load (GL) value, which can better quantify the glycemic impact of a food (Salmeron et al. 1997). Initial studies involving potatoes were limited by the sole use of GI for their glycemic evaluation (Crapo et al. 1977; Soh and Brand-Miller 1999), which categorized them with a high GI. In contrast, potatoes have been generally noted to have a medium to low glycemic impact based on the GL estimation (Lynch et al. 2007).

2.2.2.2 Sugars

Potato tubers also contain significant quantities of free sugars with glucose and fructose as the principal monosaccharides and sucrose as the major disaccharide. Glucose, fructose, and sucrose concentrations in raw tubers of tetraploid potatoes range from 3.25 to 255 mg/100 g FW, from 2.5 to 153.7 mg/100 g FW, and from 43 to 159.7 mg/100 g FW, respectively (Amrein et al. 2003; Rodríguez et al. 2010). Higher levels of glucose, fructose, and sucrose have been recently reported for diploid potato group *Phureja* with concentrations ranging from 11.5 to 701 mg/100 g FW for glucose, from 7.25 to 605 mg/100 g FW for fructose and from 159 to 737 mg/100 g FW for sucrose (Duarte-Delgado et al. 2016). The reducing sugars glucose and fructose as well as free asparagine are acrylamide precursors. Acrylamide is formed through the Maillard reaction during high temperature cooking such as frying, roasting, or baking (Muttucumaru et al. 2008). Acrylamide has been classified as "probably carcinogenic to humans" by the WHO and the International Agency for Research on Cancer. Therefore, the reducing sugar content in potatoes has been recommended not to be greater than 100 mg/100 g FW in order to keep acrylamide formation on a low level (Kumar et al. 2004). Importantly, cold storage (2–4 °C) may induce an accumulation of reducing sugars in tuber tissue leading to undesirable browning, production of bitter flavors, and increased levels of acrylamide with cooking (Neilson et al. 2017).

2.2.3 Protein

According to Camire et al. (2009), the protein content of potatoes generally ranges from 1 to 1.5 g/100 g FW depending on the cultivar. De Haan et al. (2019) reported higher levels of protein in cooked tubers of Peruvian floury landraces (1.76–2.95 g/100 g FW). Potato protein content is generally low compared with other major staples like maize and beans although potato yields more protein per unit growing area than do cereals (Bamberg and Del Rio 2005). Also, the quality of the potato protein, which reflects its digestibility and indispensable amino acid content, is very good. The biological value of potato protein—the proportion retained for growth or maintenance divided by the amount absorbed—is high. Depending on the cultivar, the biological value of potato protein is between 90 and 100 and is very similar to the biological value of whole egg protein (100) and is higher than that of soybeans (84) and legumes (73) (Camire et al. 2009).

The levels of lysine, methionine, threonine, and tryptophan are likely to limit the protein quality of mixed diets consumed by humans. Potatoes exceed the recommended levels of these indispensable amino acids, demonstrating that potato protein is of high quality. Compared with pasta, white rice, and whole grain cornmeal, potatoes are the only staple food meeting the recommended lysine level. However, sulfur-containing amino acids (methionine + cysteine) are lower in potatoes than in the other common plant staple foods (King and Slavin 2013).

2.2.4 Lipids

Total lipids of potatoes are low and range from 0.1 to 0.5 g/100 g FW and consist mainly of phospholipids (47%), glycol and galactolipids (22%), which are structural elements of biological membranes as well as neutral lipids (21%) such as acylglycerols and free fatty acids (Ramadan and Oraby 2016). More than 94% of the tuber lipids contain esterified fatty acids. The essential polyunsaturated fatty acids with one to three double bounds consist of mainly linoleic acid (C18:2 cis-9,12, an *n*-6 fatty acid) and linolenic acid (C18:3 cis-9,12,15, an *n*-3 fatty acid) (70–75%), precursors of a wide range of bioactive compounds generated endogenously (Galliard 1973). The composition of the fatty acids of the potato lipids is nutritionally advantageous. For example, potato consumption in the United Kingdom was estimated to provide 10 and 13% of the dietary *n*-6 and *n*-3 polyunsaturated fatty acid intake, respectively (Gibson and Kurilich 2013). In contrast, potato intake provided only 4% of saturated fatty acid and 6% trans fatty acid intake, which was largely attributed to the addition of fats and oils such as butter and margarine to potato dishes.

2.2.5 Fiber

Dietary fiber represents the undigested and unabsorbed carbohydrate part in the diet. These resistant carbohydrates may be fermented in the large intestine. Soluble fibers lower serum lipids, whereas insoluble fibers increase stool weight (Slavin 2008). Potatoes contain dietary fiber in their cell walls, especially in the thickened cell walls of the peel (Camire et al. 2009). Cooked potatoes without the skin provide 1.8 g fiber/100 g, FW, whereas cooked potatoes with the skin provide 2.1 g fiber/100 g FW. Potatoes contain less fiber than whole-grain cornmeal (7.3 g/100 g), but more fiber than white rice (0.3 g/100 g) or whole-wheat cereal (1.6 g/100 g). Although potatoes cannot therefore be considered a high-fiber food, they can be a significant source of fiber for individuals regularly eating potatoes, particularly in developed countries where fiber intake is generally far below recommended levels (Auestad et al. 2015). In that regard, potatoes have been indicated to contribute 14.4–26.2% of daily fiber intake in men and women living in the USA based on the National Health and Nutrition Examination Survey (NHANES) data (2009–2010).

2.2.6 Minerals

Potassium is the most abundant mineral in potato with concentrations varying from 150 to 1386 mg/100 g FW (Nassar et al. 2012). Potassium functions as an important electrolyte in the nervous system. High intake levels of potassium can help control high blood pressure and may decrease the risk of stroke (Bethke and Jansky 2018). One hundred grams of boiled potatoes can contribute up to 16% of the Adequate Intake (AI) of potassium recommended for adults (4700 mg per day).

Phosphorus and magnesium are also present in potato in moderate quantities ranging from 42 to 120 mg/100 g FW and from 16 to 40 mg/100 g FW, respectively (Bonierbale et al. 2010). One hundred grams of boiled potatoes can contribute up to 11% of the Estimated Average Requirement (EAR) of phosphorus and magnesium for adults (42–120 and 265–340 mg per day, respectively). Calcium is present in minor quantities in potato ranging from 2 to 20 mg/100 g FW; contributing no more of 2% of the EAR of calcium for adults (800–1100 mg per day).

Iron and zinc concentrations from raw potatoes range from 0.25 to 0.83 mg/100 g FW and from 0.23 to 0.39 mg/100 g FW, respectively (Burgos et al. 2007). Iron and zinc concentrations are significantly affected by the growing environment. Interestingly, Lombardo et al. (2013) reported that soil composition affects the mineral concentration of crops, with a sandy texture of the soil favoring the iron oxidation processes to insoluble polymers and consequently reducing iron availability to the plant.

Burgos et al. (2007) reported iron and zinc concentration in cooked potatoes ranging from 0.29 to 0.69 mg/100 g FW and from 0.29 to 0.48 mg/100 g FW, respectively. These values are lower than iron and zinc concentrations reported for cereals and legumes but bioavailability of iron and zinc from potatoes may be higher due to the presence of high levels of ascorbic acid—which facilitates iron absorption in the human body—and low levels of phytic acid, an inhibitor of iron and zinc absorption. It has been recently demonstrated that the bioaccessibility of iron in potato is higher than that reported in crops such as wheat and beans. Approximately 63–79% of the potato iron is released from the food matrix after in vitro gastrointestinal digestion, and therefore available for intestinal absorption (Andre et al. 2015).

In the Andean highlands, where there is little access to meat and the levels of anemia and malnutrition are high, potatoes are an important dietary source of iron due to their high consumption. For example, in Huancavelica, in the Peruvian highlands, women and children consume on average 840 and 200 g of potato per day, respectively (De Haan et al. 2019). Similarly, in parts of Rwanda and other African countries, women consume an average of 400 g of potatoes per day. Therefore, improving the iron and zinc concentrations of potato and their bioavailability would have a real impact to contribute to reduce malnutrition and improve life quality in these and other areas where anemia and/or stunted growth are still pervasive.

The International Potato Center (CIP) has been working for the past 15 years on potato mineral biofortification to increase the concentration of iron and zinc in this crop. The CIP Biofortification Potato Program started from a baseline of 0.48 mg/100 g FW for iron and 0.35 mg/100 g FW for zinc. After three cycles of breeding and recurrent selection, concentrations of the first biofortified potatoes reach 0.73 mg iron and 0.63 mg zinc/100 g FW. Considering 400 g of potato consumption for women of the Andes, the consumption of biofortified potatoes would cover 41 and 37% of the EAR of iron and zinc in women.

Presently, CIP is combining the first products of its biofortification breeding program with advanced breeding lines to release new varieties that will be able to withstand major potato pests and diseases, tolerate heat and drought, providing high yields, and respond to preferences of farmers and consumers.

2.2.7 Vitamins

Potatoes are a good source of ascorbic acid (vitamin C) and pyridoxine (vitamin B_6). Vitamin C as an antioxidant plays an important role in protection against oxidative stress. Vitamin C is an important free radical scavenger of reactive oxygen species such as hydroxyl radicals, superoxide anions, singlet oxygen, and hydrogen peroxide that can cause tissue damage resulting from lipid peroxidation, DNA breakage or base alterations, which may contribute to degenerative diseases such as heart disease or cancer (Bates 1997). In addition, due to its participation in the oxidation of transition metal ions, vitamin C also plays an important role in enhancing the bioavailability of non-haem iron (Teucher et al. 2004) and serves as a cofactor in the synthesis of collagen needed to support cardiovascular function, maintenance of cartilage, bones, and teeth, as well as wound healing (Naidu 2003).

Fresh potatoes have varying concentrations of vitamin C, which can reach 50 mg/100 g FW (Han et al. 2004) when they are freshly harvested. Significant variation in vitamin C concentrations of potatoes occur due to genotype and environment and genotype by environment interactions (Andre et al. 2007; Burgos et al. 2009).

Cooking and storage reduce the concentration of vitamin C in potato tubers. In addition, there are differences in the degree of reduction of vitamin C content depending on the cooking types. Retention levels of vitamin C after boiling in 20 native landraces varied between 50 and 90%. The losses may be caused by: (1) leaching into cooking water, (2) destruction by heat treatment, and (3) oxidation. It is interesting to note that the peel forms a barrier preventing loss of nutrients during cooking. As a consequence, boiling potato when it is peeled results in 10% more loss of vitamin C or phenolic compounds than if it is cooked with the peel (Woolfe and Poats 1987). Retention levels after storing under farming conditions has been shown to vary between 22 and 62%, depending on the variety (Burgos et al. 2009). Retention levels of vitamin C in 12 genotypes grown in Colorado state in the USA after 7 months of cold storage was less than 50% (Külen et al. 2012).

One hundred grams of cooked potatoes with vitamin C levels around 20 mg/100 g FW can provide between 27 and 33% of the EAR of vitamin C for an adult (75 for males and 60 for females, according to FAO/WHO 2001). One hundred grams of cooked potatoes contains lower concentrations of vitamin C than 100 g of cooked broccoli (68–108 mg/100 g FW; depending on the way of cooking; Yuan et al. 2009), 100 g of cooked spinach (44–79 mg/100 g FW; depending on the way of cooking; Zeng 2013) and 100 g of raw red pepper (up to 200 mg/100 g; Wahyuni et al. 2011). However, it is noteworthy that the final contribution of a particular food to the total intake of vitamin C depends on the total amount consumed in the diet and so potatoes may therefore contribute to a significant extent to the total dietary intake of vitamin C (Love and Pavek 2008). In that respect, potatoes have been estimated to provide over 50% of the daily vitamin C requirement in the USA and approximately 20% of the dietary vitamin C intake in Europe (Love and Pavek 2008).

Vitamin B_6, also called pyridoxine, is a versatile cofactor for key metabolic processes (Hellmann and Mooney 2010) that plays a major role in various cellular reactions and also confers several health benefits for humans, which may be partly attributed to its antioxidant capabilities (Fitzpatrick et al. 2012). It helps in maintaining normal nerve function and plays a crucial role in the synthesis of neurotransmitters such as dopamine and serotonin. Vitamin B_6 also assists normal nerve cell communication and acts as a coenzyme in the breakdown and utilization of carbohydrates, fats and proteins. In plant, it is a potent antioxidant, critical for plant pathogen resistance (Spinneker et al. 2007).

Potatoes are considered to be a good dietary source of vitamin B_6, with concentrations ranging from 0.450 mg/100 g FW to 0.675 mg/100 g FW (Moonney et al. 2013). Physical and chemical factors such as heat, light exposure, and pH also influence vitamin B_6 content, but this vitamin is relatively stable during storage (Fitzpatrick et al. 2012).

The mean concentration of vitamin B_6 in cooked potatoes (0.299 mg/100 g FW) is higher than the mean concentration of other staple crops such as maize (0.139 mg/100 g FW), rice (0.050 mg/100 g FW), cassava (0.051 mg/100 g FW), and wheat (0.034 mg/100 g FW) (Fudge et al. 2017). One hundred grams of cooked potatoes can provide between 17 and 23% of the Recommended Dietary Allowance (RDA) of B_6 from an adult (1.3–1.7 mg per day).

Potato tuber contains also moderate amount of vitamin E (Chitchumroonchokchai et al. 2017). Vitamin E is the collective name for a set of eight related tocopherols and tocotrienols, characterized by a hydrophobic isoprenoid tail and a more hydrophilic chromanol head (Bramley et al. 2000). In potato, significant amount of α-tocopherol has been found in raw tubers of Andean genotypes, ranging from 68 to 517.5 μg/100 g FW (recalculated from Andre et al. 2007), whereas amounts in commercial varieties varied between 15 and 75 μg/100 g FW (recalculated from Andre et al. 2007 and Chun et al. 2006).

In humans, as in plants, vitamin E is located primarily within the phospholipid bilayer of cell membranes. It reacts with and quenches free radicals in cell membranes, preventing polyunsaturated fatty acids from damage by lipid oxidation. Vitamin E deficiency has been associated with an elevated risk of artherosclerosis and other degenerative diseases. It is generally assumed that increases of α-tocopherol in the diet may contribute to a decreased risk of chronic diseases (Andre et al. 2010). The EAR for vitamin E is of 15 mg for women and men (Otten et al. 2010). The consumption of high vitamin E containing potato tubers, such as the Andean varieties, could therefore significantly increase the dietary vitamin E intake.

2.2.8 *Antioxidants*

Potato is one of the most important sources of antioxidants in the human diet (Lachman and Hamouz 2005). As such, it supports the antioxidant defense that reduces cellular and tissue toxicities that result from free radical-induced protein, lipid, carbohydrate, and DNA damage (Andre et al. 2010). In this way, potato antioxidants may reduce the risk for cancers, cardiovascular diseases, and type 2 diabetes.

Based on metabolic relationships and structural composition, there are three major groups of antioxidants present in potato, as in most plants. The first group consists in the aromatic phenolic compounds, which encompasses flavonoids including anthocyanins and flavonols produced by the flavonoid pathway, hydroxycinnamic acids and their derivatives produced by the phenylpropanoid pathway, and the amino acids tyrosine, phenylalanine, and tryptophan produced by the shikimate pathway. The second group encompasses the isoprenoid antioxidants such as the carotenoids and tocopherols; and the third group includes antioxidants related to ascorbate and glutathione functions in a redox system of compound-recycling that include ascorbic acid (Lovat et al. 2016).

2.2.9 *Phenolics*

Phenolic compounds, also known as polyphenols, constitute one of the most widely distributed group of dietary antioxidants in the plant kingdom, presenting more than 10,000 different structures, ranging from relatively simple phenols to complex polymers such as lignans and suberins. Phenolic compounds are produced in the cytoplasm and are subsequently transported in the vacuole or deposited in the cell wall. Routes to the major classes of phenolic compounds involve: (1) the core phenylpropanoid pathway from phenylalanine to an activated (hydroxy)cinnamic acid derivative, as well as specific branch pathways for the formation of (2) simple phenolic acids, lignins and lignans, (3) flavonoids, (4) tannins, and (5) stilbenes (Andre et al. 2009). Their aromatic cycles can be further modified through hydroxylations, methylations, glycosylations, acylations, or prenylations, extending their variability and complexity (Winkel-Shirley 2001). Phenolic acids include chlorogenic, caffeic, ferulic, and sinapic acids. Among flavonoids, anthocyanins are natural pigments, responsible for the red-blue color of many fruits and vegetables. Anthocyanins can also impact the organoleptic characteristics of foods, which may influence their technological behavior during food processing and also have implications in the field of human health (Pascual-Teresa and Sanchez-Ballesta 2008). Flavonols represent one of the most widespread flavonoid classes in plant and include compounds like quercetin and kaempferol that are most commonly found in their glycosylated form, i.e., linked with glucose or rutinose. As compared to other phenolic compounds, flavonol concentrations are known to be largely influenced by the environmental conditions during plant growth (Lancaster et al. 2000).

Phenolic compounds are considered to be health-promoting phytochemicals as they have shown in vitro antioxidant activity and have been reported to exhibit beneficial anti-bacterial, hypoglycemic, anti-viral, anti-carcinogenic, anti-inflammatory and vasodilatory properties (Duthie et al. 2000; Mattila and Hellstrom 2006).

2.2.9.1 Chlorogenic Acid

Chlorogenic acid has been reported as the predominant phenolic acid in raw and boiled potato tubers (Burgos et al. 2013b). The main function of chlorogenic acid in the plant appears to defend against pathogens. Concentrations of chlorogenic acid as well as other hydroxycinnamic acids are significantly induced following pathogen invasion, and deposited to enforce the cell walls to arrest pathogen development (Yogendra et al. 2015). In humans, these compounds consumed through diet are increasingly considered as effective protective agents against reactive oxygen species (ROS), which are known to be involved in aging and many degenerative diseases (Liang and Kitts 2015).

The isomers of chlorogenic acid, neo-chlorogenic acid, and crypto-chlorogenic acid, as well as caffeic acid are also found in potato tubers (Andre et al. 2007). Potatoes contain three isomers of chlorogenic acid depending on whether the hydroxycinnamate is attached to 3-, 4-, or 5-position of the quinic acid moiety with 5-*O*-caffeoylquinic acid as the principal chlorogenic acid. The 5-*O*-caffeoylquinic acid (CQA) isomer is also the principal chlorogenic acid component of coffee and apples (Stalmach et al. 2010; Clifford 1999). In vitro and ex vivo studies have demonstrated a reduction in oxidation of human LDL following the consumption of coffee suggesting that 5-*O*-CQA protects against in vitro oxidation of human LDL, a key step in the formation of atherosclerotic plaques (Natella et al. 2007; Richelle et al. 2001). 5-*O*-CQA has also been shown to exert anti-carcinogenic effects in animal models (Stalmach et al. 2010).

Lachman et al. (2013) have reported chlorogenic acid concentrations of raw potatoes ranging from 7.87 to 60.07 mg/100 g FW in nonpeeled potatoes and from 5.11 to 46.13 mg/100 g FW in their peeled counterparts, while Burgos et al. (2013b) report a wider range of variation in the chlorogenic acid concentration of raw purple potatoes (ranging from 63 to 329.75 mg/100 g FW). Boiling, baking, and microwaving reduce the chlorogenic acid concentration of potatoes with boiled tubers having a higher retention of chlorogenic acid than baked and microwaved ones (Lachman et al. 2013). In a recent study conducted by Piñeros-Niño et al. (2017), the chlorogenic acid concentration of cooked tubers from 193 potato varieties ranged from 19.25 to 399 mg/100 g FW. In a previous study by Burgos (2014), the chlorogenic acid concentration in cooked tubers of purple-fleshed cultivars ranged from 36.17 to 395.73 mg/100 g FW and in red-fleshed cultivars from 14.45 to 48.60 mg/100 g FW.

The highest concentration of chlorogenic acid reported in 100 g of cooked potato tubers is similar to the maximum provided by a single cup of coffee (350 mg chlorogenic acid; Clifford 1999) and is tenfold higher than the maximum amount provided by whole apples (38.5 mg/100 g FW, Spanos and Wrolstad 1992).

Chlorogenic acid is only partially bioavailable and its bioactivity may be modulated by the gut microbiota that can generate bioactive secondary microbial phenolic metabolites such as caffeic acid that have much greater bioavailability (Tomas-Barberan et al. 2014; Olthof et al. 2003). Chlorogenic acid may also promote a healthy gut microbiome. In a batch culture fermentation model of the colon, chlorogenic acid was found to promote growth of Bifidobacterium bacterial species that could be beneficial for gut health (Mills et al. 2015).

2.2.9.2 Anthocyanins

Anthocyanins are a class of water-soluble flavonoids, which show a range of pharmacological effects, such as prevention of cardiovascular disease, obesity control, and anti-tumor activity. Their potential anti-tumor effects are reported to be based on a wide variety of biological activities including antioxidant, anti-inflammation, anti-mutagenesis, induction of differentiation, inhibiting proliferation by modulating signal transduction pathways, inducing cell cycle arrest, and stimulating apoptosis or autophagy of cancer cells; anti-invasion; anti-metastasis; reversing drug resistance of cancer cells and increasing their sensitivity to chemotherapy (Lin et al. 2017).

Anthocyanins are present in the flesh and skin of several purple- and red-fleshed potatoes such as those landraces found in the Andes, which show a wide range of anthocyanin structures and concentrations that are largely cultivar-dependent (Brown et al. 2003) and location-dependent (Ieri et al. 2011). Increased height above sea level, higher annual sum of precipitation, and lower annual average temperatures cause higher anthocyanin concentrations (Lachman et al. 2009).

The total anthocyanin concentration of raw and cooked purple-fleshed potatoes ranges from 63 to 588 mg/100 g FW and from 71 to 453 mg/100 g FW, respectively (Burgos et al. 2013a, b). Total anthocyanin concentration of cooked red-fleshed potatoes ranges from 8.2 to 55.3 mg/100 g FW (Burgos 2014).

Giusti et al. (2014) identified five major anthocyanidins (cyanidin, petunidin, pelargonidin, peonidin, and malvidin) in extract from purple potato and three major anthocyanidins (cyanidin, pelargonidin, and peonidin) in extracts of red potatoes. The extract of purple potatoes contained four major anthocyanins: cyanidin-3-rutinoside-5-glucoside, petunidin-3-rutinoside-5-glucoside, pelargonidin-3-rutinoside-5-glucoside, and peonidin-3-rutinoside-5-glucoside, with petunidin and peonidin glycosides being the most predominant. The extract of red-fleshed potatoes contained four major anthocyanins: cyanidin-3-rutinoside-5-glucoside, pelargonidin-3-rutinoside-5-glucoside, peonidin-3-rutinoside-5-glucoside, and pelargonidin-3-rutinoside, with pelargonidin-3-rutinoside-5-glucoside being the most predominant.

Burgos (2014) characterized the anthocyanin profile of 12 purple-fleshed accessions and 6 red-fleshed accession from CIP's genebank and found that in purple-fleshed accessions the predominant anthocyanin is petunidin-3-(coumaroyl) rutinoside-5-glucoside (petanin), representing from 37 to 78% of the total anthocyanins. It is followed by peonidin-3-(coumaroyl) rutinoside-5-glucoside, cyanidin-3-(coumaroyl) rutinoside-5-glucoside, and minor proportions of malvidin 3-(coumaroyl) rutinoside-5-glucoside and pelargonidin-3-(coumaroyl) rutinoside-5-glucoside. In red-fleshed accessions, the predominant anthocyanin is pelargonidin-3-(coumaryl) rutinoside-5-glucoside, representing 41–75% of the total anthocyanins. It is followed by peonidin-3-(coumaroyl) rutinoside-5-glucoside, pelargonidin-3-rutinoside, and cyanidin-3-(coumaroyl) rutinoside-5-glucoside in various proportions and then by pelargonidin-3-(coumaryl) rutinoside in minor proportions. Figure 2.2 shows as an example the anthocyanin profile for two purple-fleshed accessions (CIP 705534 and CIP 702363) and two red-fleshed accessions (CIP 703625 and CIP 702453). Pt3(c)R5G represented by the purple bar is dominant in the purple-fleshed accessions while PI3R(c)R5G represented by the pink bar is dominant in the red-fleshed potatoes.

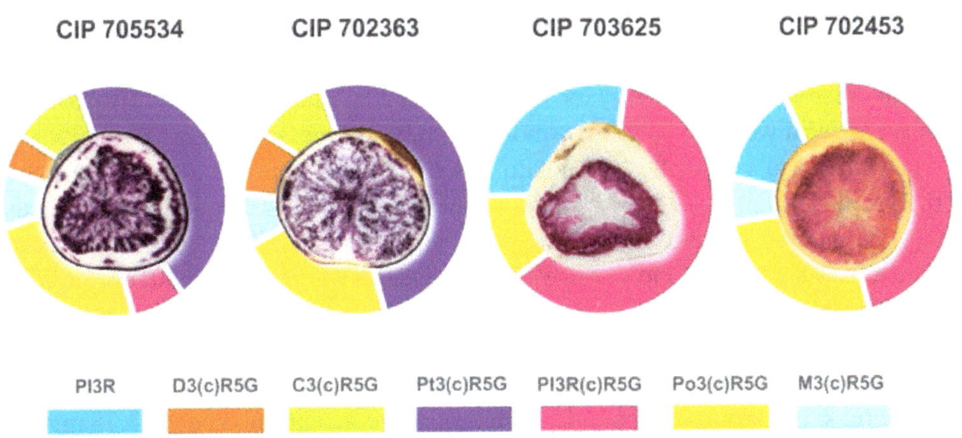

Fig. 2.2 Anthocyanin profile in purple-fleshed and red-fleshed potatoes. (PI3R: pelargonidin-3-rutinoside, PI3R(c)R5G: pelargonidin-3-(coumaroyl) rutinoside-5-glucoside, Pt3(c)R5G: petunidin-3-(coumaroyl) rutinoside-5-glucoside, Po3(c)R5G: peonidin-3-(coumaroyl) rutinoside-5-glucoside, C3(c)R5G: cyanidin-3-(coumaroyl) rutinoside-5-glucoside, M3(c)R5G: malvidin 3-(coumaroyl) rutinoside-5-glucoside)

The most prominent anthocyanins present in the red- and purple-fleshed accessions are acylated with hydroxycinnamic acid (Fossen and Andersen 2000). Three different cinnamic acids were found acylating the anthocyanins in the extract of purple and red potatoes: caffeic, *p*-coumaric and ferulic acid (Giusti et al. 2014). Acylated anthocyanins are known to be stable and hence can be considered as promising natural colorants for the food industry.

The highest anthocyanin concentration reported in a dark purple-fleshed potato (above 400 mg/100 g FW; Andre et al. 2007) is lower than in blueberries (558 mg/100 g, FW; Hosseinian and Bea 2007), cranberries (589 mg/100 g FW; Wada and Ou 2002), eggplant (750 mg/100 g FW; Wu et al. 2006), and purple corn (1642 mg/100 g FW; Cevallos-Casals and Cisneros Zevallos 2003). However, the contribution of purple-fleshed potatoes to the diet can be considerably higher considering the high mean intake of potatoes in some areas like the Andean highlands where consumption may reach 500 g per day, as compared to the mean intake of blueberries, cranberries, and eggplant (1 g per day in the United States; Wu et al. 2006).

Kubow et al. (2017) studied the biotransformation of anthocyanins from cooked purple-fleshed potatoes using a dynamic human gastrointestinal (GI) model that includes stomach, small intestine, and colonic vessels. After 24 h digestion, liquid chromatography-mass spectrometry identified 15–36 anthocyanin species throughout the GI vessels. Genetic background of the purple potato cultivars led to major variances in the pattern of anthocyanin breakdown and release during digestion composition. Diminished concentrations of several anthocyanin species in the colonic vessels indicated microbial biotransformation which is, in turn, associated to increased bioaccessibility.

The cytotoxicity and cell viability of colonic Caco-2 cancer cells and nontumorigenic colonic CCD-112CoN cells after 24 h exposure to colonic fecal water of purple-fleshed potato digests has been also tested by Kubow et al. (2017). The cultivar Leona showed a significant potency to induce cytotoxicity and decrease viability of Caco-2 cells. The differing microbial anthocyanin metabolite profiles in colonic vessels between cultivars were indicated to play a significant role in the impact of fecal water toxicity on tumor and nontumorigenic cells.

In white- and yellow-fleshed potato tubers, flavonols predominate in the flavonoid profile (Andre et al. 2007). Flavonols have been extensively studied in the past 10 years as they present a range of putative health-promoting effects, including reduced risk of cancer and cardiovascular diseases (Wang et al. 2016). Rutin in particular has shown strong antioxidative and anti-inflammatory effects at the cellular level (Habtemariam and Lentini 2015).

In potato tubers, rutin (quercetin-3-*O*-rutinoside) and kaempferol-3-*O*-rutinoside are the most important compounds, with reported concentrations of 0–4.78 mg and 0–5.68 mg/100 g FW, respectively, in Andean raw potato tubers (Andre et al. 2007). The influence of various cooking methods on potato flavonols has been investigated, which revealed the stability of the concentrations through treatment (Navarre et al. 2010). The bioaccessibility of these compounds was also high (close to 100% on average) when evaluated in a collection of 12 Andean potato genotypes (Andre et al. 2015).

2.2.10 Carotenoids

Potatoes contain lipophilic phytonutrients in the form of carotenoids that have numerous health-promoting properties including decreasing risk of several chronic diseases (Gammone et al. 2015; Wu et al. 2015). Carotenoids have been reported to exhibit chemoprevention by a variety of mechanisms including immune system activation, protection against oxidative stress, promotion of gap junction communication, inhibition of DNA damage, enhanced metabolic detoxification, and tumor suppressor action and inhibition of oncogene expression (Khachik et al. 1999; Fiedor and Burda 2014).

Potato carotenoid concentrations and profiles are related to the flesh color with dark yellow cultivars showing approximately tenfold higher concentrations of total carotenoids than white-fleshed varieties (Brown et al. 2005). Significant and predominant amounts of zeaxanthin and antheraxanthin are found in deep yellow-fleshed potatoes while the carotenoid profile of yellow potatoes is composed of violaxanthin, antheraxanthin, lutein, and zeaxanthin and that of cream-fleshed potatoes of violaxanthin, lutein, and β-carotene (Burgos et al. 2009).

The violaxanthin, antheraxanthin, lutein, and β-carotene concentration of raw tubers of potatoes from the *Tuberosum* group ranged from 1.5 to 87.8 μg/100 g FW, 0.6 to 15.8 μg/100 g FW; 1.6 to 35.1 μg/100 g FW; and 0.1 to 2.1 μg/100 g FW, respectively (Fernandez-Orozco et al. 2013), while the concentration of these carotenoids in tubers from the *Phureja* group ranged from 20.0 to 410 μg/100 g FW; 9.3 to 503 μg/100 g FW; 55 to 211 μg/100 g FW and 4.8 to 27 μg/100 g FW, respectively (Burgos et al. 2009), and in tubers from the Andigenum group from 14.3 to 173 μg/100 g FW, 7 to 16 μg/100 g FW; 43.3 to 442 μg/100 g FW and 10.5 to 54.8 μg/100 g FW, respectively (Andre et al. 2007).

Boiling does not affect the lutein and zeaxanthin concentration of potato; however, violaxanthin and antheraxanthin concentrations of potatoes are significantly reduced after boiling. Lutein and zeaxanthin concentrations of cooked yellow-fleshed potatoes ranged from 73 to 253 μg/100 g FW and from 0 to 1048 μg/100 g FW, respectively (Burgos et al. 2012) with deep yellow-fleshed potatoes being a significant source of zeaxanthin (above 500 μg/100 g FW).

Lutein and zeaxanthin are important dietary carotenoids that are selectively taken up into the macula of the eye, where they protect against development of age-related macular degeneration and cataracts (Wu et al. 2015). Moreover, these compounds have been reported to have other health-promoting effects, including immune-enhancement and reduction of the risk of developing degenerative diseases such as cancer and cardiovascular diseases (Krinsky and Johnson 2005). There is no recommended daily intake for lutein and zeaxanthin, but many studies show a health benefit for lutein supplementation at 10 mg per day and zeaxanthin at 2 mg per day (American Optometric Association 2009).

The highest values of lutein and zeaxanthin reported in 100 g of yellow-fleshed potatoes are lower compared to the amount of lutein provided by 100 g of lettuce (540 μg; Kimura and Rodriguez-Amaya 2003), broccoli (3250 μg; Khachick et al.

1992), parsley (5800 μg; Hart and Scott 1995), or spinach (4180 μg; Tee and Lim 1991); and lower than the amount of zeaxanthin provided by 100 g of maize at its highest zeaxanthin concentration (3800 μg/100 g) (Brenna and Berardo 2004) and of red paprika (2200 μg/100 g) (Müller 1997; Mínguez Mosquera and Hornero-Méndez 1994). However, it is important to consider that potato consumption can be as high as 500 g per day whereas the mean intake of the above-mentioned vegetables is less than 50 g per day; hence, the overall contribution of potato-based carotenoids to the dietary intake can be higher. Furthermore, the contribution of a food source to lutein and zeaxanthin intake depends on their digestive stability, bioaccessibility, and bioavailability in the respective food matrix. Bioaccessibility refers to the proportion of ingested carotenoid that is released from the food matrix and incorporated into micelles in the gastrointestinal tract, and thus available for intestinal absorption (Rodriguez-Amaya 2015). Bioavailability refers to the portion of the carotenoid that is absorbed in the body, enters in systemic circulation and becomes available for utilization in normal physiological functions or for storage in the human body (van Het Hof et al. 2000). The bioavailability of carotenoids from plant foods is influenced by the species and structure of carotenoids present in the food, composition, and release of carotenoids from the food matrix, absorption in the intestinal tract, transportation within the lipoprotein fractions, biochemical conversions, and tissue-specific depositions, as well as by the nutritional status of the ingesting consumer (Bohn 2017). Research on bioavailability of potato carotenoids is required to have more useful and complete information regarding their nutritional and health benefits.

Burgos et al. (2013a) evaluated the in vitro digestive stability and the efficiency of micellarization or bioaccessibility of lutein and zeaxanthin in yellow-fleshed potatoes. The gastric and duodenal digestive stability of lutein and zeaxanthin in boiled tubers ranged from 70 to 95% while the bioaccessibility ranged from 33 to 71% for lutein and from 51 to 71% for zeaxanthin. A more recent study has reported that bioaccessibility of lutein and zeaxanthin in yellow-fleshed clones range from 76 to 82% for lutein and from 24 to 55% for zeaxanthin (Andre et al. 2015).

The maximum bioaccessible lutein concentration reported in yellow-fleshed potatoes is around 300 μg/100 g FW while the maximum bioaccessible zeaxanthin concentration is around 600 μg/100 g FW. Considering the mean potato intake in the Andes of Peru, Ecuador, and Bolivia (500 g per day), potato tubers from the variety with the highest bioaccessible lutein could provide 14% of the suggested level of lutein intake for having health benefit (10 mg per day). Likewise, potato tubers of the variety with the highest bioaccessible zeaxanthin concentration could provide 50% more than the suggested level of zeaxanthin intake (2 mg per day). In Spain, potato was shown to contribute 13–20% towards the total dietary intake of zeaxanthin and was ranked as the third main contributor after citrus fruits and green vegetables (Garcia-Closas et al. 2004). More population studies are needed, however, regarding the nutritional and health contributions of carotenoids as provided by potatoes.

2.3 Antioxidant Activity

Antioxidant activity (AA) describes the capacity of redox molecules in foods and biological systems to scavenge free radicals considering the additive and synergistic effects of all antioxidants rather than the effect of single compounds, and may, therefore, be useful to study the potential health benefits of antioxidants on oxidative stress-mediated diseases (Puchau et al. 2010). In that context, mixtures of phytochemicals found in plant foods are more effective in improving antioxidant status than isolated phytochemicals (DeGraft-Johnson et al. 2007). The antioxidant activity of foods as assessed by indices such as ferric reducing ability of plasma (FRAP) has been indicated to be a valid and reproducible determinant of human plasma AA measurements (Rautiainen et al. 2008). Antioxidant capacities of food staples such as potatoes could thus potentially affect the antioxidant status of that population.

The antioxidants in potato are mainly hydrophilic (polyphenols, ascorbic acid, anthocyanins, and flavanols) (Fig. 2.3) (Reyes et al. 2005). In white- or yellow-fleshed potatoes, prevalent contributors of AA are chlorogenic acid, gallic acid, caffeic acid, and catechin (Reddivari et al. 2007a), while in purple- and red-fleshed potatoes the major contributors to AA are anthocyanins and chlorogenic acid (Lachman et al. 2009). Potatoes also contain lipophilic antioxidants (carotenoids and vitamin E) (Fig. 2.4).

Because antioxidant activity of potato anthocyanins results from the synergistic effect of each anthocyanin pigment (Hayashi et al. 2003), it is important to assess different pigmented potato cultivars for individual anthocyanidin content, as well as the contribution of the anthocyanidin composition to their antioxidant activity. A high degree of hydroxylation and/or methoxylation of individual anthocyanidins could contribute in conjunction with other phenolics to high AA (Lachman et al. 2009).

Burgos et al. (2013b) reported that boiled potatoes of purple-fleshed potato varieties have an AA ranging from 4017 to 17,304 mg Trolox equivalents (TEq)/g, FW as determined by the 2,2-azino-bis-3-ethylbenzthiazoline-6-sulfonic acid (ABTS) antioxidant capacity measure and from 2369 to 9754 mg TEq/g, FW as determined by 1,1-diphenyl-2-picryl-hydrazyl (DPPH) antioxidant capacity assay. Compared to other sources of antioxidants, potato has lower AA than strawberry, blackberry, and blueberry as determined by the ABTS assay (around 25,030–50,000 mg TEq/g, FW, Garcia-Alonso et al. 2004). However, as indicated above the overall contribution of potato to the antioxidant intake of a population will finally depend in the amount of potatoes typically consumed.

Ombra et al. (2015) reported after simulated gastrointestinal digestion, the extracts from purple potato have high in vitro antioxidant, antimicrobial, and anti-proliferative activities against the colon cancer cells Caco-2 and SW48 and the breast cancer cells MCF-7 and MDA-MB-231.

After digestion of cooked tubers from purple-fleshed potatoes using a dynamic human gastrointestinal model, Kubow et al. (2017) found an increased FRAP anti-

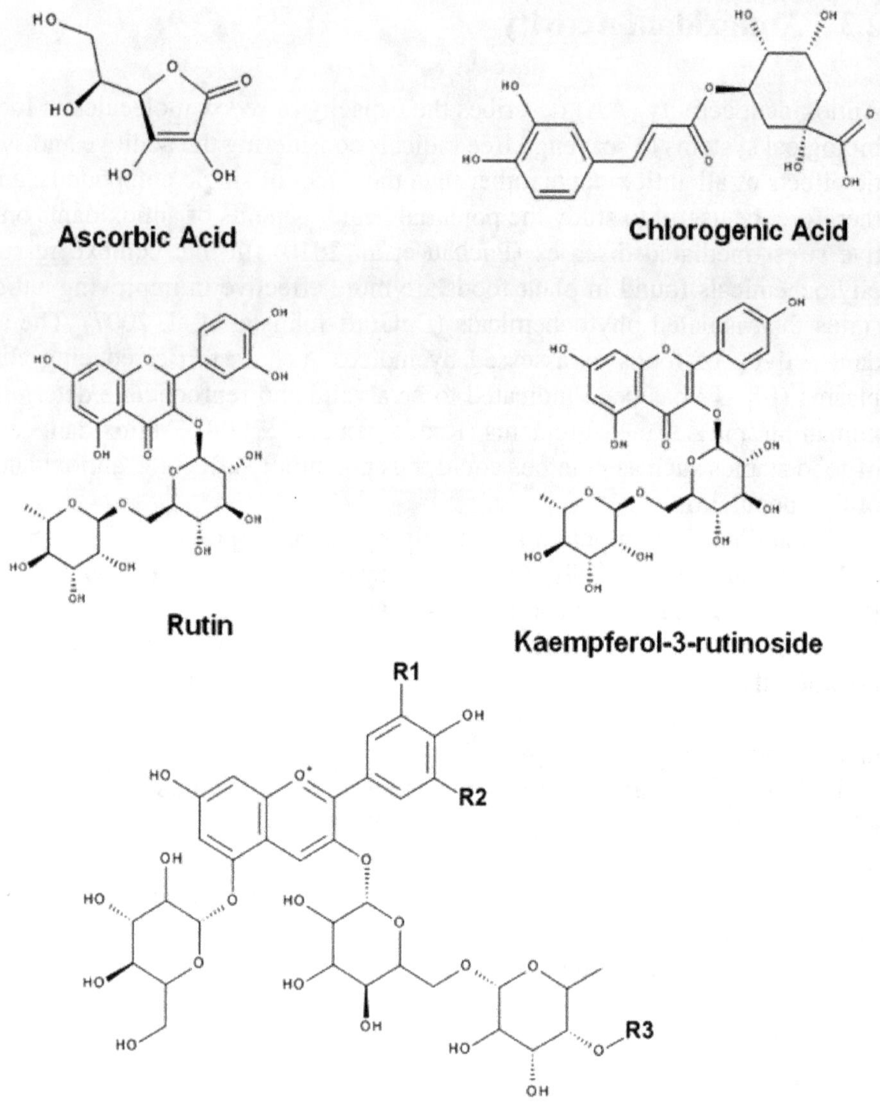

Anthocyanins	R1	R2	R3
pet-3-coum-rut-5-glc	OMe	OH	p-coumaric acid
peo-3-coum-rut-5-glc	OMe	H	p-coumaric acid
cyan-3-coum-rut-5-glc	OH	H	p-coumaric acid
mal-3-coum-rut-5-glc	OMe	OMe	p-coumaric acid
pel-3-coum-rut-5-glc	H	H	p-coumaric acid

Fig. 2.3 Chemical structure of hydrophilic antioxidants in potato

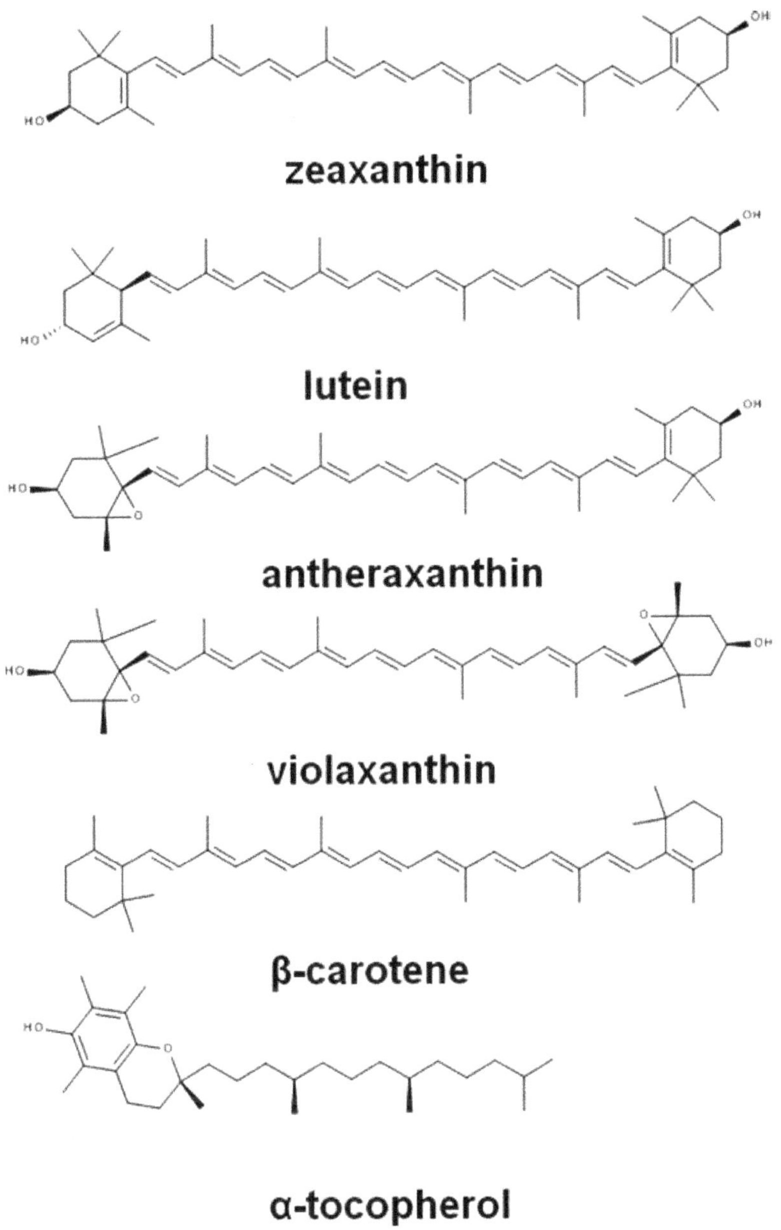

Fig. 2.4 Chemical structure of lipophilic antioxidants in potato

oxidant activity in the colonic reactors. Metabolic microbial breakdown of anthocyanins over a 24 h period appeared to generate sufficient amounts of microbial metabolites to produce an improvement in antioxidant capacity. Anthocyanins and their metabolites can, via antioxidant activity, provide protection for intestinal cells against oxidative stress in the gut, and hence alleviate gut inflammation, protect against colorectal cancer, and generally enhance colorectal health.

2.4 Glycoalkaloids

Glycoalkaloids are secondary plant metabolites that serve as natural defenses against bacteria, fungi, viruses, and insects (Friedman 2004). They can be toxic for humans when present in high concentrations, and can impart a bitter taste to potatoes. However, glycoalkaloids and hydrolysis products without the carbohydrate side chain (aglycones) also have beneficial effects that include: lowering of cholesterol (Friedman et al. 2003) and inhibition of the growing of cancer cells in culture as well as tumor growth in vivo (Friedman 2015).

Although there are many glycoalkaloids, α-chaconine and α-solanine make up 95% of the total glycoalkaloids present (Friedman et al. 1997); α-solanine is found in greater concentrations than α-chaconine, and α-solanine has only half as much specific toxic activity as a α-chaconine (Lachman et al. 2001).

Experiments with human taste panels revealed potato varieties with glycoalkaloid levels exceeding 14 mg/100 g FW tasted bitter (Friedman 2006). Those in excess of 22 mg/100 g FW also induced mild to severe burning sensations in the mouths and throats of panel members.

Glycoalkaloid levels vary greatly in different potato varieties and may be influenced by factors such as light, mechanical injury, and storage. They are also influenced by stress such as heat and drought during production. This raises concern for maintaining the quality of potatoes under climate change (Andre et al. 2009), and suggests increased attention may be needed to glycoalkaloid concentrations of potato varieties bred for or grown in warm environments.

Glycoalkaloid concentration of raw potatoes ranges from 0.7 to 18.7 mg/100 g FW (Friedman et al. 2003). Peeling significantly reduced the glycoalkaloid levels in the tubers: solanine to 43.6% and chaconine to 31% (Lachman et al. 2013). Cooking also significantly reduced the levels of glycoalkaloids (Tajner-Czopek et al. 2008), with boiling reducing the levels of glycoalkaloids more than baking and microwaving (Lachman et al. 2013).

Glycoalkaloid content in potato tubers should not exceed 20 mg/100 g FW, because this level is dangerous for human health (Ruprich et al. 2009). The toxicity of glycoalkaloids at appropriate high levels may be due to adverse effects such as anticholinesterase activity on the central nervous system and to disruption of cell membranes adversely affecting the digestive system and general body metabolism (Friedman et al. 2003). The toxicity of glycoalkaloids is associated with the synergistic interaction between two main components of glycoalkaloids: α-solanine and α-chaconine.

However, glycoalkaloids also have anti-carcinogenic properties. Exposure of cancer cells to glycoalkaloids produced potatoes (α-chaconine and α-solanine) or their hydrolysis products (mono-, di-, and trisaccharide derivatives and the aglycones solasodine, solanidine, and tomatidine) inhibits the growth of the tumor cells in culture as well as in vivo tumor growth (Friedman 2015). On the basis of the anti-carcinogenic properties of these potato components, it is conceivable that the levels typically noted in commercial potatoes might help to protect against multiple cancers. Epidemiological studies, however, are needed to substantiate this possibility.

2.5 Contribution to Health

Population-based epidemiological studies have emphasized the importance of nutrition to combat metabolic disorders emerging worldwide that have been associated with diet, such as diabetes, cancer, and cardiovascular diseases. In that regard, higher intakes of fruits and vegetables have been consistently indicated to exert protective effects against such chronic diseases (Dragsted et al. 2006). Potato has been underappreciated relative to other vegetables as it has been subject to controversy such as being labeled as a contributor to development of obesity and diabetes (Burlingame et al. 2009). On the other hand, potatoes contain relatively high concentrations of key phytonutrients that have shown bioactivities that could counteract chronic disease development (Ezekiel et al. 2013). Potatoes have shown promising health-promoting effects in human cell culture, experimental animals, and human clinical studies, including anti-cancer, hypocholesterolemic, anti-inflammatory, anti-obesity, and anti-diabetic properties. Nutritional compounds of potatoes such as phenolics, anthocyanins, fiber, starch as well as compounds considered anti-nutritional such as glycoalkaloids, lectins, and proteinase inhibitors are believed to contribute to the health benefits of potatoes (Fig. 2.5). As there are many biological activities attributed to the compounds present in potato, some of which could be beneficial or detrimental depending on specific circumstances, long-term studies investigating the association between potato consumption and diabetes, obesity, cardiovascular disease, and cancer while controlling for fat intake are needed (Visvanathan et al. 2016).

2.5.1 Anticancer Effect

Several studies have shown a reduction in proliferation of cancer cells when treated with potato extracts. Potato antioxidants such as phenolic acids and anthocyanins, glycoalkaloids, fiber, and proteinase inhibitors identified in potatoes have been implicated in the suppression of cancer cell proliferation in vitro and in vivo.

2.5.1.1 Role of Potato Antioxidants

Phenolic acids and anthocyanins are potato antioxidants that have reported anti-carcinogenic activity. Hayashi et al. (2006) reported that anthocyanins in steamed purple and red potatoes suppressed the growth of benzopyrene-induced stomach cancer in mice. Reddivari et al. (2007b) found that the anthocyanin fractions from potato extracts were cytotoxic to prostate cancer cells through activation of caspase-dependent and caspase-independent pathways. Madiwale et al. (2011) reported that purple flesh potatoes rich in anthocyanins suppressed proliferation and elevated

Fig. 2.5 Health benefits of potatoes

apoptosis of colon cancer cells compared with white and yellow flesh potatoes. In a more recent study, Charepalli et al. (2015) found that extracts of purple-fleshed potatoes suppress colon tumorigenesis via elimination of colon cancer stem cells. Chlorogenic acid, the main phenolic acid of potato, is effective against human liver, colon, and prostate cancer cells (Wang et al. 2011) and inhibits significantly the proliferation of colon cancer and prostate cancer cells.

2.5.1.2 Role of Potato Glycoalkaloids

α-Solanine and α-chaconine, the main steroidal glycoalkaloids in potatoes, are well studied for their antitumor properties (Friedman 2015). Lee et al. (2004) found that α-solanine exhibited growth inhibition and apoptosis induction in multiple cancer cells such human colon (HT29) and liver (HepG2) cancer cells. Friedman et al. (2005) evaluated the anti-carcinogenic effect of α-solanine and α-chaconine extracted from five fresh potato varieties (Dejima, Jowon, Sumi, Toya, and Vora Valley) and found that glycoalkaloids exerted anti-proliferative effects of the following human tumor cell lines: cervical (HeLa), liver (HepG2), lymphoma (U937), stomach (AGS and KATO III) cells, and on normal liver (Chang) cells. Friedman et al. (2005) also reported that the anti-proliferative effects of the glycoalkaloids were concentration dependent and that α-chaconine was more bioactive than α-solanine. Yang et al. (2006) found that α-chaconine induced the apoptosis of HT-29 human colon cancer cells through caspase-3 activation and inhibition of extracellular signal-regulated kinase phosphorylation.

Reddivari et al. (2010) showed that α-chaconine exhibited potent anti-proliferative properties and increased cyclin-dependent kinase inhibitor p27 levels in two prostate cancer cell lines, LNCaP and PC3. More recently, it has been reported that α-solanine, has a positive effect on the inhibition of pancreatic cancer cell growth in vitro and in vivo. Sun et al. (2014) demonstrated that α-solanine inhibited cancer cell growth through caspase 3-dependent mitochondrial apoptosis and that the expression of tumor metastasis-related proteins, MMP-2 and MMP-9, was also decreased in the cells treated with α-solanine. Lv et al. (2014) reported that α-solanine inhibited proliferation of PANC-1, sw1990, MIA PaCa-2 cells in a dose-dependent manner, as well as cell migration and invasion with a toxic dose and that the administration of α-solanine during 2 weeks in a xenograft model reduces the tumor volume and weight by 43–61%. These studies showed beneficial effects on pancreatic cancer in vitro and in vivo, which may be mediated via suppressing pathways involving proliferation, angiogenesis, and metastasis.

2.5.1.3 Role of Potato Fiber

Langner et al. (2009) reported that commercially available potato fiber extract (Potex) exhibited anti-proliferative effects in several tumor cell cultures. The fiber extract decreased cancer cell motility, induced apoptosis, and also caused morphological changes in tumor cells.

2.6 Anti-diabetic and Anti-obesity Effects

Potato consumption has often been associated in cohort studies with elevated risk of type 2 diabetes (Muraki et al. 2016) and obesity (Borch et al. 2016), which has been attributed to a relatively high glycemic index in some potato varieties and processed potato products containing added saturated and trans fats. A major confounding factor in such studies is typical Western dietary patterns associated with increased disease risk typically include potato consumption along with high intake of red and processed meat, refined grains, high-fat dairy products, fried foods and sugar (Pastorino et al. 2016). More research is needed to adjust association of food items such as potatoes in such dietary patterns (Hu 2002). Moreover, RDS present in cooked potatoes (especially amylose) tends to retrograde upon cooling generating appreciable amounts of slowly digestible starch (SDS) or resistant starch (RS) that contribute to dietary fiber content (Sajilata et al. 2006) and potentially positively impact health by slowing postprandial glucose release from cooked potatoes (King and Slavin 2013).

GI values below 56 are considered as low glycemic index while values above 74 are considered to indicate a high glycemic index. GI values in potato ranged from 56 to 94 for eight British cultivars (Henry et al. 2005) and from 53 to 103 for seven Australian cultivars (Wang et al. 2014). When boiled red potatoes were served hot to volunteers, a GI of 89.4 was found (Fernandes et al. 2005). When cooking is followed by cooling, amylose retrogrades to produce resistant starch. The GI response was only 56.2 when cooking was followed by refrigeration of 12–24 h.

Potato chips and French fries have been implicated by some nutrition researchers as major contributors to obesity risk as these products contain a high fat and caloric content. Potato servings, however, are not likely in themselves to promote obesity as potatoes are considered to have a low energy density as they are a low-fat food with a high-water content (Anderson et al. 2013). Potato-based foods with high calorie fat additions have been considered as a major culprit towards obesity risk (Camire et al. 2009).

Conversely, potatoes may have a role in controlling appetite and therefore weight gain, by contributing to satiety. Satiety is the feeling of fullness and the loss of hunger that occur after eating. Many factors influence satiety, including the rate of gastric emptying and the proportion of macronutrients in the food. Foods that increase satiety are thought to promote weight control by delaying subsequent meals and total calories consumed (Camire et al. 2009). Compared to rice and pasta, adult feeding studies have shown that satiating amounts of potatoes co-ingested with meat resulted in lower energy intake and postprandial insulin concentrations; and higher levels of ghrelin, which is a gastric orexigenic appetite-stimulating hormone that contributes to feeding regulation (Erdmann et al. 2007). Likewise, studies involving children showed that meat co-ingestion with boiled mashed potato resulted in an approximate 40% lower energy intake as compared to meat consumed together with either pasta or rice (Akilen et al. 2016). The stronger satiety of boiled mashed potato for the calories consumed was related to similar suppression of ghrelin postprandially relative to the other carbohydrate-rich foods despite the lower potato meal intake. Short-term intervention studies have generally indicated that high GI meals decrease satiety, and an increase in the return of hunger and energy intake at a later meal as opposed to low glycemic index meals containing potatoes (Roberts 2000).

2.7 Anti-hyperlipidemic, Anti-hypertensive and Anti-inflammatory Effects

A variety of animal feeding studies have shown cholesterol-lowering properties from potato intake that have been related to its content of protein, resistant and phosphorylated starch, fiber, glycoalkaloids (Friedman 2006), and phenolic compounds (Friedman 1997). Robert et al. (2006) found that consumption of cooked potatoes (consumed with skin) improved lipid metabolism in cholesterol-fed rats. Rats fed a potato-enriched diet for 3 weeks had lower concentrations of plasma cholesterol and triglycerides and reduced liver cholesterol content. Hashimoto et al. (2006) showed that retrograded starch from two varieties of potato pulp lowered serum total cholesterol and triglyceride concentrations. The authors indicated that the retrograded starch promoted the excretion of bile acids resulting in a low concentration of serum cholesterol; and that retrograded starch inhibited the synthesis of fatty acids at the mRNA levels of fatty acid synthase (FAS) and SREBP-1c, which might be related to the observed reduction of the serum triglyceride concentrations. Kanazawa et al. (2008) reported that gelatinized potato starch containing a high level of phosphate reduced concentrations of serum-free fatty acids and triglycerides and liver triglycerides.

Liyanage et al. (2008) have demonstrated the hypocholesterolemic effect of potato peptides. Rats fed a cholesterol-free diet containing 20% (w/w) potato peptides showed greater concentrations of serum high-density lipoprotein (HDL) cholesterol and increased fecal steroid output and lesser non-HDL cholesterol concentrations than rat fed diets containing 20% casein peptides. The results were attributed to inhibition of cholesterol absorption, possibly via suppression of micellar solubility of cholesterol. In a follow-up study, Liyanage et al. (2009) found that potato peptides reduced the serum non-HDL cholesterol concentrations by stimulating fecal steroid excretion, accelerated by cecal short-chain fatty acids in a hypercholesterolemic rat model. There is a lack of data, however, from randomized controlled trials to demonstrate a relationship between potato consumption and blood lipid parameters in humans.

Vinson et al. (2012) showed a significant lowering of systolic blood pressure in humans after supplementation to hypertensive subjects in a 4-week cross-over trial involving consumption of six to eight small purple potatoes twice daily versus no potato intake. The blood pressure lowering effect was related to high intake of polyphenols associated with the pigmented potatoes. This latter intervention trial is contrasted by an analysis from three large prospective cohort studies indicating increased hypertension risk in association with potato intake of four or more servings per month as opposed to one serving per month (Borgi et al. 2016). A major limitation of such trials is that co-ingestion of salt, high-salt foods, saturated or trans fats with potatoes could have contributed to the hypertension risk as opposed to potato per se, particularly since potatoes are typically eaten in a meal context (Miller and Stanner 2016). In support of this contention, a 3-year longitudinal study of Japanese people showed that adherence to a traditional Japanese dietary pattern exerted favorable effects on blood pressure that was partly associated with potato intake (Niu et al. 2016).

Relatively high intake of potassium is needed to counteract the blood pressure raising effects of a high sodium diet and so protect against hypertension (Camire et al. 2009). An increase in consumption of potassium-rich foods has been promoted to combat hypertension and cardiovascular disease (WHO 2012). In that regard, intake of potassium-rich foods has been indicated to protect against stroke risk (Adebamowo et al. 2015). As potatoes are rich in potassium and are naturally very low in sodium content, this food could counter development of hypertension-associated diseases. Additionally, Makinen et al. (2008) reported that a protein isolated from vascular bundle and inner tuber tissues of potato enhanced the inhibition of the angiotensin converting enzyme I, a biochemical factor affecting blood pressure that contributes to hypertension.

Kaspar et al. (2011) found anti-inflammatory effects in healthy men consuming white and pigmented potatoes with greater effects from pigmented potatoes. Potato phenolics and glycoalkaloids have shown evidence of anti-inflammatory activities (Kenny et al. 2013). Indigestible carbohydrates including resistant starch and fiber have demonstrated the ability to modulate inflammatory markers in both animal models (Vaziri et al. 2014) and human clinical trials (Jiao et al. 2015). Hence, the contribution of resistant starch or fiber from select potato products may have also direct impact on inflammatory stress in humans.

2.8 Potato and Its Relationship with Cardiovascular Diseases

As a key dietary source of potassium, vitamin C, and dietary fiber, potatoes contribute significantly to nutrients with defined roles in promoting cardiovascular health (McGill et al. 2013). Boiled potatoes have been shown to have favorable impact on several measures of cardiometabolic health in animals and humans, including lowering blood pressure, improving lipid profiles, and decreasing markers of inflammation (McGill et al. 2013). When eaten as a regular food item and consumed with skin, potato intake can significantly enhance cardioprotective fiber intake that is generally lacking in Western-type diets (Lockyer et al. 2016). Large prospective studies in Sweden involving a 13-year follow-up showed no adverse relationship of higher potato intake with cardiovascular risk for either morbidity or mortality (Larsson and Wolk 2016). Similarly, a systematic review of five observational studies carried out by Borch et al. (2016) showed no convincing evidence to support an adverse association between unprocessed potato intake and the risk of developing metabolic disorders including obesity, type 2 diabetes, and cardiovascular disease. On the other hand, processed potato products like French fries and potato crisps, with high lipid and trans fats content and added sodium can have adverse effect of the heart health and so should be minimized in the diet. In that regard, despite the lack of a relationship between chronic disease risk and potato intake in the above comprehensive review by Borch et al. (2016), they showed that French fries and fried potato were associated with an increased risk for obesity and type 2 diabetes. Likewise, a longitudinal study involving 4440 subjects with an 8-year follow-up showed no association between higher potato intake and mortality risk, whereas participants who consumed fried potatoes two to three times/week had an increased mortality risk (Veronese et al. 2017). Camire et al. (2009) have recommended preparing potatoes with minimum lipid addition and consume potatoes with peels to conserve their cardiovascular health promoting properties.

2.9 Concluding Remarks

Potato is an important source of carbohydrates, resistant starch, quality protein, vitamins C and B6 as well as potassium. Potatoes are also a source of antioxidants. Chlorogenic acid and glycoalkaloids are present in all potatoes independently of the flesh color while deep yellow-fleshed potatoes contain high amounts of lutein and zeaxanthin; and purple-fleshed potatoes contain high amounts of anthocyanins. Potatoes glycoalkaloids in high concentrations can be toxic to humans but in low concentrations can have beneficial effects such as inhibition of the growth of cancer cells.

The contribution of potato to the diet is affected by cooking, potato intake, and the bioavailability of potato nutrients. Potato vitamins are significantly reduced after cooking. However, 100 g of cooked potatoes provide around 30% of the requirement of vitamin C and 20% od the requirement of vitamin B6. Potato carotenoids and anthocyanins show high recoveries after cooking due to an improved release of these antioxidants. In vitro studies demonstrate that potato lutein and zeaxanthin have a high bioaccessibility and that potato phenolics undergo microbial transformation in the intestinal tract producing metabolites that may also promote a healthy gut microbiome. Further research in humans is needed to confirm the beneficial effect of potato phenolics in the gut.

In areas where potato is consumed in large quantities like in the highlands of Latin American countries, the potato contribution to the energy, protein, iron, and zinc intake is significant. In those areas, iron and zinc biofortified potatoes are expected to contribute to reduce malnutrition and anemia. Nevertheless, to assess the full potential of the biofortified potatoes, human studies are required to gain insight on how much of the iron from biofortified potatoes are absorbed by the human body.

Regarding its contribution to human health, potatoes have shown promising health-promoting effects in human cell culture, experimental animals, and human clinical studies. Potato compounds such as phenolic acids and anthocyanins, glycoalkaloids, fiber, and proteinase inhibitors have been implicated in the suppression of cancer cell proliferation in vitro and in vivo and are believed to contribute to the hypocholesterolemic, anti-inflammatory, anti-obesity and anti-diabetic properties of potato.

References

Adebamowo SN, Spiegelman D, Willett WC (2015) Association between intakes of magnesium, potassium, and calcium and risk of stroke: 2 cohorts of US women and updated metaanalyses. Am J Clin Nutr 101:1269–1277

Akilen R, Deljoomanesh N, Hunschede S, Smith CE, Arshad MU, Kubant R, Anderson GH (2016) The effects of potatoes and other carbohydrate side dishes consumed with meat on food intake, glycemia and satiety response in children. Nutr Diabetes 6:e195. https://doi.org/10.1038/nutd.2016.1

American Optometric Association (2009) Diet, nutrition and eye health. http://www.aoa.org/documents/nutrition/Diet_Nutrition_Eye_Health_booklet.pdf. Accessed 28 Sept 2017

Amrein T, Bachmann S, Noti A, Biedermann M, Barbosa M, Biedermann-Brem S, Grob K, Keiser A, Realini P, Escher F, Amadó R (2003) Potential of acrylamide formation, sugars, and free asparagine in potatoes: a comparison of cultivars and farming systems. Agric Food Chem 51(18):5556–5560

Anderson GH, Soeandy CD, Smith CE (2013) White vegetables: glycemia and satiety. Adv Nutr 4(Suppl):356S–367S

Andre CM, Oufir M, Guignard C, Hoffmann L, Hausman JF, Evers D, Larondelle Y (2007) Antioxidant profiling of native Andean potato tubers (*Solanum tuberosum* L.) reveals cultivars with high levels of β-carotene, alpha-tocopherol, chlorogenic acid, and petanin. Agric Food Chem 55:10839–10849

Andre CM, Schafleitner R, Legay S, Lefèvre I, Aliaga CA, Nomberto G, Hoffmann L, Hausman JF, Larondelle Y, Evers D (2009) Gene expression changes related to the production of phenolic compounds in potato tubers grown under drought stress. Phytochemistry 70:1107–1116

Andre CM, Larondelle Y, Evers D (2010) Dietary antioxidants and oxidative stress from a human and plant perspective: a review. Curr Nutr Food Sci 6(1):2–12

Andre CM, Evers D, Ziebel J, Guignard C, Hausman JF, Bonierbale M, zum Felde T, Burgos G (2015) In vitro bioaccessibility and bioavailability of iron from potatoes with varying vitamin C, carotenoid, and phenolic concentrations. J Agric Food Chem 63(41):9012–9021. https://doi.org/10.1021/acs.jafc.5b02904

Auestad N, Hurley J, Fulgoni V, Schweitzer CM (2015) Contribution of food groups to energy and nutrient intakes in five developed countries. Nutrients 7:4593–4618

Bach S, Yada RY, Bizimungu B, Fan M, Sullivan JA (2013) Genotype by environment interaction effects on starch content and digestibility in potato (*Solanum tuberosum* L.). J Agric Food Chem 61(16):3941–3948. https://doi.org/10.1021/jf3030216. Epub 2013 Apr 12

Bamberg JB, del Rio A (2005) Conservation of potato genetic resources. In: Razdan MK, Mattoo AK (eds) Genetic improvement of solanaceous crops. Volume I: Potato. Science Publishers, Inc., Plymouth, p 476

Bates C (1997) Vitamin Analysis. Ann Clin Biochem 34(6):599–626. https://doi.org/10.1177/000456329703400604

Bethke P, Jansky S (2018) The Effects of Boiling and Leaching on the Content of Potassium and Other Minerals in Potatoes. J Food Sci 73(5):H80–H85

Bohn T (2017) Bioactivity of carotenoids–chasms of knowledge. Int J Vitam Nutr Res 10:1–5. https://doi.org/10.1024/0300-9831/a000400

Bonierbale M, Burgos G, zum Felde T, Sosa P (2010) Composition nutritionnelle des pommes de terre. Cahiers de nutrition et dietétique 45:S28–S36

Borch D, Juul-Hindsgaul N, Veller M, Astrup A, Jaskolowski J, Raben A (2016) Potatoes and risk of obesity, type 2 diabetes, and cardiovascular disease in apparently healthy adults: a systematic review of clinical intervention and observational studies. Am J Clin Nutr 104:489–498

Borgi L, Rimm EB, Willett WC, Forman JP (2016) Potato intake and incidence of hypertension: results from three prospective US cohort studies. BMJ 353:i2351. https://doi.org/10.1136/bmj.i2351

Bramley PM, Elmadfa I, Kafatos A, Kelly FJ, Manios Y, Roxborough HE, Schuch W, Sheehy PJ, Wagner JH (2000) Vitamin E. J Sci Food Agric 80:913–938

Brenna O, Berardo N (2004) Applications of near-infrared reflectance spectroscopy (NIRS) to the evaluation of carotenoids in maize. J Agri Food Chem 52:5577–5582

Brennan CS (2005) Dietary fibre, glycaemic response, and diabetes. Mol Nutr Food Res 49:560–570

Brown CR, Wrolstad R, Durst R, Yang P, Clevidence B (2003) Breeding studies in potatoes containing high concentrations of anthocyanins. Am J Pot Res 80:241. https://doi.org/10.1007/BF02855360

Brown CR, Culley D, Yang CP, Durst R, Wrolstad R (2005) Variation of anthocyanin and carotenoid contents and associated antioxidant values in potato breeding lines. J Am Soc Hortic Sc 130:174–180

Burgos G (2014) Concentración y bioaccesibilidad de carotenoides y compuestos fenólicos en papas cocidas. Tesis doctoral. Universidad de la Laguna. Departamento de Química Analítica, Nutrición y Bromatología

Burgos G, Amoros W, Morote M, Stangoulis J, Bonierbale M (2007) Iron and zinc concentration of native Andean potato varieties from a human nutrition perspective. J Sci Food Agric 87(4):668–675

Burgos G, Salas E, Amoros W, Auqui M, Munoa L, Kimura M, Bonierbale M (2009) Total and individual carotenoid profiles in the Phureja group of cultivated potatoes: I. concentrations and relationships as determined by spectrophotometry and high performance liquid chromatography (HPLC). J Food Compos Anal 22:503–508

Burgos G, Amoros W, Salas E, Muñoa L, Sosa P, Díaz C, Bonierbale M (2012) Carotenoid concentrations of native Andean potatoes as affected by cooking. Food Chem 133:1131–1137

Burgos G, Muñoa L, Sosa P, Bonierbale M, Zum Felde T, Díaz C (2013a) In vitro bioaccessibility of lutein and zeaxanthin of yellow fleshed boiled potatoes. Plant Foods Hum Nutr 68(4):385–390

Burgos G, Amoros W, Muñoa L, Sosa P, Cayhualla E, Sanchez C, Diaz C, Bonierbale M (2013b) Total phenolic, total anthocyanin and phenolic acid concentrations and antioxidant activity of purple-fleshed potatoes as affected by boiling. J Food Compos Anal 30:6–12

Burlingame B, Mouille B, Charrondiere R (2009) Nutrients, bioactive non-nutrients and antinutrients in potatoes. J Food Compos Anal 22:494–502

Camire ME, Kubow S, Donnelly DJ (2009) Potatoes and human health. Crit Rev Food Sci Nutr 49:823–840

Cevallos-Casals B, Cisneros Zevallos L (2003) Stoichiometric and kinetic studies of phenolic antioxidants from Andean purple corn and red-fleshed sweet potato. J Agric Food Chem 51:3313–3319

Champ MMJ (2004) Physiological aspects of resistant starch and in vivo measurements. J AOAC Int 87:749–755

Charepalli V, Reddivari L, Radhakrishnan S, Vadde R, Agarwal R, Vanamala JK (2015) Anthocyanin-containing purple-fleshed potatoes suppress colon tumorigenesis via elimination of colon cancer stem cells. J Nutr Biochem 26(12):1641–1649. https://doi.org/10.1016/j.jnutbio.2015.08.005

Chitchumroonchokchai C, Diretto G, Parisi B, Giuliano G, Failla ML (2017) Potential of golden potatoes to improve vitamin A and vitamin E status in developing countries. PLoS One 12(11):e0187102

Chun J, Lee J, Ye L, Exler J, Eitenmiller RR (2006) Tocopherol and tocotrienol contents of raw and processed fruits and vegetables in the United States diet. J Food Compos Anal 19:196–204

Clifford M (1999) Chlorogenic acids and other cinnamates – nature, occurrence and dietary burden. J Sci Food Agric 79:362–372

Crapo PA, Reaven G, Olefsky J (1977) Postprandial plasma-glucose and -insulin responses to different complex carbohydrates. Diabetes 26:1178–1183

DeGraft-Johnson J, Kolodziejczyk K, Krol M, Nowak P, Krol B, Nowak D (2007) Ferric-reducing ability power of selected plant polyphenols and their metabolites: implications for clinical studies on the antioxidant effects of fruits and vegetable consumption. Basic Clin Pharmacol Toxicol 100:345–352

De Haan S, Burgos G, Liria R, Rodriguez F, Creed-Kanashiro H, Bonierbale M (2019) The Nutritional Contribution of Potato Varietal Diversity in Andean Food Systems: a Case Study. Am J Potato Res 96:151. https://doi.org/10.1007/s12230-018-09707-2

Dragsted LO, Krath B, Ravn-Haren G, Vogel UB, Vinggaard AM, Jensen PB, Loft S, Rasmussen SE, Sandstrom B, Pedersen A (2006) Biological effects of fruit and vegetables. Proc Nutr Soc 65:61–67

Duarte-Delgado D, Ñústez-López C, Narváez-Cuenca C, Restrepo-Sánchez L, Melo S, Sarmiento F, Kushalappa A, Mosquera-Vásquez T (2016) Natural variation of sucrose, glucose and fructose contents in Colombian genotypes of *Solanum tuberosum* Group Phureja at harvest. J Sci Food Agric 96(12):4288–4294. https://doi.org/10.1002/jsfa.7783. Epub 2016 Jun 21

Duthie G, Duthie S, Kyle J (2000) Plant polyphenols in cancer and heart disease: implications as nutritional antioxidants. Nutr Res Rev 13:79–106

Ek KL, Brand-Miller J, Copeland L (2012) Glycaemic effect of potatoes. Food Chem 133:1230–1240

Englyst HN, Kingsman SM, Cummings JH (1992) Classification and measurement of nutritionally important starch fractions. Eur J Clin Nutr 46:S33–S50

Erdmann J, Hebeisen Y, Lippl F, Wagenpfeil S, Schusdziarra V (2007) Food intake and plasma ghrelin response during potato-, rice- and pasta-rich test meals. Eur J Clin Nutr 46:196–203

Ezekiel R, Singh N, Sharma S, Kaur A (2013) Beneficial phytochemicals in potato—a review. Food Res Int 50:487–496

FAO (2013) FAO statistical databases FAOSTAT. http://www.fao.org/faostat/en/#data/FBS. Accessed October 12th 2017

FAO/OMS/UNU (2004) Human energy requirements. Report of joint FAO/WHO/UNU expert consultation. FAO, Rome

FAO/WHO (2001) Human vitamin and mineral requirements. Report of a joint FAO/WHO expert consultation. Bangkok, Thailand

Fernandes G, Velangi A, Wolever T (2005) Glycemic index of potatoes commonly consumed in North America. J Am Diet Assoc 105:557–562

Fernández-Garcia E, Carvajal-Lérida I, Pérez-Gálvez A (2009) In vitro bioaccessibility assessment as a prediction tool of nutritional efficiency. Nutr Res 29:751–760

Fernandez-Orozco R, Gallardo-Guerrero L, Hornero-Méndez D (2013) Carotenoid profiling in tubers of different potato (*Solanum* sp.) cultivars: accumulation of carotenoids mediated by xanthophyll esterification. Food Chem 141:2864–2872

Fiedor J, Burda K (2014) Potential role of carotenoids as antioxidants in human health and disease. Nutrients 6(2):466–488

Fitzpatrick TB, Basset GJ, Borel P, Carrari F, Della Penna D, Fraser PD, Hellmann H, Osorio S, Rothan C, Valpuesta V (2012) Vitamin deficiencies in humans: can plant science help? Plant Cell 24:395–414

La Frano M, de Moura F, Boy E, Lönnerdal B, Burri B (2014) Bioavailability of iron, zinc, and provitamin A carotenoids in biofortified staple crops. Nutr Rev 72(5):289–307. https://doi.org/10.1111/nure.12108.

Fossen T, Andersen Ø (2000) Anthocyanins from tubers and shoots of the purple potato, Solanum tuberosum. J Hortic Sci Biotechnol 75(3):360–363. https://doi.org/10.1080/14620316.2000.11511251

Friedman M (1997) Chemistry, biochemistry, and dietary role of potato polyphenols. A review. J Agric Food Chem 45:1523–1540

Friedman M (2004) Analysis of biologically active compounds in potatoes (Solanum tuberosum), tomatoes (Lycopersicon esculentum), and jimson weed (Datura stramonium) seeds. J Chromatogr A 1054(1–2):143–155

Friedman M (2006) Potato glycoalkaloids and metabolites: roles in the plant and in the diet. J Agric Food Chem 54:8655–8681

Friedman M (2015) Chemistry and anticarcinogenic mechanisms of glycoalkaloids produced by eggplants, potatoes, and tomatoes. J Agric Food Chem 63:3323–3337
Friedman M, McDonald G, Filadelfi-Keszi M. (1997) Potato glycoalkaloids: chemistry, analysis, safety, and plant physiology. Critical Reviews in Plant Sciences 16(1):55–132. https://doi.org/10.1080/07352689709701946
Friedman M, Roitman JN, Kozukue N (2003) Glycoalkaloid and calystegine contents of eight potato cultivars. J Agric Food Chem 51(10):2964–2973
Friedman M, Lee KR, Kim HJ, Lee IS, Kozukue N (2005) Anticarcinogenic effects of glycoalkaloids from potatoes against human cervical, liver, lymphoma, and stomach cancer cells. J Agric Food Chem 53:6162–6169
Fudge J, Mangel N, Gruissem W, Vanderschuren H, Fitzpatrick T (2017) Rationalising vitamin B_6 biofortification in crop plants. Curr Opin Biotechnol 44:130–137. https://doi.org/10.1016/j.copbio.2016.12.004
Galliard T (1973) Lipids of potato tubers. I. Lipid and fatty acid composition of tubers from different varieties of potato. J Sci Food Agric 24(5):617–622
Gammone MA, Riccioni G, D'Orazio N (2015) Carotenoids: potential allies of cardiovascular health? Food Nutr Res 59:26762. https://doi.org/10.3402/fnr.v59.26762
Garcia-Alonso M, Pascual-Teresa S, Santos-Buelga C, Rivas-Gonzalo J (2004) Evaluation of the antioxidant properties of fruits. Food Chem 84:13–18
Garcia-Closas R, Berenguer A, Tormo MJ, Sanchez MJ, Quiros JR, Navarro C, Arnaud R, Dorronsoro M, Chirlaque MD, Barricarte A, Ardanaz E, Amian P, Martinez C, Agudo A, Gonzalez CA (2004) Dietary sources of vitamin C, vitamin E and specific carotenoids in Spain. Br J Nutr 91:1005–1011
Gibson S, Kurilich A (2013) The nutritional value of potatoes and potato products in the UK diet. Nutr Bull 38:389–399
Giusti MM, Polit MF, Ayvaz H, Tay D, Manrique I (2014) Characterization and quantitation of anthocyanins and other phenolics in native Andean potatoes. J Agric Food Chem 62:4408–4416
Habtemariam S, Lentini G (2015) The therapeutic potential of rutin for diabetes: an update. Mini Rev Med Chem 15(7):524–528
Han J, Kosukue N, Young K, Lee K, Friedman M (2004) Distribution of ascorbic acid in potato tubers and in home-processed and commercial potato foods. J Agric Food Chem 52(21):6516–6521
Hart D, Scott KJ (1995) Development and evaluation of an HPLC method for the analysis of carotenoids in foods and measurement of the carotenoid content of vegetables and fruits commonly consumed in the UK. Food Chem 54:101–111
Hashimoto N, Ito Y, Han KH, Shimada K, Sekkikawa M, Topping DL (2006) Potato pulps lowered the serum cholesterol and triglyceride levels in rats. J Nutr Sci Vitaminol 52:445–450
Hayashi K, Mori M, Matsunani Y, Suzutan T, Ogasawara M, Yoshida I, Hosokawa TA, Azuma M (2003) Anti Influenza Virus Activity of a Red-Fleshed Potato Anthocyanin. Food Sci Technol Res 9(3):242–244
Hayashi K, Hibasami H, Murakami T, Terahara N, Mori M, Tsukui A (2006) Induction of apoptosis in cultured human stomach cancer cells by potato anthocyanins and its inhibitory effects on growth of stomach cancer in mice. Food Sci Technol Res 12:22–26
Hellmann H, Mooney S (2010) Vitamin B6: A Molecule for Human Health? Molecules 15:442–459
Henry C, Lightowler H, Strik C, Storey M (2005) Glycaemic index values for commercially available potatoes in Great Britain. Br J Nutr 94(6):917–921
Holst B, Williamson G (2008) Nutrients and phytochemicals: from bioavailability to bioefficacy beyond antioxidants. Curr Opin Biotechnol 19:73–82
Hosseinian FS, Bea T (2007) Saskatoon and wild blueberries have higher antho-cyanin content than other Manitoba berries. J Agric Food Chem 55(26):10832–10838
Hu FB (2002) Dietary pattern analysis: a new direction in nutritional epidemiology. Curr Opin Lipidol 13(1):3–9
Ieri F, Innocenti M, Andrenelli L, Vecchio V, Mulinacci N (2011) Rapid HPLC/DAD/MS method to determine phenolic acids, glycoalkaloids and anthocyanins in pigmented potatoes (*Solanum tuberosum* L.) and correlations with variety and geographical origin. Food Chem 125:750–759

Jiao J, Xu J-Y, Zhang W, Han S, Qin L-Q (2015) Effect of dietary fiber on circulating C-reactive protein in overweight and obese adults: a meta-analysis of randomized controlled trials. Int J Food Sci Nutr 66:114–119

Kanazawa T, Atsumi M, Mineo H, Fukushima M, Nishimura N, Noda T (2008) Ingestion of gelatinized potato starch containing a high level of phosphorus decreases serum and liver lipids in rats. J Oleo Sci 57:335–343

Kaspar KL, Park JS, Brown CR, Mathison BD, Navarre DA, Chew BP (2011) Pigmented potato consumption alters oxidative stress and inflammatory damage in men. J Nutr 141:108–111

Kenny OM, McCarthy CM, Brunton NP, Hossain MB, Rai DK, Collins SG, Jones PW, Maguire AR, O'Brien NM (2013) Anti-inflammatory properties of potato glycoalkaloids in stimulated Jurkat and Raw 264.7 mouse macrophages. Life Sci 92:775–782

Khachick F, Goli M, Beecher G, Holden J, Lusby W, Tenorio M, Barrera M (1992) Effect of food preparation on qualitative and quantitative distribution of major carotenoid constituents of tomatoes and several green vegetables. J Agric Food Chem 40:390–398

Khachik F, Cohen L, Zhao Z (1999) Metabolism of dietary carotenoidsand their possible role in prevention of cancer and macular de-generation. In: Shibamoto T, Terao J, Osawa T (eds) Functional foods for disease prevention I, American Chemical Society symposium series. Oxford University Press, New York, pp 71–85

Kimura M, Rodriguez-Amaya DB (2003) Carotenoid Composition of Hydroponic Leafy Vegetables. J Agric Food Chem 51:2603–2607. http://dx.doi.org/10.1021/jf020539b

King J, Slavin J (2013) White potatoes, human health, and dietary guidance. Adv Nutr 4:393S–401S. https://doi.org/10.3945/an.112.003525

Krinsky NI, Johnson EJ (2005) Carotenoid actions and their relation to health and disease. Mol Aspects Med 26(6):459–516. Review

Kubow S, Iskandar MM, Melgar-Bermudez E, Sleno L, Sabally K, Azadi B, How E, Prakash S, Burgos G, Felde TZ (2017) Effects of simulated human gastrointestinal digestion of two purple-fleshed potato cultivars on anthocyanin composition and cytotoxicity in colonic cancer and non-tumorigenic cells. Nutrients 9(9):pii: E953. https://doi.org/10.3390/nu9090953

Külen O, Stushnoff C, Holm DG (2012) Effect of cold storage on total phenolics content, antioxidant activity and vitamin C level of selected potato clones. J Sci Food Agric 93:2437–2444

Kumar D, Singh B, Parveen K (2004) An overview of the factors affecting sugar content of potatoes. Ann Appl Biol 145:247–256

Lachman J, Hamouz K (2005) Red and purple coloured potatoes as a significant antioxidant source in human nutrition – a review. Plant Soil Environ 51(11):477–482

Lachman J, Hamouz K, Orsak M, Pivec V (2001) Potato glycoalkaloids and their significance in plant protection and human nutrition – review. Rosstlinna Vyrova 47(4):181–191

Lachman J, Hamouz K, Sulc M, Orsak M, Pivec V, Hejtmankova A (2009) Cultivar differences of total anthocyanins and anthocyanidins in red and purple-fleshed potatoes and their relation to antioxidant activity. Food Chem 114:836–843

Lachman J, Hamouz K, Musilová J, Hejtmánková K, Kotíková Z, Pazderu K, Domkárová J, Pivec V, Cimr J (2013) Effect of peeling and three cooking methods on the content of selected phytochemicals in potato tubers with various colour of flesh. Food Chem 138:1189–1197

Lancaster J, Reay P, Norris J, Butler R (2000) Induction of flavonoids and phenolic acids in apple by UV-B and temperature. J Hortic Sci Biotechnol 75(2):142–148. https://doi.org/10.1080/14620316.2000.11511213

Langner E, Rzeski W, Kaczor J, Kandefer-Szersze M, Pierzynowski SG (2009) Tumour cell growth-inhibiting properties of water extract isolated from heated potato fibre (Potex). J Pre-Clin Clin Res 3:36–41

Larsson SC, Wolk A (2016) Potato consumption and risk of cardiovascular disease: 2 prospective cohort studies. Am J Clin Nutr 104:1245–1252

Lee KR, Kozukue N, Han JS, Park JH, Chang EY, Baek EJ, Chang JS, Friedman M (2004) Glycoalkaloids and metabolites inhibit the growth of human colon (HT29) and liver (HepG2) cancer cells. J Agric Food Chem 52:2832–2839

Leeman M, Östman E, Björck I (2005) Vinegar dressing and cold storage of potatoes lowers postprandial glycaemic and insulinaemic responses in healthy subjects. Eur J Clin Nutr 59:1266–1271

Liang N, Kitts D (2015) Role of chlorogenic acids in controlling oxidative and inflammatory stress conditions. Nutrients 8(1). https://doi.org/10.3390/nu8010016

Lin BW, Gong CC, Song HF, Cui YY (2017) Effects of anthocyanins on the prevention and treatment of cancer. Br J Pharmacol 174:1226–1243

Liu Q, Tarn R, Lynch D, Skjodt N (2007) Physicochemical properties of dry matter and starch from potatoes grown in Canada. Food Chem 105:897–907

Liyanage R, Han KH, Watanabe S, Shimada K, Sekikawa M, Ohba K (2008) Potato and soy peptide diets modulate lipid metabolism in rats. Biosci Biotechnol Biochem 72:943–950

Liyanage R, Han KH, Shimada KI, Sekikawa M, Tokuji Y, Ohba K (2009) Potato and soy peptides alter caecal fermentation and reduce serum non-HDL cholesterol in rats fed cholesterol. Eur J Lipid Sci Technol 111:884–892

Lockyer S, Spiro A, Stanner S (2016) Dietary fibre and the prevention of chronic disease—should health professionals be doing more to raise awareness? Nutr Bull 41:214–231

Lombardo S, Pandino G, Mauromicale G (2013) The influence of growing environment on the antioxidant and mineral content of "early" crop potato. J Food Compos Anal 32(1):28–35

Lovat C, Nassar A, Kubow S, Li X, Donnelly D (2016) Metabolic biosynthesis of potato (*Solanum tuberosum* L.) antioxidants and implications for human health. Crit Rev Food Sci Nutr 56(14):2278–2303

Love SL, Pavek JJ (2008) Positioning the potato as a primary food source of vitamin C. Am J Potato Res 85:277–285

Lv C, Kong H, Dong G, Liu L, Tong K, Sun H, Chen B, Zhang C, Zhou M (2014) Antitumor efficacy of α-solanine against pancreatic cancer in vitro and in vivo. PLoS One 9(2):e87868

Lynch DR, Liu Q, Tarn TR, Bizimungu B, Chen Q, Harris P, Chik CL, Skjodt NM (2007) Glycemic index: a review and implications for the potato industry. Am J Potato Res 84:179–190

Madiwale GP, Reddivari L, Holm DG, Vanamala J (2011) Storage elevates phenolic content and antioxidant activity but suppresses antiproliferative and pro-apoptotic properties of colored-flesh potatoes against human colon cancer cell lines. J Agric Food Chem 59:8155–8166

Makinen S, Kelloniemi J, Pihlanto A, Makinen K, Korhonen H, Hopia A (2008) Inhibition of angiotensin converting enzyme I caused by autolysis of potato proteins by enzymatic activities confined to different parts of the potato tuber. J Agric Food Chem 56:9875–9883

Mattila P, Hellstrom J (2006) Phenolic acids in potatoes, vegetables, and some of their products. J Food Compos Anal 20:152–160

McGill C, Kurilich A, Davignon J (2013) The role of potatoes and potato components in cardiometabolic health: a review. J Ann Med 45(7):467–473

Miao M, Jiang B, Cui S, Zhang T, Jin Z (2015) Slowly digestible starch–a review. Crit Rev Food Sci Nutr 55(12):1642–1657. https://doi.org/10.1080/10408398.2012.704434

Miller R, Stanner S (2016) Baked, mashed, boiled or fried—can potatoes increase the risk of hypertension? Nutr Bull 41:252–256

Mills CE, Tzounis X, Oruna-Concha MJ, Mottram DS, Gibson GR, Spencer JP (2015) In vitro colonic metabolism of coffee and chlorogenic acid results in selective changes in human faecal microbiota growth. Br J Nutr 113:1220–1227

Mínguez Mosquera M, Hornero-Méndez D (1994) Comparative Study of the Effect of Paprika Processing on the Carotenoids in Peppers (Capsicum annuum) of the Bola and Agridulce Varieties. J Agric Food Chem 42(7):1555–1560

Monro J, Mishra S, Blandford E, Anderson J, Genet R (2009) Potato genotype differences in nutritionally distinct starch fractions after cooking, and cooking plus storing cool. J Food Compos Anal 22(6):539–545

Moonney S, Chen L, Kuhn C, Navarre R, Knowles NR, Hellman H (2013) Genotype-specific changes in vitamin B_6 content and the PDX family in potato. Bio Med Res Int. Article ID 389723, 7 pages. https://doi.org/10.1155/2013/389723

Müller HZ (1997) Determination of the carotenoid content in selected vegetables and fruit by HPLC and photodiode array detection. Lebensm Unters Forsch 204:88. https://doi.org/10.1007/s002170050042

Muraki I, Rimm EB, Willett WC, Manson JE, Hu FB, Sun Q (2016) Potato consumption and risk of type 2 diabetes: results from three prospective cohort studies. Diabetes Care 39(3):376–384. https://doi.org/10.2337/dc15-0547

Muttucumaru N, Elmore J, Curtis T, Mottram D, Parry M, Halford N (2008) Reducing acrylamide precursors in raw materials derived from wheat and potato. J Agric Food Chem 56(15):6167–6172. https://doi.org/10.1021/jf800279d. Epub 2008 Jul 15

Naidu K (2003) Vitamin C in human health and disease is still a mystery? An overview. Nutr J 2:7

Nassar AM, Sabally K, Kubow S, Leclerc YN, Donnelly DJ (2012) Some Canadian-grown potato cultivars contribute to a substantial content of essential dietary minerals. J Agric Food Chem 60(18):4688–4696. https://doi.org/10.1021/jf204940t

Natella F, Nardini M, Belelli F, Scaccini C (2007) Coffee drinking induces incorporation of phenolic acids into LDL and increases the resistance of LDL to ex vivo oxidation in humans. Am J Clin Nutr 86:604–609

Navarre DA, Shakya R, Holden J, Kumar S (2010) The effect of different cooking methods on phenolics and vitamin C in developmentally young potato tubers. Am J Potato Res 87(4):350–359

Neilson J, Lagüe M, Thomson S, Aurousseau F, Murphy AM, Bizimungu B, Deveaux V, Bègue Y, JME J, Tai HH (2017) Gene expression profiles predictive of cold-induced sweetening in potato. Funct Integr Genomics 17(4):459–476

Niu K, Momma H, Kobayashi Y, Guan L, Chujo M, Otomo A, Ouchi E, Nagatomi R (2016) The traditional Japanese dietary pattern and longitudinal changes in cardiovascular disease risk factors in apparently healthy Japanese adults. Eur J Nutr 55:267–279

Ombra M, Fratianni F, Granese T, Cardinale F, Cozzolino A, Nazzaro F (2015) In vitro antioxidant, antimicrobial and anti-proliferative activities of purple potato extracts (Solanum tuberosum cv Vitelotte noire) following simulated gastro-intestinal digestion. Nat Prod Res 29(11):1087–1091. https://doi.org/10.1080/14786419.2014.981183

Olthof MR, Hollman PC, Buijsman MN, van Amelsvoort JM, Katan MB (2003) Chlorogenic acid, quercetin-3-rutinoside and black tea phenols are extensively metabolized in humans. J Nutr 133:1806–1814

Otten J, Hellwig P, Meyers L (2010) Dietary reference intakes: the essential guide to nutrient requirements. The National Academies Press, Washington, DC

Pascual-Teresa S, Sanchez-Ballesta MT (2008) Anthocyanins: from plant to health. Phytochem Rev 7:281–299

Pastorino S, Richards M, Pierce M, Ambrosini GL (2016) A high-fat, high-glycaemic index, low-fibre dietary pattern is prospectively associated with type 2 diabetes in a British birth cohort. Br J Nutr 115:1632–1642

Piñeros-Niño C, Narváez-Cuenca C, Kushalappa A, Mosquera T (2017) Hydroxycinnamic acids in cooked potato tubers from Solanum tuberosum group Phureja. Food Sci Nutr 5(3):380–389

Puchau B, Zulet MA, Gonzalez de Echavarri A, Hermsdorff HH, Martinez JA (2010) Dietary total antioxidant capacity is negatively associated with some metabolic syndrome features in healthy young adults. Nutrition 26(5):534–541

Raigond P, Ezekiel R, Raigond B (2014) Resistant starch in food. J Sci Food Agric 95(10):1968–1978. https://doi.org/10.1002/jsfa.6966. Review

Ramadan M, Oraby H (2016) Fatty acids and bioactive lipids of potato cultivars: an overview. J Oleo Sci 65(6):459–470. https://doi.org/10.5650/jos.ess16015

Rautiainen S, Serafini M, Morgenstern R, Prior RL, Wolk A (2008) The validity and reproducibility of food-frequency questionnaire–based total antioxidant capacity estimates in Swedish women. Am J Clin Nutr 87:1247–1253

Reddivari L, Hale AL, Miller JC (2007a) Determination of phenolic content, composition and their contribution to antioxidant activity in specialty potato selections. Am J Potato Res 84:275–282

Reddivari L, Vanamala J, Chintharlapalli S, Safe SH, Miller J (2007b) Anthocyanin fraction from potato extracts is cytotoxic to prostate cancer cells through activation of caspase-dependent and caspase-independent pathways. Carcinogenesis 28(10):2227–2235

Reddivari L, Vanamala J, Safe SH, Miller JC (2010) The bioactive compounds α-chaconine and gallic acid in potato extracts decrease survival and induce apoptosis in LNCaP and PC3 prostate cancer cells. Nutr Cancer 62:601–610

Reyes LF, Miller JC, Cisneros-Zevallos L (2005) Antioxidant capacity, anthocyanins and total phenolics in purple- and red-fleshed potato (*Solanum tuberosum* L.) genotypes. Am J Potato Res 82:271–277

Richelle M, Tavazzi I, Offord E (2001) Comparison of antioxidant activity of commonly consumed polyphenolic beverages (coffee, cocoa, tea) prepared per cup serving. J Agric Food Chem 49:3438–3442

Robert L, Narcy A, Rock E, Demigne C, Mazur A, Rémésy C (2006) Entire potato consumption improves lipid metabolism and antioxidant status in cholesterol-fed rat. Eur J Nutr 45:267–274

Roberts SB (2000) High-glycemic index foods, hunger, and obesity: is there a connection? Nutr Rev 58:163–169

Rodríguez B, Ríos D, Rodríguez E, Díaz C (2010) Influence of the cultivar on the organic acid and sugar composition of potatoes. J Sci Food Agric 90(13):2301–2309. https://doi.org/10.1002/jsfa.4086

Rodriguez-Amaya DB (2015) Status of carotenoid analytical methods and in vitro assays for the assessment of food quality and health effects. Curr Opin Food Sci 1:56–63. https://doi.org/10.1016/j.cofs.2014.11.005

Ruprich J, Rehurkova I, Boon PE, Svensson K, Moussavian S, Van der Voet H, Bosgra S, Van Klaveren JD, Busk L (2009) Probabilistic modeling of exposure doses and implications for health risk characterization: glycoalkaloids from potatoes. Food Chem Toxicol 47:2899–2905

Sajilata MG, Singhal RS, Kulkarni PR (2006) Resistant starch—a review. Compr Rev Food Sci Food Saf 5:1–17

Salmeron J, Manson JE, Stampfer MJ, Colditz GA, Wing AL, Willett WC (1997) Dietary fiber, glycemic load, and risk of non-insulin-dependent diabetes mellitus in women. J Am Med Assoc 277:472–477

Slavin JL (2008) Position of the American Dietetic Association: health implications of dietary fiber. J Am Diet Assoc 108:1716–1731

Soh NL, Brand-Miller J (1999) The glycaemic index of potatoes: the effect of variety, cooking method and maturity. Eur J Clin Nutr 53:249–254

Spanos GA, Wrolstad RE (1992) Phenolics of apple, pear, and white grape juices and their changes with processing and storage. A review. J Agric Food Chem 40:1478–1487

Spinneker A, Sola R, Lemmen V, Castillo MJ, Pietrzik K, González-Gross M (2007) Vitamin B_6 status, deficiency and its consequences—an overview. Nutr Hosp 22(1):7–24

Stalmach A, Steiling H, Williamson G, Crozier A (2010) Bioavailability of chlorogenic acids following acute ingestion of coffee by humans with an ileostomy. Arch Biochem Biophys 501:98–105

Sun H, Lv C, Yang L, Wang Y, Zhang Q, Yu S, Kong H, Wang M, Xie J, Zhang C, Zhou M (2014) Solanine induces mitochondria-mediated apoptosis in human pancreatic cancer cells. BioMed Res Int 2014:805926

Tajner-Czopek M, Jarych-Szyszka M, Lisińska G (2008) Changes in glycoalkaloids content of potatoes destined for consumption. Food Chem 106(2):706–711

Tee E, Lim C (1991) Carotenoid composition and content of Malaysian vegetables and fruits by the AOAC and HPLC methods. Food Chem 41:309–339

Teucher B, Olivares M, Cori H (2004) Enhancers of iron absorption: ascorbic acid and other organic acids. Int J Vitam Nutr Res. 74(6):403–419

Tian J, Chen C, Ye X, Chen S (2016) Health benefits of the potato affected by domestic cooking: a review. Food Chem 202:165–175

Tomas-Barberan F, García-Villalba R, Quartieri A, Raimondi S, Amaretti A, Leonardi A, Rossi M (2014) In vitro transformation of chlorogenic acid by human gut microbiota. Mol Nutr Food Res 58(5):1122–1131. https://doi.org/10.1002/mnfr.201300441. Epub 2013 Dec 23

van Het Hof KH, West CE, Weststrate JA, Hautvast JG (2000) Dietary factors that affect the bioavailability of carotenoids. J Nutr 130(3):503–506

Vaziri ND, Liu SM, Lau WL, Khazaeli M, Nazertehrani S, Farzaneh SH, Kieffer DA, Adams SH, Martin RJ (2014) High amylose resistant starch diet ameliorates oxidative stress, inflammation, and progression of chronic kidney disease. PloS One 9:e114881

Venn B, Green T (2007) Glycemic index and glycemic load: measurement issues and their effect on diet-disease relationships. Eur J Clin Nutr 61(Suppl 1):S122–S131

Veronese N, Stubbs B, Noale M, Solmi M, Vaona A, Demurtas J, Nicetto D, Crepaldi G, Schofield P, Koyanagi A, Maggi S, Fontana L (2017) Fried potato consumption is associated with elevated mortality: an 8-y longitudinal cohort study. Am J Clin Nutr 106:162–167

Vinson J, Demkosky C, Navarre D, Smyda M (2012) High-antioxidant potatoes: acute in vivo antioxidant source and hypotensive agent in humans after supplementation to hypertensive subjects. J Agric Food Chem 60(27):6749–6754. https://doi.org/10.1021/jf2045262

Visvanathan R, Jayathilake C, Chaminda Jayawardana B, Liyanage R (2016) Health-beneficial properties of potato and compounds of interest. J Sci Food Agric 96(15):4850–4860. https://doi.org/10.1002/jsfa.7848. Epub 2016 Jul 7. Review

Wada L, Ou B (2002) Antioxidant activity and phenolic content of Oregon cane-berries. J Agric Food Chem 50(12):3495–3500

Wahyuni Y, Ballester A, Sudarmonowati E, Bino RJ, Bovy AG (2011) Metabolite biodiversity in pepper (Capsicum) fruits of thirty-two diverse accessions: variation in health-related compounds and implications for breeding. Phytochemistry 72(11–12):1358–1370

Wang Q, Chen Q, He M, Mir P, Su J, Yang Q (2011) Inhibitory effect of antioxidant extracts from various potatoes on the proliferation of human colon and liver cancer cells. Nutr Cancer 63:1044–1052

Wang S, Copeland L, Brand-Miller J (2014) Discovery of a low-glycaemic index potato and relationship with starch digestion in vitro. Br J Nutr 111(4):699–705. https://doi.org/10.1017/S0007114513003048

Wang W, Sun C, Mao L, Ma P, Liu F, Yang J, Gao Y (2016) The biological activities, chemical stability, metabolism and delivery systems of quercetin: a review. Trends Food Sci Technol 56:21–38

WHO (World Health Organization) (2012) Guideline: potassium intake for adults and children. World Health Organization, Geneva

Winkel-Shirley B (2001) Flavonoid biosynthesis. A colorful model for genetics, biochemistry, cell biology, and biotechnology. Plant Physiol 126:485–493

Woolfe J, Poats S (1987) The Potato in the Human Diet. Cambridge University Press. 0521326699, 9780521326698

Wu X, Beecher GR, Holden JM, Haytowitz DB, Gebhardt SE, Prior RL (2006) Concentration of anthocyanins in common foods in the United States and estimation of normal consumption. J Agric Food Chem 54(11):4069–4075

Wu J, Cho E, Willett WC, Sastry SM, Schaumberg DA (2015) Intakes of lutein, zeaxanthin, and other carotenoids and age-related macular degeneration during 2 decades of prospective follow-up. JAMA Ophthalmol 133(12):1415–1424. https://doi.org/10.1001/jamaophthalmol.2015.3590

Yang SA, Paek SH, Kozukue N, Lee KR, Kim JA (2006) Alpha-chaconine, a potato glycoalkaloid, induces apoptosis of HT-29 human colon cancer cells through caspase-3 activation and inhibition of ERK 1/2 phosphorylation. Food Chem Toxicol 44:839–846

Yogendra KN, Kushalappa AC, Sarmiento F, Rodriguez E, Mosquera T (2015) Metabolomics deciphers quantitative resistance mechanisms in diploid potato clones against late blight. Funct Plant Biol 42:284–298

Yuan G, Sun B, Yuan J, Wang Q (2009) Effects of different cooking methods on health-promoting compounds of broccoli. J Zhejiang Univ Sci B 10(8):580–588. https://doi.org/10.1631/jzus.B0920051

Zeeman SC, Kossmann J, Smith AM (2010) Starch: its metabolism, evolution, and biotechnological modification in plants. Annu Rev Plant Biol 61:209–234

Zeng C (2013) Effects of different cooking methods on the vitamin C content of selected vegetables. Nutr Food Sci 43(5):438–443. https://doi.org/10.1108/NFS-11-2012-0123

3

Collective Health and Value Chain Development

André Devaux, Claudio Velasco, Miguel Ordinola, and Diego Naziri

Abstract The development community has shown increasing interest in the potential of innovation systems and value chain development approaches for reducing poverty and stimulating greater gender equity in rural areas. Nevertheless, there is a shortage of systematic knowledge on how such approaches have been implemented in different contexts, the main challenges in their application, and how they can be scaled to enable large numbers of poor people to benefit from participation in value chains. This chapter provides an overview of value chain development and focuses on the International Potato Center's experiences with the Participatory Market Chain Approach (PMCA), a flexible approach that brings together smallholder farmers, traders, processors, researchers, and other service providers in a collective process to explore potential business opportunities and develop innovations to exploit them. The PMCA is an exemplary case of South–South knowledge exchange: it was first developed and implemented in the Andes, but has since been introduced, adapted, and applied to different market chains in Africa and Asia, where it has contributed to improved rural livelihoods. The experiences of adjusting and implementing the approach in these different contexts and the outcomes of those interventions, and complementary approaches, are examined in this chapter. Lessons learned from these experiences are shared with a goal of informing the promotion, improvement, and scaling of value chain approaches in the future.

A. Devaux (✉) · C. Velasco
International Potato Center, Quito, Ecuador
e-mail: a.devaux@cgiar.org; c.velasco@cgiar.org

M. Ordinola
International Potato Center, Lima, Peru
e-mail: cip-incopa@cgiar.org

D. Naziri
International Potato Center, Hanoi, Vietnam, and Natural Resources Institute, University of Greenwich, Chatham, UK
e-mail: d.naziri@cgiar.org

3.1 Introduction

Agriculture is rapidly changing in developing countries in response to a variety of factors that include enhanced access to technologies, increased urbanization, shifting in diets, improvements in farmers' education and health, institutional and policy reforms, and investments in rural infrastructure. At the same time, the growing number of supermarkets—referred to as "supermarket revolution" (Reardon and Hopkins 2006)—and the smallholder agriculture integration in changing food markets are helping to create new income opportunities and a more dynamic environment; one in which new technologies and agricultural practices can contribute to helping smallholder farmers respond to these changes and associated challenges. Facilitating sustainable access to high-value markets can enable poor farmers to increase their incomes, making it an effective strategy for reducing poverty (Wiggins and Keats 2013). But those farmers are often at a disadvantage when it comes to producing for and doing business in high-quality food chains and ensuring compliance to both public and private standards (Henson and Humphrey 2010), given their limited access to financial and other services, and their largely poor organizational capacity for collective marketing. Furthermore, their market connections are mostly informal and characterized by high levels of distrust, uncertainty, and transaction costs. This is particularly true for perishable crops, such as potatoes, which are grown on small farms in mountainous areas. For agricultural research to benefit such farmers and make value chains more inclusive, it must be complemented by other efforts to improve the regulatory environment, alleviate resource constraints, and build local capacity to respond to evolving technological and economic challenges and opportunities. It may also be necessary to act to influence the incentives and constraints faced by market actors, so that they can communicate more effectively with farmers and establish mutually beneficial and enduring business relationships.

There are clear signs that agro-industries are having a significant impact on economic development and poverty reduction globally, both in urban and rural communities (FAO 2013). Despite the risks associated with high-value markets, changes in and around the agri-food sector can contribute to the development of better support services for farmers, such as technology, extension, and financial products. There is evidence that small producers with access to technical support services are more willing to adopt new technologies and make investments to take advantage of emerging market opportunities (Royer et al. 2016).

Traditionally, different organizations have designed and implemented different types of interventions in agriculture and associated markets. While public agricultural research and extension programs have focused mainly on increasing agricultural production and productivity, nongovernmental organizations (NGOs) and other entities have focused on commercialization and the development of inclusive value chains. The impact of interventions in these areas has primarily been constrained by the lack of holistic approaches that address the challenges and opportunities along the value chain while taking into consideration the needs and capacities of different value chain actors, from input suppliers to farmers, agro-industries, and consumers. Accordingly, development practitioners have progressively introduced the so-called innovation system approaches which bring together different value chain stakeholders in recognition of the fact that systematic change occurs through the interactions of multiple actors—both individual and institutional—and sources of agricultural knowledge and innovation (Biggs 2008; Schut et al. 2016). However, the practical application of agricultural innovation systems and inclusive value chain approaches—and in particular the integration of these two approaches—is a challenge. This has been documented by studies, such as those presenting the results of recent work by the CGIAR[1] consortium and partners in Africa, Asia, and Latin America, in which the opportunities arising from new and expanding markets for agricultural products are analyzed and the challenges for smallholder participation in these markets, and benefits derived from participation, are identified (Devaux et al. 2016). In Latin America, Devaux et al. (2009) present the case of the Papa Andina Initiative in the Andes, which used collective action to promote innovation in the market chain through two approaches: The Participatory Market Chain Approach (PMCA) (Devaux et al. 2009) and Innovation Platforms (Thiele et al. 2011b). Both approaches sought to promote the interaction of small-scale potato producers with market actors and agricultural service providers, with a goal of establishing alliances and contractual agreements in response to new market opportunities.

This chapter offers some perspectives on value chain development and presents the experiences of the International Potato Center (CIP) with the PMCA. This approach, originally developed to increase competitiveness and improve the livelihoods of small-scale potato producers in the Andes, has proved useful in other market chains and in other parts of the world, such as East Africa and Asia. The chapter unfolds as follows. We first present the experiences of implementing the approach in the Andes, and the adjustments that have been made to it, while analyzing factors that have influenced its implementation. We then describe and analyze the experiences of replicating and validating the approach in different contexts and regions. Finally, we discuss the lessons learned in order to inform the design of interventions that use the PMCA approach as a research for development tool and provide insights for its replication and adaptation elsewhere.

3.2 General Concepts of Value Chain Development

The term "value chain" is used in different ways in literature. In this chapter, by "value chain" we refer to a set of actors that interact to transform inputs and services into products with attributes that consumers are willing to buy. The debate surrounding value chains in recent years has been about understanding the changes that

[1] CGIAR is a global research partnership for a food-secure future (http://www.cgiar.org/)

take place in some rapidly evolving markets for agricultural products, the consequences of these changes for the poorest actors in the market (specifically small producers and small-to-medium-sized companies), and effective alternatives that governments, development organizations, and the private sector can provide to support those value chain actors. Millions of people with low-incomes—a large portion of them women—participate in agricultural value chains as producers, traders, processors, and retailers (Fig. 3.1). Many millions more, including the majority of the poor in the developing world, participate in agricultural value chains as workers or consumers. Improving the performance of agricultural value chains can thus benefit a large number of people (Reardon and Timmer 2012; Reardon et al. 2012).

We refer to "value chain development" (which we abbreviate as VCD hereafter) as a type of intervention aimed at reducing poverty through improved links between companies, urban dwellers and rural producers. It has been defined as "A positive or desirable change in a value chain to extend or improve productive operations and generate social benefits: poverty reduction, income and employment generation, economic growth, environmental performance, gender equity and other sustainable development goals" (UNIDO 2011). From this perspective, many development agencies, donors, and governments have adopted the value chain approach as a key element of their strategies to reduce rural poverty (Humphrey and Navas-Alemán 2010). In contrast to traditional agricultural research and development (R&D) approaches, which focus on improving the capacities of small producers to increase their productivity or better manage natural resources, the VCD approach challenges R&D organizations to work with diverse actors to understand how a value chain functions and identify mutually beneficial options to improve its efficiency.

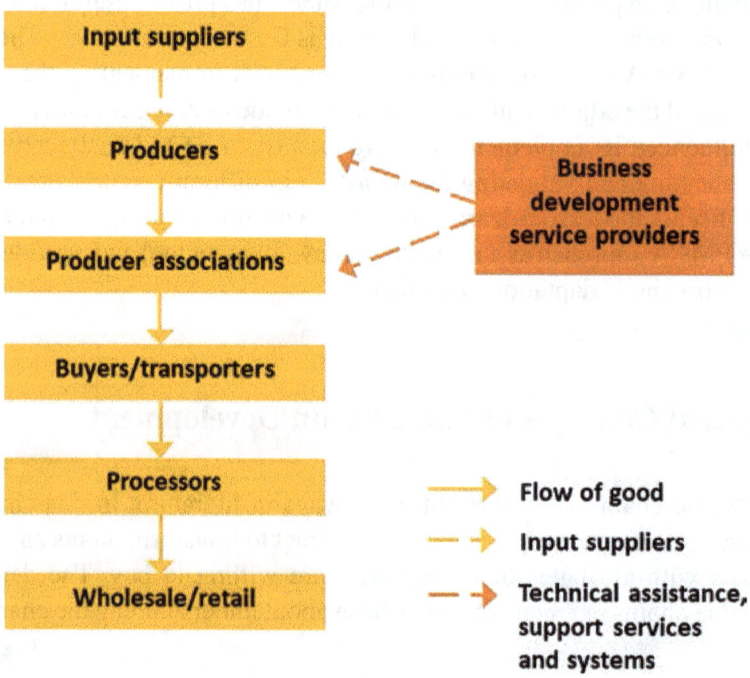

Fig. 3.1 Simple value chain (Devaux et al. 2016)

While the globalization of markets offers opportunities to market higher value products that simply did not exist before, these markets generally demand much more in terms of business acumen, efficiency and attention to quality and food safety standards than traditional product markets (Reardon et al. 2009). Participation in value chains for more demanding markets requires small-scale producers to deliver a regular supply of products of consistent quality in sufficient quantity at the right time and price. The fulfillment of these conditions requires access to land, inputs, technology, knowledge, organizational capacity, ability, and infrastructure, which may be lacking in some communities or groups of resource-poor farmers. Public policies are also needed to adapt government strategies to different situations and support the participation of smallholder farmers in more dynamic value chains. Smallholders are at a disadvantage compared to medium- and large-scale producers when it comes to taking advantage of these transformations, since they are often located in areas with less private and public infrastructure, are further away from markets, produce comparatively little amounts of products than need consolidation, and have less favorable conditions for high-yielding agriculture.

In this context, CIP, within the framework of its activities in the Andean region, developed a more integrated, participatory approach that combines agricultural, institutional, and value chain innovations while seeking synergies. Called the Participatory Market Chain Approach (PMCA), it is a flexible approach that tries to involve small-scale farmers, market agents (traders, companies, and processors among others), researchers, and service providers in a collective process to identify and explore potential business opportunities that can equitably benefit the diverse actors of a selected value chain (Bernet et al. 2006). It was developed and applied for the first time in Peru, to increase competitiveness in potato market chains, which are important components of local agri-food systems, and to contribute to improving the livelihoods of small farmers. Subsequently, thanks to CIP's global presence, the approach was introduced and adapted together with local organizations that applied it to different market chains in other countries in the Andes, Africa, and Asia, in an example of South–South knowledge exchange. The PMCA is most effective when it is implemented as part of a comprehensive strategy that includes support to farmers' organizations, business development, policy change, and public advocacy. CIP spearheaded the development of the PMCA and has supported the development of local capacities needed to facilitate successful innovation processes (Devaux et al. 2013).

3.3 The Participatory Market Chain Approach: Origin and Characteristics

The regional initiative known as *Papa Andina* (Andean Potato) was implemented from 1999 to 2010 to strengthen the capacity of R&D organizations in Bolivia, Ecuador and Peru with the overall objective of increasing competitiveness and improving the livelihoods of small-scale potato producers (Devaux et al. 2011).

In 2002, CIP social scientists working in the Papa Andina program and the Potato Innovation and Competitiveness Project in Peru (INCOPA)[2] began to experiment with a participatory approach known as Rapid Appraisal of Agricultural Knowledge Systems (RAAKS) (Engel and Salomon 1997). This approach brings together diverse stakeholders in a participatory process to stimulate collective learning, foster trust, and promote innovation. Papa Andina used RAAKS to identify market opportunities involving small-scale farmers, together with other value chain actors, researchers, and service providers.

The participation of traders, supermarkets, food processors, and chefs in a process of research for development was a radical break from previous participatory R&D efforts, which had been limited to researchers and farmers. Many researchers felt—and some still do—that working with market agents could distract research for development scientists from their focus, which was solving farmers' production problems. When the interventions were implemented, additional steps were added to RAAKS for the development of new products, resulting in a new approach, the PMCA, which was implemented in Peru and then validated in Bolivia in 2003. In subsequent years, this approach was applied in different contexts and thoroughly documented (Ordinola et al. 2009; Devaux et al. 2011).

The PMCA involves the individuals participating in a market chain and service providers, public and private, that support the chain—such as NGOs, financial services providers, development professionals, and researchers—in a facilitated process for identifying and developing innovations to exploit market opportunities. Those innovations can be technological, commercial, or institutional. The PMCA was conceived to be implemented in three phases that last between 9 and 18 months, according to the public and private context and the actors involved (Bernet et al. 2006; Antezana et al. 2008). However, when applied in different contexts, it became apparent that various activities might be required to consolidate progress in strengthening the relationships between the market chains' actors and follow up on technical and commercial innovations after the conclusion of the three phases.

The PMCA's phases (Fig. 3.2):

- **Phase 1. Getting to know market chain actors and their activities through an assessment**. The PMCA is initiated by an R&D organization that takes the lead in selecting the market chain in which it will work, identifying potential R&D partners and conducting exploratory and participatory diagnostic research on the chain. This phase, which can last from 2 to 4 months, concludes with a public event to discuss the assessment's results, generate ideas for possible innovations and motivate the actors of the market chain and service providers to participate in Phase 2.

[2] The INCOPA project, managed by CIP with support from the Swiss Agency for Development and Cooperation, was created to improve potato value chain competitiveness in Peru, with an emphasis on native potatoes grown by smallholder farmers in Andean highlands.

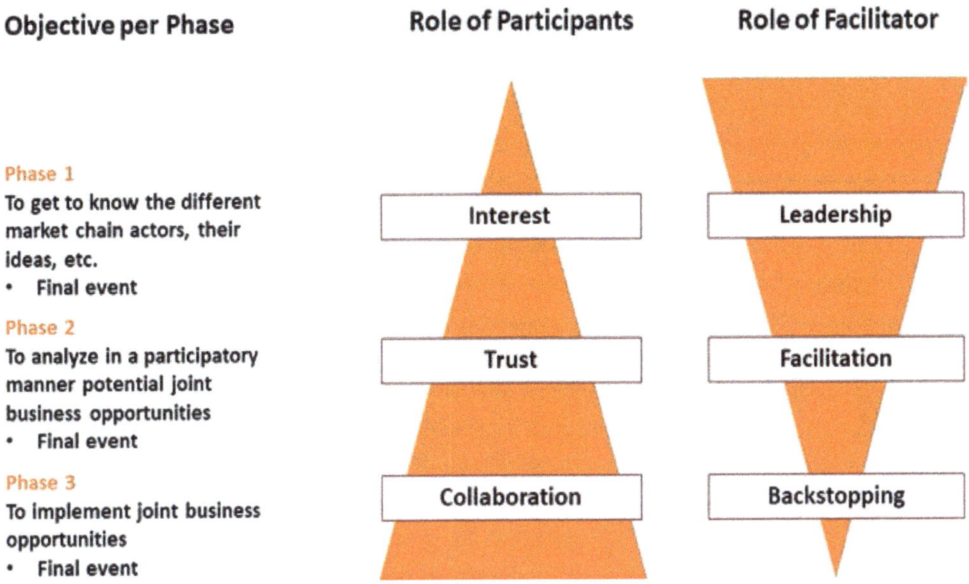

Fig. 3.2 The structure of the methodology: the three phases of the PMCA (Source: Bernet et al. 2006)

Phase 2. Joint analysis of potential market opportunities and innovations needed to take advantage of them. Representatives of participating R&D organizations facilitate the exploration and analysis of potential market opportunities. Actors of the value chain and service providers are organized in thematic groups that meet between 6 and 10 times to analyze potential opportunities, including thorough technical or market studies. A second public event takes place at the end of Phase 2 to discuss prioritized business opportunities and to encourage new stakeholders, with the appropriate knowledge and complementary experience, to join Phase 3.

Phase 3. Joint development of innovations. During this final phase, the group focuses on the development, market testing, and launch of specific innovations. This may require work in specialized areas such as processing, packaging, labeling or branding tests. Companies increasingly play a leading role in this phase. The PMCA exercise closes with a third public event in which the market innovations developed are presented to a wide selection of stakeholders such as public officials, potential donors, policy makers, decision makers from the private sector (supermarkets, processors, etc.), the public and representatives of the media.

Experiences show that the implementation of the PMCA can deviate from the orderly sequence of phases described above. While some groups have dissolved before producing any useful innovation, others have advanced and launched cost-effective innovations during Phase 2. Some groups have also continued to interact and generate innovations years after the conclusion of the PMCA exercise.

3.4 Examples of Value Chain Intervention Implemented in Several Geographies

Since the PMCA was developed, it has been adapted and applied to foster innovation in varied value chains in different countries, including the following:

- Potatoes in Bolivia, Peru, Ecuador, and Indonesia
- Dairy products in Bolivia
- Organic coffee in Peru
- Yams in Colombia
- Sweetpotatoes, tomatoes, and peppers in Uganda
- Potatoes, sweetpotatoes, bananas, and cassava in Uganda
- Aquaculture in Nepal and Bangladesh
- Amazonian fruits in Bolivia
- Fruits and vegetables in Nepal
- Organic and traditional products (e.g., home-made pasta, goat cheese, nuts, honey, mushrooms, tea, and dried fruits and vegetables) in Albania

Only a few of these applications have been well documented and evaluated. In the following sections, we analyze concrete examples of applications of the PMCA in interventions for which there is sufficient published information (Table 3.1).

3.4.1 Peru's Native Potato Revolution

From 2001 to 2010, more than 20 public organizations, NGOs and companies collaborated on the INCOPA project, within the framework of the Papa Andina regional initiative. The project focused on both commercial potato varieties and the

Table 3.1 Experiences with the PMCA around the globe

Experience	Product/s	Country/ies	Principal sources
(1) Peru's native potato revolution	Potato	Peru	Ordinola et al. (2009, 2013, 2014), Horton and Samanamud (2013)
(2) PMCA experiences in different value chains in the Andes	Coffee, milk, yam, and potato	Peru, Bolivia, Colombia	Horton et al. (2013c)
(3) The revalorization of native potatoes in Ecuador	Potato	Ecuador	Montesdeoca et al. (2013), Montesdeoca (2016)
(4) Building capacity for innovation in Ugandan value chains	Potato, sweetpotato, tomato, and hot peppers	Uganda	Mayanja et al. (2012, 2013), Horton et al. (2010)
(5) The PMCA and Farmer Business Schools in Indonesia	Potato, sweetpotato, vegetables, coffee, and cattle	Indonesia, Philippines	Horton et al. (2013a, b), International Potato Center (2017a)

native potato varieties that indigenous farmers in the Andes have traditionally grown and consumed. Its objective was to promote innovations in native potato production and commercialization in order to improve the competitiveness of this value chain for the benefit of the small-scale farmers that produce them. To this end, CIP and partners developed and used the PMCA as an applied research and development approach. INCOPA worked to link small-scale producers of native potatoes with researchers, development professionals, and a range of actors in potato market chains (supermarkets, processors, etc.) to capitalize on the biodiversity of native potatoes and their culinary, nutritional, and cultural attributes. The implementation of the PMCA was complemented by efforts to promote alliances and coordination between value chain actors through innovation platforms, to increase public awareness of the virtues of native potatoes and to support the formulation of policies in support of these chains. The PMCA acted as a catalyst for innovations that—together with different interventions in the technical, institutional and political spheres—triggered commercial, productive, and institutional processes that continue to this day, and contributed to what has been dubbed "Peru's native potato revolution" (Horton and Samanamud 2013).

In Peru, two cycles of the PMCA were implemented. The first application focused on the commercialization of improved potato varieties, whereas the second cycle focused on native potatoes. The exercises included the participation of not only researchers, Ministry of Agriculture officials, and typical market chain actors but also chefs and supermarket managers. This was the first time that such a diverse group met to collaborate on options for promoting the potato in Peru. The "new actors" brought new energy, perspectives, and ideas to discussions that had previously been dominated by R&D professionals.

The first application of the PMCA, which started in 2002 in Peru, gave rise to the country's first brand of selected fresh potatoes, *Mi Papa* (My Potato), which was distributed through the wholesale market in Lima. Subsequently, a new organization, CAPAC Peru, was created with the participants to promote the commercialization of high-quality local agricultural products, including potatoes marketed under the *Mi Papa* brand. The second application of the PMCA resulted in two new products made with Peru's native potatoes. *T'ikapapa* was launched as the first brand of gourmet, fresh native potatoes sold in supermarkets in the Peruvian capital, Lima. Shortly after that, *Jalca Chips*, an innovative potato chip product made of naturally colored (blue, red, yellow) native potatoes, was launched. Jalca Chips, sold in duty-free shops in Lima's international airport, opened the way for native potatoes into high-value, processed products markets. Both products were sold in grocery stores but were eventually replaced in the market by other brands of better quality using native potatoes, thanks to a process of creative imitation and improvement. Since then, more than 20 comparable native potato products have appeared on the market ranging from the ones produced by small provincial companies to the products of multinational companies, some of which are exported, as illustrated in Fig. 3.3 (Ordinola et al. 2009).

In addition to the new products, important ideas for policy initiatives, advocacy, and awareness campaigns to promote potato consumption emerged during the

Fig. 3.3 The PMCA as a trigger for innovation in the development of new potato products in Peru. (Source: Updated based on Ordinola et al. 2009)

second cycle of the PMCA. Perhaps the most remarkable one was the creation of a National Potato Day in Peru. It was established in 2005 and has been held annually since then, in both Lima and the provinces. The Minister of Agriculture (MINAGRI) established a special commission to organize the first event, which included organizations that participated in the PMCA. Today this commission continues to function as an institutionalized network and MINAGRI allocates an annual budget for the celebration of National Potato Day, which promotes potatoes to the Peruvian people. This commission supported the efforts of Peruvian government officials to get the United Nations to declare 2008 as the International Year of the Potato. After that proposal was accepted, the commission led numerous outreach activities to promote the celebration of the International Year of the Potato in Peru throughout 2008. The combination of innovations in the market chain, changes in supporting policies, and enhanced public awareness have contributed to improving the image and perceptions of native potatoes in Peru.

Encouraging the big players in the market During the application of the PMCA in Peru, some of the large players in the Peruvian market participated in the initial meetings, but then abandoned the process because they perceived the process time-consuming and saw few benefits in the short term. As a consequence, the first innovations involved mostly small processing companies and not all of their efforts were sustainable for different reasons, such as quality issues, lack of continuity and failure to position their products. The INCOPA project and its network of public-private collaborators backed the initiatives of the most innovative small entrepreneurs. As these products began to penetrate the markets, the large market players returned to learn about opportunities for developing new potato-based products. They supported the creation of institutional arrangements to organize an orderly supply of quality potatoes through commercial arrangements such as contracts to promote good practices that were backed by the project.

These efforts were supported by NGOs that worked in the regions where native potatoes are grown, and that helped producers to organize themselves so to be to supply a product that met the market's quality requirements. The CAPAC Peru platform,[3] which emerged as an institutional innovation through the PMCA, has supported the native potato value chain by promoting interactions between producers and market agents (Ordinola et al. 2009). With the revamped interest of bigger market players led to the creation of an array of new potato-based products and different brands of native potato chips of superior quality, which ranged from the products of multinationals such as Frito Lay and Grupo Gloria to those produced by small provincial companies, in response to market's diverse demands. The PMCA thus "triggered" a process whereby the second and third generation of innovations are often more important than the ones developed during the approach's initial application (Ordinola et al. 2013; Proexpansión 2011). Figure 3.3 illustrates this process.

Measuring impact In the province of Andahuaylas, in Peru's Apurímac Region, a study was implemented to evaluate the initial scope and impact of the INCOPA project's intervention through the CAPAC Peru platform, using the "impact pathway" methodological framework (Maldonado et al. 2011). In this region, INCOPA used the PMCA to promote the exploitation of market opportunities for native potato biodiversity and improvements in the competitiveness of the potato value chain. For this study, 80 producers in the intervention area were selected through stratified sampling (sampling units are grouped by geographical location), and additional 66 farmers in the same areas were identified as a counterfactual group.

The main conclusions of this case study include:

1. Potato producers in the studied area expanded and diversified their commercial relationships, mainly due to greater demand for native potatoes locally and from supermarkets and agribusinesses in Lima;
2. Small producers in the study area managed to develop business skills and improved their post-harvest management to the degree that they enjoyed increased demand from the new markets that they accessed;
3. Potato producers in the target group belong to organizations that maintained relationships and links with institutions which were engaged through the PMCA process and had continued to provide technical assistance and other services that allowed more efficient business operations;

[3] *Cadenas Productivas Agrícolas de Calidad en el Perú* (Quality Agricultural Production Chains), known as CAPAC Perú, was a second-level, social, economic, and technology promotion organization that provided specialized services for the development of production chains for potato and other tubers in Peru.

4. Farmers involved in the project attained higher average incomes through better prices (26% higher than the control group) and higher sales volumes of potatoes, especially native potatoes.

In a subsequent case study on the participation of potato growers in Peru's central highlands in native potato value chains, Tobin et al. (2016) studied the social differentiation between participants and nonparticipants within communities and the reasons for the inclusion or exclusion of households in the value chain. His findings indicate that participation in value chains is not necessarily beneficial for all small-scale farmers. For such programs to have broad social benefits, policies and other institutional arrangements are needed to minimize the risks associated with participation in such value chains and to provide support for participation in other types of less demanding markets. Since high-value markets often require more investment and assets, they are probably not the most appropriate option for lower-income households. Horton and Samanamud (2013) indicate that families with less land, less education, less access to credit, and less developed social networks have benefited less from new market opportunities. But despite these shortcomings, the same authors found that the innovation process generated by the PMCA helped to improve the image of native potatoes and link small-scale farmers with sufficient capacity to dynamic urban markets for potato-based products.

If we analyze the development of native potato value chains in Peru, it is clear that the country's economic and policy contexts have been favorable, including gov-ernment support and beneficial policies, the participation of a committed private sector, a gastronomic sector willing to promote native products, and the support of international donors such as the Swiss Agency for Development and Cooperation, which strengthened the innovation process through training and coaching in the implementation of the PMCA and the development of innovations. This process made it possible to transform a situation in which native potatoes were nothing more than a subsistence food for poor farmers in the Andean highlands, with little prospect of market-oriented agriculture, to one in which they are now recognized as a noteworthy and nutritious Peruvian product that deserves a price premium in urban markets and gourmet restaurants.

More recently, CIP has built upon this achievement with support from the International Fund for Agricultural Development (IFAD) in a project to strengthen innovation to improve the incomes, food security, and resilience of potato producers in Bolivia, Ecuador, and Peru. This initiative complements IFAD's public invest-ment projects in these three countries and has contributed to expanding the PMCA approach in the region.

3.4.2 *Analysis of PMCA Experiences in Different Value Chains in the Andes*

From 2007 to 2010, the Andean Change Alliance (*Alianza Cambio Andino*)[4] evaluated the processes and results of PMCA implementation in VCD interventions in the Andes. Eight applications of the PMCA were initiated under the leadership of professionals at agricultural R&D organizations in Bolivia, Colombia, Ecuador, and Peru. None of them had previously implemented the approach, but the Alliance provided them with training and backstopping during implementation. Out of the eight cases initiated, five were completed, four of which were analyzed based on the greatest learning potential with the resources available for the study (Horton et al. 2013c). Those cases are:

Case 1. Marketing high-quality coffee in San Martin, Peru
Case 2. Developing and marketing of new dairy products in Oruro, Bolivia
Case 3. Development of new markets for yams in Northern Colombia
Case 4. Conserving and marketing native potatoes in Northern Potosí, Bolivia

A summary of the results achieved and analysis of each case, according to Horton et al. (2013c), follows.

3.4.2.1 Case 1. Marketing of High Quality Coffee in San Martín, Peru

This case study focused on a women's group that took the lead in the development of a local market for locally produced coffee. Members of the group acquired knowledge and skills in coffee processing and marketing and established a new brand of coffee for the local market. Innovations included more careful selection of coffee beans and improvements in roasting, grinding, and packaging. The application of the PMCA motivated the creation of more networks and relationships among the different stakeholders. They participated in a public event to promote the region's coffee in 2010 that attracted local authorities, private sector players, media, and about 500 members of the public. Success in the commercialization of coffee helped to consolidate the women's group and raise its visibility in public and political circles, as well as in fairs and markets for organic products. They now play a more prominent role in the local agricultural system.

[4] Alianza Cambio Andino (Andean Change Alliance) was a regional program of cooperation among organizations and businesses in Bolivia, Colombia, Ecuador, and Peru that contributed to sustainable livelihoods in poor communities by improving the participation of small-scale farmers in innovation processes.

3.4.2.2 Case 2. Development and Marketing of New Dairy Products in Oruro, Bolivia

A local foundation, SEDERA, directed and facilitated implementation of the PMCA with the aim of diversifying the production of community dairy plants. SEDERA and a group of local farmers were successful in producing mozzarella cheese that met local quality requirements and was marketed under the brand *Vaquita Andina*. The PMCA motivated local dairy producers to diversify the types of cheese they produce and to improve the quality and sanitary standards of their product. While the original goal was to sell the new mozzarella cheese to local pizzerias, this did not materialize, mainly due to its relatively high price. Instead, the main consumers were high-income households willing to pay a premium for a locally produced cheese. The product has been sold in a store operated by SEDERA and in some high-end food markets, including a supermarket in Oruro. Although the economic benefits for small producers were limited, SEDERA gained experience in analyzing market chains and facilitating innovation processes and is now using a more market-oriented approach to its development work.

3.4.2.3 Case 3. Development of New Markets for Yams in Northern Colombia

In April 2008, the PBA Foundation facilitated the PMCA's implementation in 7 market chains over a period of 13 months. Some progress was made in improving the marketing of yams, but no new distinctive yam product was developed or marketed. To sell higher quality yams at better prices, small farmers increased plant density and improved the selection and cleaning of harvested tubers. Some shipments of fresh yams to the US were made, but the Colombian farmers faced strong competition from other Caribbean suppliers in that market. Commercial trials of high-quality yam fiber were hampered by a lack of resources for the construction of a pilot plant. In light of the small size of local farmers' organizations, the PBA Foundation worked to establish a regional network of local associations to improve their marketing performance. An unexpected result was the organization of suppliers within the local market to better coordinate the flow of products and stabilize price. The PBA Foundation has now incorporated elements of the PMCA into its portfolio of participatory methods.

3.4.2.4 Case 4. Conservation and Commercialization of Native Potatoes in Northern Potosí, Bolivia

PROINPA and the Center for Agricultural Development (CAD) have worked for years to conserve the biodiversity of potatoes and other Andean crops and to reduce rural poverty in northern Potosí. They facilitated the development of a new product called *Miskipapa*, which consisted of selected and washed native potatoes sold in meshes. To market Miskipapa, farmers had to improve the sorting and grading of their harvested potatoes. Although they expressed interest, there was little support from local government agencies. Miskipapa was sold for 3 years in supermarkets in La Paz and Cochabamba, in the shop of a mining union, two tourist hotels and farmers' markets. However, due to limitations in both supply and demand of native potatoes, the economic benefits for farmers were limited. Nevertheless, greater awareness of the crop's value contributed to renewed efforts to conserve the region's native potato biodiversity. The most significant result was the experience acquired by CAD, which changed its emphasis from enhancing production to promoting market chain innovation.

To analyze the factors that influenced PMCA implementation and results in these interventions, Horton et al. (2013c) used the "Institutional Analysis and Development (IAD) Framework" developed by Ostrom (2005) and modified Devaux et al. (2009) and Thiele et al. (2011a). This framework considers four main groups of factors that can influence the implementation and results of the PMCA:

- Macro context: which includes government policies, socio-economic conditions, and the agro-ecological characteristics of the region that can influence VCD.
- Market chain: biophysical and technological characteristics of the market chain in which the PMCA is being applied.
- Principal actors: attributes of the relevant market chain actors and service providers involved in the PMCA process.
- Rules in use: formal and (mainly) informal norms and customs that govern the behavior of participants.

The macro context The pro-market policies of Colombia and Peru provided a more favorable environment for the use of the PMCA than the policies of the Bolivian government, which emphasize the role of the state and "communitarian socialism." Agro-ecological environments can also have an effect on the implementation processes and results. In the Bolivian highlands, where poverty is very high and production conditions are adversely affected by climatic risks and the reduced use of inputs, there are severe limitations on the implementation of VCD approaches for the reduction of rural poverty.

The attributes of the chain Successful innovation is more likely in some market chains than in others. In the cases involving coffee in Peru and, to a lesser extent, dairy products in Bolivia, it was possible to bring in outside expertise to improve processing. On the contrary, in the cases of native potato and yam, the knowledge base for commercialization and processing was more restricted. One of the reasons was eventually because the participatory process did not involve the required diversity of participants because of the location and the opportunities linked to the value chain selected. Coffee and dairy products also presented more opportunities for processing, branding and product differentiation than potatoes and yams. As mentioned in the previous section, the processing of native potatoes into colorful potato chips has emerged as a viable enterprise that can respond to the demands of urban consumers in Peru and, to a lesser extent, in Bolivia. However, this type of industry is typically based in urban areas and was not considered to be a viable option for the potato farmers in northern Potosi, Bolivia.

The main actors The cases analyzed enabled the identification of three types of "champions" that can be crucial for the successful implementation of the PMCA, and for the approach's integration into the work of R&D organizations. The first type of champion is the facilitator who coordinates the group and supports the innovation processes; the second type is a manager/decision maker, who coordinates the VCD intervention and the implementation of the approach, including mobilization of resources; the third is a recognized leader within the market chain. In the first case study, for example, the facilitator, from the NGO Practical Action in Peru, played a key role in identifying and supporting local actors and facilitating the change processes. A senior manager of the NGO also provided strong institutional support. And the leader of the women's group led the development of the new coffee brand and the creation of networks with other actors in the local coffee sector. The leadership and investment capacity of the private sector are also crucial for the ultimate success of efforts to stimulate business innovations.

The rules in use This refers to social customs, norms and rules, both formal and informal, that guide human behavior on a day-to-day basis. It is an important component in the value chains that were analyzed, which were generally characterized by distrust and limited communication and interaction between the different actors in the chain (such as producers, intermediaries, processors and retailers), which of course limits coordination and collaboration. Ethnic and racial divisions and discrimination were notable in the value chains for potato in Bolivia and yam in Colombia, as compared to the coffee chain in Peru, where such divisions were not observed.

The rules in use (or "standard operating procedures") of the R&D organizations are also important. The PMCA is facilitated by individuals based at R&D organizations that have their own mandates, program structures, cultures, norms and external relationships. The mandate and culture at public agricultural research organizations can pose challenges for the successful implementation of the PMCA, as these organizations may be reluctant to work with the private entrepreneurs who process and market agricultural products. For example, the implementing organization in Peru, Practical Action, had a strong tradition of working with all sectors, so it easily incorporated the PMCA into its program to develop coffee markets in Peru. This is not always the case with public entities that are more focused on productive and technological issues.

Another important factor for the implementation of the PMCA is the process of training the people who will implement the approach, especially the group facilitators. If not done properly, deviations or failure to implement the approach may occur, as Horton et al. (2013c) documented. To support the training process, a guide for trainers was produced (Antezana et al. 2008).

3.4.3 The Revalorization of Native Potatoes in Ecuador

In November of 2008, through the Andean Change Alliance, CIP began collaborating with the Potato Program at Ecuador's National Agricultural and Livestock Research Institute (INIAP), a local NGO, Fundación Marco, and the Small-Scale

Potato Producers' Consortium (CONPAPA) on the implementation of PMCA in Ecuador, as part of an effort to create market opportunities based on the country's wealth of native potato varieties.

The qualitative assessment in the first phase of the PMCA involved 29 chefs and administrators of restaurants and hotels in different provinces of the country, in order to determine their knowledge and attitudes towards native potatoes and identify market niches. The results showed that these varieties were largely unknown in restaurant and hospitality sectors of the cities, where they could not be found in markets. Nevertheless, the participants agreed that they had an interesting business potential.

The final event of Phase 1 was thus dubbed as a "Meeting to seek for business opportunities in native potatoes," an activity attended by 35 participants that included chefs, farmers, processors, representatives of local NGOs and public officials. Two interest groups were formed in the event: one on culinary research with native potatoes and the other, considering the requirements of hotels and restaurants, focused on the supply of raw materials.

The participation of the chefs was central to the second phase of this PMCA experience to promote and give visibility to native potatoes in Ecuador. That phase culminated in a massive event organized by the Ecuador Gourmet Cooking School to present and share variations of some of the most traditional dishes in Ecuadorian cuisine made with native potatoes. This event raised the visibility of those varieties and subsequent efforts focused on taking advantage of opportunities in niche markets linked to gastronomy.

In the third phase, other restaurants joined the process in order to promote the consumption of native potatoes and visits by chefs to the farms of native potato growers were encouraged. Those who visited farms understood for the first time the reality of the cultivation of those potatoes, in mountainous areas with difficult access, and accepted that native potatoes should fetch a higher price in the market than commercial varieties. Farmers, for their part, had access for the first time to information concerning the quality and timeliness required for supplying potatoes to this type of market.

A year after initiating the PMCA, the *Papa Nativa* (Native Potato) brand was launched at the final event of the process with the slogan "Discovering the Andean flavor." The product consisted of selected and washed native potatoes in a mesh. The launch included press conferences and food fairs in the framework of the celebration of Ecuador's National Potato Day.

Developing the product as a commercial innovation triggered technological innovations at the farm level (fertilizer application, better seed management, pest management) and in the presentation of potatoes in the market (moving from an unsorted product to a selected one with a brand name). For the CONPAPA farmers' association, this meant raising members' capacities for postharvest handling and commercialization, since they began selling native potatoes in diverse presentations and linked with consumers such as restaurants and supermarkets in urban areas for the first time.

Although the "*Papa Nativa*—Discovering the Andean Flavor" brand has not prospered in the market due to supply problems, high transaction costs, and limited demand, the results of the PMCA were important because it showcased native potatoes in markets other than the local highland markets that they had traditionally been limited to. This result captured the attention of the company INALPROCES, which had been promoting high-quality products with social responsibility criteria for several years. Managers of INALPROCES were inspired by CIP's Papa Andina experience with native potatoes in Peru and foresaw interesting business opportunities in the production and sale of potato chips made with colorful native varieties in the Ecuadorian market and abroad.

INALPROCES partnered with INIAP, the Minga Foundation for Rural Action and Cooperation, CONPAPA and CIP's Ecuador office to identify native potatoes adequate for processing that could be produced and supplied by CONPAPA (Montesdeoca et al. 2013). Two colorful varieties obtained by INIAP by crossing selected native potatoes were chosen for the production of quality potato chips: Puca Shungo (Red Heart) and Yana Shungo[5] (Black Heart). Their main characteristic is the intense reddish or purple coloration in their flesh, which means they are rich in antioxidants, vitamins, and proteins, and differentiates them from traditional potatoes. In 2011, the brand of colorful native potato chips Kiwa was launched, promoted with support from international cooperation programs as a business model for corporate social responsibility. This business model included working in collaboration with a number of local actors to maintain consistent production levels and to ensure a supply of quality seed potatoes. The model also included training services for farmers based on the Farmers' Field Schools methodology. The business model promoted by INALPROCES was so successful from the beginnings that, in 2011, the company's work with Andean potato farmers was recognized as the best Corporate Social Responsibility project in Ecuador by the Ecuadorian-German Chamber of Industry and Commerce. It also won an international Award "Taste 11 Award for Top Innovation of Anuga—2011" in Germany for the native potato chips.

Since its launch on the market, the product has increased its sales in the domestic and export markets, including United States, Europe, and the Middle East (Fig. 3.4). This growth has promoted collaboration between CIP, INIAP, and CONPAPA farmers to carry out activities aimed at promoting the production of quality seed of the two selected varieties, increasing the volumes of their production and planning sales by CONPAPA to INALPROCES.

The development of Kiwa native potato chips is one of the most noteworthy examples of the evolution of the innovation process that, as we noted, continues and evolves long after the PMCA has formally ended. This commercial innovation has also catalyzed a range of technological innovations. For example, to satisfy the demand for seed potatoes of the varieties used in the production of the Kiwa product, these varieties were included in the seed multiplication program of the INIAP

[5] See technical data sheet: http://repositorio.iniap.gob.ec/handle/41000/3267

Fig. 3.4 Kiwa potato chips displayed at a supermarket in Saudi Arabia

high-quality seed production greenhouse. However, there are other challenges at the production level, mainly with changes in climate and frost in the Ecuadorian Andes that affected up to 80% of the production in some years. It is necessary to continue with the selection of varieties that are appropriate for quality processing and more tolerant to weather shocks. In the processing area, it is necessary to find more uses for these potatoes, such as mashed potatoes or frozen French-fry-cut native potatoes for export (Martin Acosta, Kiwa CEO, personal communication).

The Kiwa case is a good example of how market opportunities and desire of social responsible business can be driving forces in the establishment and operation of public–private partnerships for rural development. In particular, the partnership between INIAP, INALPROCES, and CONPAPA generated concrete research demands for the selection of varieties and specific requirements for the provision of high-quality seed and technology transfer services. Through this alliance, CONPAPA has strengthened the business management capacity of associated farmers and their ability to meet the demand of INALPROCES, access technical assistance services from public and private suppliers, and respond to other business opportunities. Currently CONPAPA, renamed "AGROPAPA" since 2014, produces certified seed under contract with the Ministry of Agriculture and has increased its customer base of fresh, quality ware potatoes in provincial markets.

Conclusions from This Case

If we apply the Institutional Analysis and Development (IAD) Framework as used by Horton et al. (2013c) to this case, we gain several insights. Regarding the macro context, it is clear that Ecuador's pro-market policies were not as favorable as Peru's in the case of the Native Potato Revolution. One of the great challenges has been to inspire entrepreneurship in face of a rather limited business mindset. Although Ecuador is one of the countries with the highest records of trademarks and inventions in the Andean region, these rarely translate into actual businesses and commercial uptake. As a result, fewer companies are initiated and fewer employment opportunities are created than in other countries (Wong and Padilla 2017). Furthermore, farmers have limited access to credit for both investment and working capital. As a matter of fact, none of the farmers involved in the PMCA had access to credit and, when necessary, they invested their limited family savings or got personal loans.

Regarding the value chain, native potatoes were unknown by urban consumers and it took a considerable effort to position them in that market as a fresh product and to connect producers with restaurants or potential buyers because there was no appropriate distribution system. Promoting native potatoes as colorful chips has been more successful. Given the smaller size of the Ecuadorian national market compared to Peru, the fact that the product is being placed in international markets has contributed to its success.

In terms of the principal actors, there was support from service providers and chefs in the initial implementation of the PMCA, but apparently, a recognized leader in the native potato chain was missing. When INALPROCESS launched the idea of producing colorful native potato chips, the leadership role was assumed by that company.

If we analyze the rules in use, the native potato value chain was not well organized; there was little communication and limited interaction between the different actors in the chain and service providers. The alliance between the private sector company INALPROCESS and the government institute INIAP was a new experience and it needed to be constructed giving consideration to the norms at INIAP, which has a more technological mandate than marketing and processing. Once the Kiwa native potato chips product was launched, the working model for providing training services and technical support by the company to help farmers respond to market requirements did not work systematically and was not connected with other service providers. For example, there was an unmet demand for a manual of procedures and best practices in crop management, post-harvest treatments, and potato selection, as well as clear and understandable rules and parameters to define the price paid to the farmer for his products.

Results: The implementation of the PMCA served to showcase native potatoes with the collaboration of the chefs and make them known to a broader public than the rural markets they have traditionally been sold in. The producers were able to meet the quality and supply schedule required by the processing company responding to demanding export potato markets, which has helped them access new market options for their potatoes.

Thanks to INALPROCESS, a colorful native potato chip product was developed which now has a presence in national and international markets. The product accounts today for 20% of the company's sales. They have been able to sell the Kiwa Native Andean Potato Chips (or Crisps) to retailers all over the world in presentations ranging from vending machine size to Club sizes, to retailers such as Costco in Canada and SnR in Philippines. The company is planning to expand mainly in the USA, the Middle East and Eastern Europe and to move into the organic-certified market (Martin Acosta, CEO Kiwa, personal communication).

3.4.4 Building Capacity for Innovation in Ugandan Value Chains

As part of CIP's efforts to facilitate South–South collaboration, the experiences of the PMCA in the Andes were shared with R&D professionals and actors in the value chains of various agricultural products in Uganda between 2005 and 2007.

The capacity development strategy implemented in Uganda included a series of complementary components. Two cross-learning visits were organized in the Andes for specialists from that country. This exchange generated enthusiasm and interest to apply the methodology in Uganda. Training workshops were organized at the beginning of each phase of the PMCA implementation in Uganda, which allowed members of the facilitating team to practice using tools in real-world situations. Participatory learning and joint decision-making by the facilitating team and thematic teams strengthened teamwork and empowered the participants.

The approach was subsequently validated in potato, sweetpotato, tomato, and hot pepper value chains in Uganda. Eight women from R&D organizations in the country led the application of the PMCA and facilitated the various meetings of the participants. Specialists from CIP and national entities from Peru and Bolivia provided capacity building and backstopping for the Ugandan team throughout the process (Mayanja et al. 2013; Horton et al. 2010).

Many organizations and individuals played roles in the introduction, validation and refinement of the PMCA in Uganda, including representatives of academic and research institutions, governmental and non-governmental organizations, the private sector and business organizations. Institutions in each of these categories had clear functions. The R&D institutions were responsible for introducing and facilitating the process of implementing the methodology. They also identified the actors, including service providers, in the market chain that ended up forming the thematic

Table 3.2 Participants in the first application of the PMCA in Uganda

R&D actors	Market chain actors
Potato work group	
Ssemwanga Center	Tom Cris (processor)
Africa 2000 Network—Uganda	Nyamarogo potato farmers
Agribusiness Initiative Trust	Uganda National Sweet Potato Producers Association (UNSPPA)
International Institute of Tropical Agriculture (IITA)/Foodnet	Merchants from Owino market

Adapted from Mayanja et al. (2012)

groups during the implementation of the approach. These different actors worked together to identify and exploit market opportunities. Through this process, innovations were generated.

The potato chain group, for example, was led by the Ssemwanga Center for Agriculture and Food Ltd., a consulting company with members who are also team leaders in the Uganda branch of the Africa 2000 network, the International Institute of Tropical Agriculture (a CGIAR center) and the Agribusiness Initiative Trust. The core team facilitated two thematic groups that were comprised of key actors, such as representatives of farmers' associations, processors and traders, who worked together in an attempt to address the challenges and opportunities identified in the market chain (Table 3.2). The diversity of organizations involved reflects the important role of partnerships in the promotion of innovations for market-driven development. Similarly, different organizations participated in the thematic groups that worked in sweetpotato and vegetable market chains.

The PMCA experience generated a series of results for participants that included the acquisition of new knowledge, skills, social networks and the ability to innovate. The market chain actors generated a series of viable commercial, technological and institutional innovations. The R&D actors (outside the core team members) and other service providers provided technical assistance, guidance and important links to the industry that contributed to the generation of innovations. Innovations that resulted from the PMCA included packaging and a better brand design for potato chips destined for high value markets, orange-fleshed sweetpotato flour for the local market, and sliced and dried hot pepper for export (Mayanja et al. 2012), some of these market innovations such as the potato chips continue to be produced.

It is important to mention that after the conclusion of the PMCA cycle, several activities were carried out to consolidate the progress in strengthening the relationships between the market chains' actors, follow up on commercial innovations, and promote and institutionalize use of the PMCA by R&D organizations in Uganda and other countries in the region. All these activities were carried out by the original facilitators, who voluntarily supported the effort after the initial funding for the PMCA had been exhausted. Advice and support was given to market chain actors to help them develop and market new products and present successful proposals for financing. Some of the original facilitators also served as trainers in workshops organized by other development programs in Uganda (Mayanja et al. 2013).

The PMCA was institutionalized by different entities through the implementation of projects that included the approach and promoted private sector participation. One example is the Mukono Zonal Institute for Agricultural Research and Development, whose director used the PMCA in other value chains, such as pineapple. The NGO VEDCO (Volunteer Efforts for Development Concerns) and local universities also used the approach for innovation in other market chains.

Development of Gender Guidelines for the PMCA in Uganda

During the different applications of the PMCA in Uganda, CIP researchers came to realize that the approach lacked ways to identify, analyze and address gender and generational differences among value chain actors, which reduced the approach's potential for having a positive impact in terms of ensuring that participants of both genders and all ages have equitable access to opportunities and benefits along value chains. To that end, CIP researchers and partners in Africa and the Andes developed a series of practical tools in recent years to integrate the gender approach into the PMCA's different phases of analysis and intervention. These tools are being tested and validated in projects both in Africa and the Andes Region, and there is an online guide available for their application (Mayanja et al. 2016). The guide seeks to create capacities among the facilitators of the PMCA to carry out a gender analysis to generate a better understanding of the different problems that men and women face in value chain interventions. The results of this analysis guide the development of gender strategies to promote equal opportunities for men and women to benefit from the PMCA.

Conclusions of PMCA Application in Uganda

The context in Uganda was favorable to the development of the market chain approach. Academic and research institutions, governmental and nongovernmental organizations and the private sector were actively involved in the concrete implementation of the PMCA.

From the beginning, the approach was applied to different chains, due to stakeholder interest, and its implementation was supported by different international cooperation projects. But despite the institutionalization of the approach by some organizations, it is important to recognize the need for adequate financing and institutional arrangements to accompany the work of the facilitators of the innovation process and to be able to follow up on it. After completing the PMCA, it is necessary to continue supporting the development process of emerging innovations and prototypes through business development services to entrepreneurs. Unfortunately, financing mechanisms for this type of services were not considered in the programs of the participating institutions.

The systematic process of training representatives of the R&D organizations that supported PMCA implementation was key to its development, and subsequent promotion in different value chains. The fact that women from different organizations led the process contributed to adjusting the approach to pay greater attention to gender issues along the chain. A recurring theme in the application of the PMCA in different contexts was the need to provide support to small-scale farmers to improve their productive and business capacity to respond to changing market demands. This is part of a development strategy that should be taken into account in national rural policies.

3.4.5 *The PMCA and Farmer Business Schools in Indonesia*

Between 2008 and 2009, CIP implemented the PMCA in fresh and processed potato value chains in West Java, Indonesia. Training and support were provided for a team of local implementers. The process resulted in 13 different, duly documented innovation processes, most of which generated new or improved processed products (mainly chips and sandwiches) rather than marketing fresh potatoes (Horton et al. 2013a, b).

During a project review, PMCA implementers identified limited business skills and inefficient farmer organizations as the main limitations that prevented the establishment of more effective links to the market. Farmers had little understanding of market opportunities and inadequate access to much needed market information as price trends and actual demand.

To address such shortcomings, CIP researchers integrated in the PMCA some elements of the Farmers Field School approach (FFS)—a group-based process that had primarily been used to promote integrated pest management by smallholder farmers (Orrego et al. 2009). This resulted in a new VCD approach that was named Farmer Business School (FBS). The FBS is a participatory action-learning process that aim to link groups of farmers to agricultural value chains (International Potato Center 2017a). It differs from other agricultural development approaches in that it focuses on the equitable and effective inclusion of small-scale farmers in value chains, instead of focusing only on farm production activities. The FBS builds farmers' capacity through a series of group learning activities based on real experiences during a farm production/marketing cycle while promoting interaction with other value chain actors. As a tangible outcome of FBS, participants are expected to have actual micro/small enterprises and business initiated or strengthened upon completing the FBS learning process. A FBS guide developed by (International Potato Center 2017b) covers five key issues:

1. Identification of market opportunities
2. Evaluation of market chains
3. Development of market-oriented innovations
4. Development of business plans
5. Provision of business support services

The FBS experience that CIP presented in Indonesia proved useful for strengthening farmers' organizations and building business skills and capacity. It was subsequently adapted by the FoodSTART project to the context of the highlands of the Cordillera region, in the Philippines, primarily to integrate it within the framework of a broader rural development investment project of the International Fund for Agricultural Development (IFAD) called CHARMP2. Between 2012 and 2013, the FBS approach was piloted with 6 groups of farmers (approximately 120 people) dedicated to commercial enterprises based on root and tuber crops, vegetables, coffee and livestock. Following the initial piloting, FBS was scaled to additional 66 groups (approximately 1600 people) between 2014 and 2015, through the collaboration between FoodSTART and CHARMP2. It is encouraging that, even after the formal partnership (and any financial and technical support from FoodSTART) ceased, CHARMP2 continued scaling the FBS approach with additional 25 farmers groups (approximately 1000 people). FBS is still the main VCD approach implemented by the Department of Agriculture through CHARMP2 and being FBS graduated has become a pre-requisite for accessing the livelihood assistance from the project (DA-CHARMP2 and CIP-FoodSTART+ 2018).

Based on the successful experience with CHARMP2, the FBS approach is currently being scaled in Asia by FoodSTART second phase (FoodSTART+) through partnerships with four IFAD investment projects in the Philippines, India and Indonesia. In the framework of FoodSTART+, FBS is being used as catalyst for introducing a number of farming and postharvest innovations to participant farmers groups: these include crop varieties, agronomic practices, processing technologies and marketing strategies needed to develop the new food products demanded by the market.

Comparison of the FFS, FBS and PMCA approaches Farmer field schools, farmer business schools and the participatory market chain approaches all employ the method of learning by doing to improve the well-being of smallholder farming families, but in different ways. The following is a comparative analysis of the three approaches that considers the scope and extent of implementation (Table 3.3).

This comparative analysis indicates that according to results achieved with the FBS in Asia, it could be an appropriate approach when the main challenge is for small-scale farmers' groups to exploit an existing well-defined and clear market

Table 3.3 Comparison of the FFS, FBS, and PMCA approaches

Criteria	FFS	PMCA	FBS
Types of actors involved	Smallholder farmers' groups	A diverse set of actors along the value chain	Focus on smallholder farmers' groups with interaction with other stakeholders in the chain
Themes covered	Agricultural production techniques	Participation to value chains and the promotion of interaction and collaboration among actors	Farmer organization and market development themes
System of focus	Crop system	Innovation with value chain actors to promote inclusive businesses	Agricultural business and value addition
Area of intervention	Strengthening farmer organization	Building "social capital bridges" to strengthen interaction and collaboration among multiple actors in different areas of the value chain	Strengthening farmer organization and innovation processes for enhanced participation in value chain
Implementation time frame	One crop cycle	Lasts as long as necessary to develop successful innovations in a value chain	Covers one farming and commercialization cycle

opportunity. PMCA, on the other hand, has demonstrated to be a preferred approach when the market opportunity is less defined or understood and where there are significant potential benefits for research and innovation in the processing or marketing stages (Devaux et al. 2011). PMCA needs commitment from research organizations while in the case of FBS it is less the case since the engagement with other chain actors and stakeholders is more sporadic. That said, the two approaches can be complementary when combined as the PMCA would contribute to conduct market research and to identify potential market opportunities to be developed. Then FBS can be run to help farmers group to exploit these market opportunities reinforcing farmers' business capacities and developing technological innovations required to respond to these opportunities.

The experience demonstrated the capacity of the actors who led the PMCA implementation process to adapt the approach to contexts and needs and adjust it. In Asia, the implementers developed the Farmer Business School approach to improve farmer's capacities to benefit from market opportunities. As indicated above, the FBS has been integrated into the framework of broader research for development projects, such as FoodSTART, as a tool for developing the farmers' capacity to access more dynamic markets, with also a gender focus, and considering the challenges of climate change. CIP and local partners are now validating the FBS in the Andes in combination with the application of PMCA. It is a process that is still in full development and needs to be further documented.

3.5 Lessons from Value Chain Approach Applications in Different Contexts

From the market chain experiences in different contexts described here we can draw lessons that can contribute to improving the design of future VCD interventions that use the PMCA as a collective research and innovation approach. It should be noted that in all of these cases, creative adaptations were made to the protocols set out in the implementation guide (Bernet et al. 2006). Following the analytical framework developed by Horton et al. (2013c), we can highlight the main factors that influenced these adaptations:

- the attributes of the context, which includes the policy environment and institutional framework;
- the attributes of the value chain;
- the characteristics of the participants and the facilitation role of the leading organization;
- the "rules in use" governing behavior and relationships between the value chain actors;
- and the importance of building the capacities of the professionals from R&D institutions responsible for implementing the approach.

The interaction of these variables and, especially the external environment of each context determined the implementation strategy and the forces that drove each implementation of the PMCA in the Andes, Asia, and Africa. Each of them was different, as were their scope and size.

Agricultural change creates opportunities for smallholders An important element for the promotion of the PMCA is the ever-growing integration of the agricultural sector into markets and value chains, which increases market opportunities for small-scale farmers. There are more opportunities for them to move beyond the practice of selling directly to rural consumers or traders in local markets to access modern value chains that can open doors to other buyers, including urban traders, processors and supermarket chains. When they have alternatives and the ability to participate in more competitive value chains, few smallholder farmers will limit themselves to local markets; the majority will try to sell to these new players, in order to reap greater benefits (Reardon et al. 2012).

By applying the PMCA, facilitating interactions between farmers and other value chain actors, and conducting market analyses, it is possible to identifying new commercial opportunities and market niches in which small-scale farmers can have a comparative advantage. In the PMCA user guide (Bernet et al. 2006), there is a tool called "impact filter" that provides a rapid, qualitative assessment of the expected impact of different market opportunities for the benefit of small-scale producers, while considering social and environmental factors. This tool allows R&D organizations to plan and guide interventions more effectively.

A holistic approach like the PMCA represents a new way of doing agricultural R&D Instead of undertaking research and then trying to transfer the results to farmers, this approach brings together a range of relevant actors—farmers, traders, processors, researchers and service providers—to set priorities and jointly develop innovations that respond better to the demands of the value chain. The approach has made it possible to achieve concrete results under different socioeconomic conditions, despite the fact that most development support for farmers continues to focus on agricultural production. This situation is reflected in the experiences analyzed in this chapter, which demonstrate a shortage of available or accessible services to accompany value chain innovation processes. The inclusive value chain development approach makes it clear that, in the debate on food security, more attention should be paid to post-harvest and value chain interventions, as opposed to focusing only on the farming sector. In the quest to improve food security, value chain efficiency, processing productivity and post-harvest management deserve almost as much attention as approaches to improve farm production.

Innovations take time and cannot always be programmed The PMCA triggers innovation processes that often continue long after the PMCA has formally ended. Second and third generation innovations tend to be even more important than the first ones, developed during the PMCA exercise, as is confirmed by the experiences in Peru, Ecuador and Uganda. However, this requires an investment in services and institutional mechanisms that allow monitoring and supporting innovations that arise beyond the life of a specific project. The lack of accompaniment of value chain innovations can be a limitation that reduces the scope of the innovation process. In the case of Peru, for example, the continued support of the Swiss Agency for Development and Cooperation allowed monitoring and accompaniment of innovation process beyond the formal conclusion of PMCA implementation.

Monitoring and evaluation challenges Most of the cases studied lacked a systematic process of monitoring and evaluating value chain innovations. The fact that these processes can take a long time, and that many innovations are developed after the PMCA implementation, with unexpected results, makes it more challenging to do impact analysis. In many cases, the optimal time for evaluating impacts is well after project financing has concluded, when it is not feasible to obtain specific financing for evaluation processes. An implementation guide and impact pathway can be useful guidelines for the implementation and evaluation of participatory approaches such as the PMCA and can serve as a basis for reflection and learning. In research and development projects with limited resources and tight deadlines, it is difficult to justify the time and resources necessary to develop actions and models of change. For this reason, when planning participatory interventions, specific resources should be allocated for the development of actions and models of change that can then be refined and tested by local implementers as the innovation process progresses.

Holistic and participatory approaches such as the PMCA are not easily "scaled" or "transferred" How can the initiatives described here be scaled or replicated better? The answer is not found in a particular or specific arrangement. The processes and approaches, and not so much the forms, are what lead to effective interactions, collaboration between actors and development of a market chain (Wiggins and Keats 2013). What needs to be scaled, in the sense of replication and adaptation, are the habilitation, facilitation and learning processes, supported by a necessary architecture that includes catalysts and leadership, forums and services to analyze and respond to specific problems and, especially, support for the farmers' groups that are organized and trained. The catalysts can be private companies and/or non-governmental organizations (NGOs). Both have their advantages and disadvantages. In the case of the private sector, the challenge is how to establish an equitable relationship with small-scale producers. In the case of NGOs, they will play the role of catalyst, based on their mission, only as long as they can be financed. To achieve the replication of successful approaches, it is important to promote innovation, learning and dissemination of experiences. To date, investment in learning and dissemination of these experiences has not reached a level comparable to that of the promotion of practical initiatives in the field. The experiences of collaboration with IFAD's investment projects, first in Asia, with FoodSTART, and more recently in South America, with the FIDA-Andes project, coordinated by CIP, has facilitated learning from and diffusion of these experience. In that sense the experience of FoodSTART in the Philippines is innovative and particularly encouraging as, following successful piloting and strengthening of the capacities of the partner's development project staff, the FBS approach has been institutionalized by the Department of Agriculture that is replicating and scaling it throughout the region. Similar outcomes are expected from the ongoing second phase of FoodSTART.

The PMCA has proven to be adaptable to a range of contexts Although the PMCA was developed to respond to a specific set of problems with a specific crop in a specific place, subsequent efforts of promotion, exchange of experiences and capacity building have contributed to its successful application to other crops that face different challenges and opportunities in other countries and regions. In the trajectory of its applications in the Andes, Uganda and Indonesia, the PMCA was not simply "transferred" from one place to another. In each case, it was necessary to adapt the approach to local circumstances and needs and to strengthen capacities to promote and accompany innovations.

There is no "one-size-fits-all" solution to the challenges of value chains There is no single factor of success. In each case presented here, a combination of interventions at different levels was needed to address the value chain development processes. A set of policy measures, development programs and technical capacity building for value chain actors are needed to facilitate innovation processes and strengthen effective links to promote the participation of small-scale producers in expanding markets. A comparative study of the variables that influence and determine the adaptation of participatory methodologies such as PMCA in different contexts would allow a better definition of the conditions needed for the replication, adaptation and scaling of these methodologies.

References

Antezana I, Bernet T, López G, Oros R (2008) Enfoque Participativo en Cadenas Productivas (EPCP): Guía para capacitadores. International Potato Center, Lima, 189 pp

Bernet T, Thiele G, Zschocke T (2006) Participatory market chain approach (PMCA)—user guide. International Potato Center, Lima. http://cipotato.org/publications/pdf/003296.pdf

Biggs S (2008) Learning from the positive to reduce rural poverty and increase social justice: institutional innovations in agricultural and natural resources research and development. Exp Agric 44(1):37–60

DA-CHARMP2 and CIP-FoodSTART+ (2018) Most significant change in the Cordilleras: a pilot evaluation. Department of Agriculture-Second Cordillera Highland Agricultural Resource Management Project and International Potato Center-Food Resilience Through Root and Tuber Crops in Upland and Coastal Communities of the Asia-Pacific, Manila, Philippines

Devaux A, Horton D, Velasco C, Thiele G, López G, Bernet T, Reinoso I, Ordinola M (2009) Collective action for market chain innovation in the Andes. Food Policy 34:31–38

Devaux A, Ordinola M, Horton D (eds) (2011) Innovation for development: the Papa Andina Experience. International Potato Center (CIP), Lima, Peru, 314 pp

Devaux, A., Ordinola, M., Mayanja, S., Campilan, D. and Horton, D. (2013) The participatory market chain approach: from the Andes to Africa and Asia. Papa Andina innovation brief 1. International Potato Center, Lima, Peru, 4 pp

Devaux A, Torero M, Horton D, Donovan J (eds) (2016) Innovation for inclusive value chain development: successes and challenges. International Food Policy Research Institute (IFPRI), Washington, DC, 529 pp

Engel P, Salomon M (1997) Facilitating innovation for development: a RAAKS resource box. Royal Tropical Institute, Amsterdam. https://www.kit.nl/sed/publications/?search=raaks&q_author=0&q_year=0&display=5

FAO (2013) Agroindustrias para el desarrollo. Roma. http://www.fao.org/3/a-i3125s.pdf

Henson S, Humphrey J (2010) Understanding the complexities of private standards in global agri-food chains as they impact developing countries. J Dev Stud 46(9):1628–1646. https://doi.org/10.1080/00220381003706494

Horton D, Samanamud K (2013) Peru's native potato revolution. Papa Andina innovation brief 2. International Potato Center, Lima, Peru, 6 pp

Horton D, Akello B, Aliguma L, Bernet T, Devaux A, Lemaga B, Magala D, Mayanja S, Sekitto I, Thiele G, Velasco C (2010) Developing capacity for agricultural market chain innovation: experience with the 'PMCA' in Uganda. J Int Dev 22:367–389

Horton D, Campilan D, Prasetya B, Gani H, Pakih M, Kusmana (2013a) The PMCA, business development services and farmer business schools in Indonesia. Papa Andina innovation brief 5. International Potato Center, Lima, Peru, 6 pp

Horton D, Campilan D, Prasetya B, Gani H, Pakih MR, Kusmana (2013b) Agricultural market chain development in Indonesia: experiences with the 'Participatory Market Chain Approach', 'Farmer Business Schools', and 'Business development services'. Social Science working document 2013-1. International Potato Center (CIP), Lima

Horton D, Rotondo E, Paz Ybarnegaray R, Hareau G, Devaux A, Thiele G (2013c) Lapses infidelities, and creative adaptations: lessons from evaluation of a participatory market development approach in the Andes. Eval Program Plann 39:28–41

Humphrey J, Navas-Alemán L (2010) Value chains, donor interventions and poverty reduction: a review of donor practice. IDS research report 63. Institute of Development Studies, Brighton

International Potato Center (2017a) Farmer Business School, a climate smart and gender responsive approach. Brief. The International Potato Center-food security through Asian Roots and Tubers (CIP-FoodSTART+), IFAD funded Project, The Philippines, 4 pp

International Potato Center (2017b) Farmer Business Schools in a changing world: a gender responsive and climate smart manual for strengthening farmer entrepreneurship. International Potato Center, Lima. 2 Volumes

Maldonado L, Ordinola M, Manrique K, Fonseca C, Sevilla M, Delgado O (2011) Estudio de caso: Evaluación de impacto de la intervención del proyecto INCOPA/CAPAC en Andahuaylas. CIP, Lima

Mayanja S, Akello B, Horton D, Kisauzi D, Magala D (2012) Value chain development in Uganda: lessons learned from the application of the participatory market chain approach. BANWA 9:64–96

Mayanja S, Akello B, Horton D, Kisauzi D, Magala D (2013) Building capacity for market-chain innovation in Uganda. Papa Andina innovation brief 4. International Potato Center, Lima, 4 pp

Mayanja S, Barone S, McEwan M, Thomas B, Amaya N, Terrillon J, Velasco C, Babini C, Thiele G, Prain G, Devaux A (2016) Prototype guide for integrating gender into participatory market chain approach. International Potato Center, Lima

Montesdeoca L (2016) Reporte trianual 2014–2016 de TRIAS—AGROPAPA-CONPAPA, 14 pp

Montesdeoca L, Acosta M, Quishpe C, Monteros C, Andrade-Piedra J, Pavez I (2013) Rescatando variedades ancestrales: Innovación de las papas nativas en Ecuador. En: "Innovaciones de Impacto, lecciones de la agricultura familiar en América Latina y el Caribe"/Editado por Priscila Enriquez, Hugo Li Pun, San José de Costa Rica. IICA, BID, 224 pp

Ordinola M, Devaux A, Manrique K, Fonseca C, Thomann A (2009) Generando innovaciones para el desarrollo competitivo de la papa en el Perú. Centro Internacional de la Papa (CIP), Lima

Ordinola M, Devaux A, Manrique K, Fonseca C (2013) Innovaciones en la cadena de la papa en el Perú: El valor de la biodiversidad. Leisa, Revista Agroecológica, Volumen 29, número 2, Nuevos mercados, nuevos valores

Ordinola M, Devaux A, Bernet T, Manrique K, Lopez G, Fonseca C, Horton D (2014) The PMCA and potato market chain innovation in Peru. Papa Andina innovation brief 3. International Potato Center, Lima, 8 pp

Orrego R, Ortiz O, Pradel W, Arévalo A, Barrantes C, Macedo O (2009) Sistematización de la implementación de las Escuelas de Campo de Agricultores (ECAs) en Andahuaylas. Lima (Perú). Centro Internacional de la Papa, Cooperación al Desarrollo—CESAL, Junta de Comunidades de Castilla—La Mancha, 42 pp. ISBN 978-92-9060-371-9

Ostrom E (2005) Understanding institutional diversity. Princeton University Press, Princeton

Proexpansión (2011) Cambios del sector papa en el Perú en la última década: Los aportes del proyecto Innovación y Competitividad de la Papa (INCOPA). Centro Internacional de la Papa, Lima

Reardon T, Hopkins R (2006) The supermarket revolution in developing countries: policies to address emerging tensions among supermarkets, suppliers and traditional retailers. Eur J Dev Res 18:522. https://doi.org/10.1080/09578810601070613

Reardon T, Timmer P (2012) The economics of the food system revolution. Ann Rev Resour Econ 4:225–264

Reardon T, Barrett CB, Berdegué JA, Swinnen JFM (2009) Agrifood industry transformation and small farmers in developing countries. World Dev 37(11):1717–1727

Reardon T, Chen K, Minten B, Adriano L (2012) The quiet revolution in staple food value chains: enter the dragon, the elephant, and the tiger. Asian Development Bank/International Food Policy Research Institute, Mandaluyong City/Washington, DC

Royer A, Bijman J, Bitzer V (2016) Linking smallholder farmers to high quality food chains: appraising institutional arrangements. In: Bijman J, Bitzer V (eds) Quality and innovation in food chains. Wageningen Academic Publishers, Wageningen, pp 33–62. Chapter 2

Schut M, Klerkx L, Sartas M, Lamers D, McCampbell M, Ogbonna I, Kaushik P, Atta-Krah K, Leeuwis C (2016) Innovation platforms: experiences with their institutional embedding in agricultural research for development. Exp Agric 52(4):537–561

Thiele G, Quirós CA, Ashby J, Hareau G, Rotondo E, López G, Paz Ybarnegaray R, Oros R, Arévalo D, Bentley J (eds) (2011a) Métodos participativos para la inclusión de los pequeños productores rurales en la innovación agropecuaria: Experiencias y alcances en la región andina 2007–2010. Programa Alianza Cambio Andino, Lima, 197 pp

Thiele G, Devaux A, Reinoso I, Pico H, Montesdeoca F, Pumisacho M, Andrade-Piedra J, Velasco C, Flores P, Esprella R, Thomann A, Manrique K, Horton D (2011b) Multi-stakeholder platforms for linking small farmers to value chains: evidence from the Andes. Int J Agric Sustain 9(3):1–11

Tobin D, Glenna L, Devaux A (2016) Pro-poor? Inclusion and exclusion in native potato value chains in the central highlands of Peru. J Rural Stud 46:71–80

UNIDO (2011) Pro-poor value chain development: 25 guiding questions for designing and implementing agroindustry projects. United Nations Industrial Development Organization, Vienna

Wiggins S, Keats S (2013) Leaping and learning: linking smallholders to markets in Africa. Agriculture for Impact, Imperial College and Overseas Development Institute, London

Wong N, Padilla M (2017) Emprendimiento social para la recuperación de la producción agrícola de patatas nativas de los Andes de Ecuador: Caso KIWA chips. En Experiencias de emprendimiento social en Iberoamérica. Editorial Universidad de Almería. Colección: Libros Electrónicos No 67, pp 77–81

Part II
Potato and its Genetic Improvement

4

Genetic Diversity of Potato and its Conservation

David Ellis, Alberto Salas, Oswaldo Chavez, Rene Gomez, and Noelle Anglin

Abstract Conserving the genetic diversity of potato is critical for the long-term future of potato improvement programs. Further, it is the social and ethical responsibility of the present generation to ensure future generations have the same opportunities to use, exploit, and benefit from the genetic diversity that exists today. Genebanks and the ex situ conservation of potato genetic resources are the only way to ensure this happens; in situ conservation plays a complementary role, but it can never ensure that the vast diversity that exists on earth today is still there for use in the future. Material in ex situ genebanks not only serve as a reservoir of ready-to-use genetic material when needed but also provide invaluable tools for research now and in the future of cultivated potato and its wild relatives.

4.1 Ex Situ Conservation of Potato

The conservation of crop diversity, such as potato diversity, outside of its natural habitat, or ex situ, is held in botanical gardens and genebanks throughout the world. While botanical gardens conserve diversity and display this diversity for the public to enjoy and learn from, as well for scientific research, the overall mandate of botanical gardens is not generally to promote the use of and share the plant diversity they hold. In contrast, the mandate of genebanks *is* to provide access to the genetic resources they hold and to promote the use of these genetic resources for training, breeding, and research. Unfortunately, many national genebanks holding potato germplasm do not have the capacity, resources, nor a national mandate to widely distribute their holdings, and thus, access to potato germplasm is mostly left to the genebanks in developing countries, United States Department of Agriculture (USDA), and the potato genebank of the CGIAR, which holds potato germplasm at the International Potato Center, or CIP after its Spanish acronym located in Lima, Peru.

D. Ellis (✉) · A. Salas · O. Chavez · R. Gomez · N. Anglin
International Potato Center, Lima, Peru
e-mail: d.ellis@cgiar.org

Potato genetic resources held ex situ are divided into potato wild relatives (maintained as seed populations) and cultivated varieties which are maintained either in field plantings (where tubers are planted and harvested annually), seed, in vitro/clonal material, or cryopreserved material. The same genetic material could be held in two or more different forms in a single genebank (in vitro and cryopreserved) or between two genebanks (seed and in vitro). What is critical for a genebank, is that the material it holds is available in a form that can be readily used by all who need it, including breeders, large and small-holder farmers, industry, researchers from developing or developed countries, and for teaching and educational purposes. Thus, the wealth of potato diversity (Figs. 4.1, 4.2, 4.4, and 4.5) held and made available by genebanks is an asset for humanity, as it safeguards known genetic traits which are useful today, along with a myriad of known and unknown traits ready to be deployed to meet the needs for the threats and challenges potato farmers face in the future.

The more uniform the crop in general, the easier it is to manage as an ex situ collection. Unfortunately, cultivated potato is an incredibly diverse crop, and while virtually all the potato planted and used commercially worldwide are tetraploid potato varieties, diploid, triploid, tetraploid, and even pentaploid landrace potatoes are still grown throughout the South American Andes. This diversity has been collected, maintained and is available for use in breeding and research programs worldwide from numerous genebanks. The diversity of cultivated potato is reflected not only in its vast ploidy range but also at the species level where over 30 years ago cultivated

Fig. 4.1 Diversity of flower color and shape in cultivated potato from CIP germplasm collection

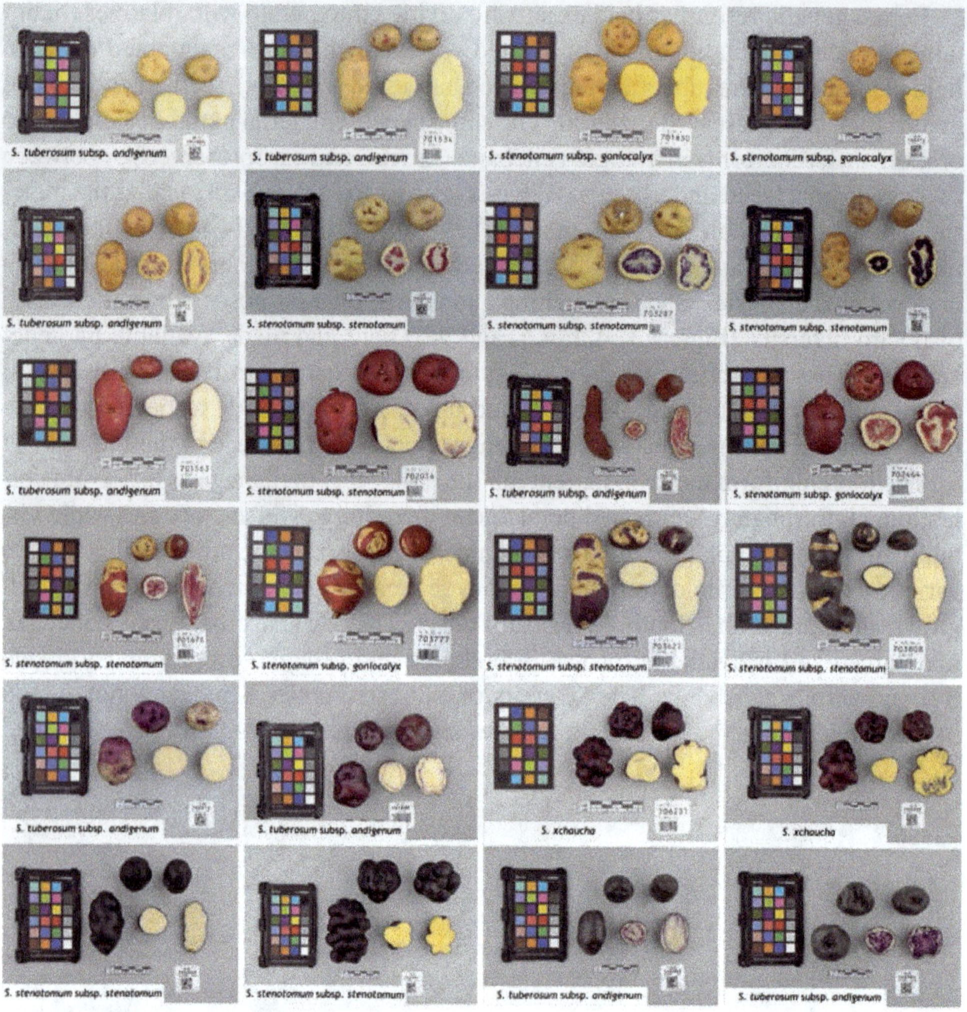

Fig. 4.2 Diversity of tuber shape and morphology and skin and flesh color in cultivated potato from the CIP germplasm collection. The color card in each photograph is used for standardizing measurement of colors of the tubers

potato was divided into seven species and nine taxa [*Solanum tuberosum* (C. Linneo) *subsp. andigenum* (Juz. & Bukasov), *S. tuberosum* (C. Linneo) subsp. *tuberosum, S. xchaucha* (Juz & Bukasov), *S. stenotomum* subsp. *goniocalyx* (Juz. & Bukasov), *S. stenotomum* subsp. *stenotomum* (Juz. & Bukasov), *S. xjuzepczukii* (Bukasov), *S. phureja* (Juz. & Bukasov), *S. xajanhuiri* (Juz. & Bukasov), and *S. xcurtilobum* (Juz. & Bukasov); (Hawkes 1990)] (Figs. 4.1 and 4.2). More recently the taxonomy of potato has been revised and cultivated potato has been regrouped into four species (*Solanum tuberosum, S. xjuzepczukii, S. xajanhuiri*, and *S. xcurtilobum*; Spooner et al. 2014). A similar taxonomic revision has occurred in the taxa of potato wild relatives with the previous taxonomic classification of Hawkes (1990) recognizing 228 wild species in 21 series. However, the recent taxonomic classification by Spooner et al. (2014) reduced the wild relatives of potato (*Solanum* section *Petota*) to 107 species.

The change in taxonomy has created a split in the ex situ potato world with some genebanks (such as the International Potato Center (CIP)] still organizing their collection based on the taxonomy of Hawkes (1990) while other major potato genebanks [such as the United States Department of Agriculture-Agricultural Research Service (USDA-ARS)] adopting the taxonomy of Spooner et al. (2014). Potato curators in all genebanks acknowledge the revision by Spooner et al. (2014) as needed and an advance in the field, yet the groupings by Hawkes (1990) make sense for managing an ex situ collection. Spooner et al. (2014) foresaw such challenges in their revision of the potato taxonomy and therefore stated that due to the "*complicating biological factors in section Petota*", they consider their "*taxonomy to be subject to critique and modification*" (Spooner et al. 2014, p 325). Two final comments are that there is consensus that ploidy in potato is not a good factor in delimitating species boundaries and that the use of different taxonomic names for the same material in different major potato genebanks is a detriment to use of the materials for the potato research community. That said, harmonizing the taxonomy used in global potato genebanks will need to be a focus in the next decade along with an extensive genetic comparison of potato collections to identify unique and redundant material between genebanks.

Taxonomic differences aside, ex situ collections and genebanks continue to play an ever increasingly dynamic role in providing the tools needed for food security in the future. The Second Report on The State of the World's Plant Genetic Resources for Food and Agriculture (FAO 2010) estimates that there are over 1750 genebanks worldwide holding approximately 7.4 million accessions; however, the report estimates that only 25–30% of these accessions are genetically unique. The report lists 98,285 potato accessions conserved in 174 genebanks around the world, and using the estimate above of the percentage of unique accessions, we can assume that there are at most an estimated 24,500–29,500 *unique* potato accessions are conserved worldwide in genebanks. In the case of potato, the accuracy of such an estimate of unique accessions is difficult to access until a thorough rationalization-comparison amongst the potato ex situ collections is carried out. According to this same report, 6 genebanks hold 41% of the global potato accessions: The French National Institute for Agricultural Research (INRA) in France (11%), Vavilov Institute in Russia (9%), The International Potato Center (CIP) in Peru (8%), The Leibniz Institute of Plant Genetics and Crop Plant Research (IPK) in Germany (5%), USDA-ARS in USA (5%), and The National Institute of Agrobiological Sciences (NIAS) in Japan (3%) and 20 global genebanks hold over 1000 potato accessions each. The listed global potato holdings collectively consist of 15% wild relatives of potato, 20% cultivated potato, 16% research and breeding materials, 14% advanced breeding lines, and 35% uncategorized accessions in these genebanks.

The last Global Strategy for Ex Situ Conservation of Potato (2006) analyzed 23 global potato collections which collectively maintain "*nearly 59,000*" potato accessions. This summary states that the genebanks in Latin America contain principally native cultivars while those in Europe and North America contain modern cultivars, breeding materials, and wild relatives. Such generalizations oversimplify the collections and indeed, although there is specialization among the potato genebanks, most major collections have a good representation of both wild and cultivated accessions.

Information about the particular attributes for each individual accession is critical for users to select which one or two of the thousands of accessions contain the trait(s) the user needs. Unfortunately, the information on accessions in global potato genebanks is generally incomplete and lacking the trait information users want and need most. Further, information on qualitative, complex traits (i.e., drought, frost tolerance, or yield) can be very specific to the physiological age at which the plant is subjected to the stress, the location, soil type, and degree of stress and, thus, is not always easy to list in a genebank database. Further, these complex traits require multiyear, multilocation evaluations in order to understand the complexities of the traits and genes associated with the trait. The USDA-ARS collection contains a large list of descriptors and traits recorded with 137 morphological, biochemical, nutritional, or physiological traits listed for potato (https://npgsweb.ars-grin.gov/gringlobal/cropdetail.aspx?type=descriptor&id=73), while CIP lists 72 descriptors or traits for their potato holdings (http://genebank.cipotato.org/gringlobal/search.aspx) and the European Union Genetic Resources of Potato lists 58 descriptors or traits (http://ecpgr.cgn.wur.nl/eupotato/). All these data are accessible in English on public internet-accessible databases, as is information from other global potato genebanks; although far too much of the information of the potato ex situ collections is not available publicly and outside of a few genebanks, much of the available listings from genebanks contain little information about the accessions. Fortunately, efforts such as Genesys, the global Gateway to Genetic Resources (https://www.genesys-pgr.org/welcome), provides descriptor and trait information from multiple genebanks in a single database, which facilitates use by facilitating the location and finding information on potato accessions worldwide.

4.2 Collection of New Potato Germplasm

There are estimations that 20% of plant species are in danger of extinction (Jansky et al. 2013) and such losses will be accelerated by climate change and habitat destruction. Further, for some plants, a large amount of their diversity may already only be able available from genebanks (Jarvis et al. 2011), which might be the future for potato where the wild relative species exist only in the fragile ecosystems of the Andes (Hijmans and Spooner 2001), where a changing climate is already having a substantial impact. Lingering questions for ex situ and in situ management of potato revolve around how to define diversity there is/was in cultivated and the large secondary genepools (wild potato relatives) of potato. For conservationists, the big question is how much of the diversity in potato is securely conserved, what diversity has been lost or is in imminent danger of being lost, and what is the economic value, as well as potential future value, of the potato diversity that is not securely conserved.

A benchmark study (Castañeda-Álvarez et al. 2015) evaluated the status of ex situ collections for 73 species (according to Spooner et al. 2014) of potato wild relatives and used environmental niche modelling (ENM) techniques to estimate potential geographic ranges of each species. Their data assigned a high priority status for collecting 32 species of wild potato, 43.8% of the species studied, due to severe gaps in the ex situ collections. Further, 20 more species were assigned medium priority for collection and only three species were determined to have good diversity representation in the ex situ collections. As part of the study, they also looked at in situ sites which were potentially threatened including that of *S. rhomboideilanceolatum* whose native habitat in Peru is increasingly threatened by road building and overgrazing. Potato wild relatives have made critical contributions to potato disease resistance, enhanced yield, and improved quality in past 50 years (Jansky et al. 2013). Therefore, it should be the obligation of present day potato researchers and breeders to do all we can to leave future potato scientists the same opportunities to explore, use, and reap benefits from the diversity of wild potato that we have today.

The geographic range of wild potato species is broadly spread out in 16 countries confined to the Americas (Mexico, United States, Costa Rica, Guatemala, Honduras, Panama, Argentina, Bolivia, Brazil, Chile, Colombia, Ecuador, Paraguay, Peru, Uruguay, and Venezuela (Fig. 4.3). For thousands of years, the ecological niches of the Andes have been the natural habitat and the center of the genetic diversity of wild potato species and innumerable native varieties. Long evolutionary processes have allowed the accumulation of genetic components that are valuable resources for the improvement of potato cultivation, such as the ability to survive adverse biotic conditions (diseases, pests) and abiotic (drought, frost, climate change), as well as, vast morphological differences (Figs. 4.4 and 4.5). However, this valuable genetic diversity is threatened by "genetic erosion." The main threats are urbanization, alternative land use, and the destruction of natural areas. Examples include overgrazing, the construction of new routes and villages, and the destruction of forests.

The domestication of potato is thought to have occurred ~10,000 years ago in the Lake Titicaca Basin on the border between Peru and Bolivia, most likely from the *S. brevicaule* complex with *S. candolleanum* as a potential ancestor (Spooner et al. 2014). The use of these wild potato species for food is likely, yet the high glycoalkaloid content needed to be dealt with in some fashion to avoid toxicity. Rumold and Aldenderfer (2016) studied early potato starch grains from tools found at a village site in the Titicaca basin (1700–3500-year-old site) which resembled freeze-dried samples rather than starch from fresh tubers. They hypothesize that a process such as drying and grinding may have played a role in the detoxification of the wild potato tubers, as in other root crops. This theory is consistent with the natural freeze drying of tubers in the Andes today known as "chuño." Recent supporting evidence for use of wild potatoes for human consumption comes from *S. jamesii* starch grains found among other 6000-year-old early human ruins in the Southwest US (Louderback and Pavlik 2017). It is interesting to note that *S. jamesii* was still used until recently as a food source for native communities in the area and that existing patterns of present day occurrence of this species suggests that early humans may

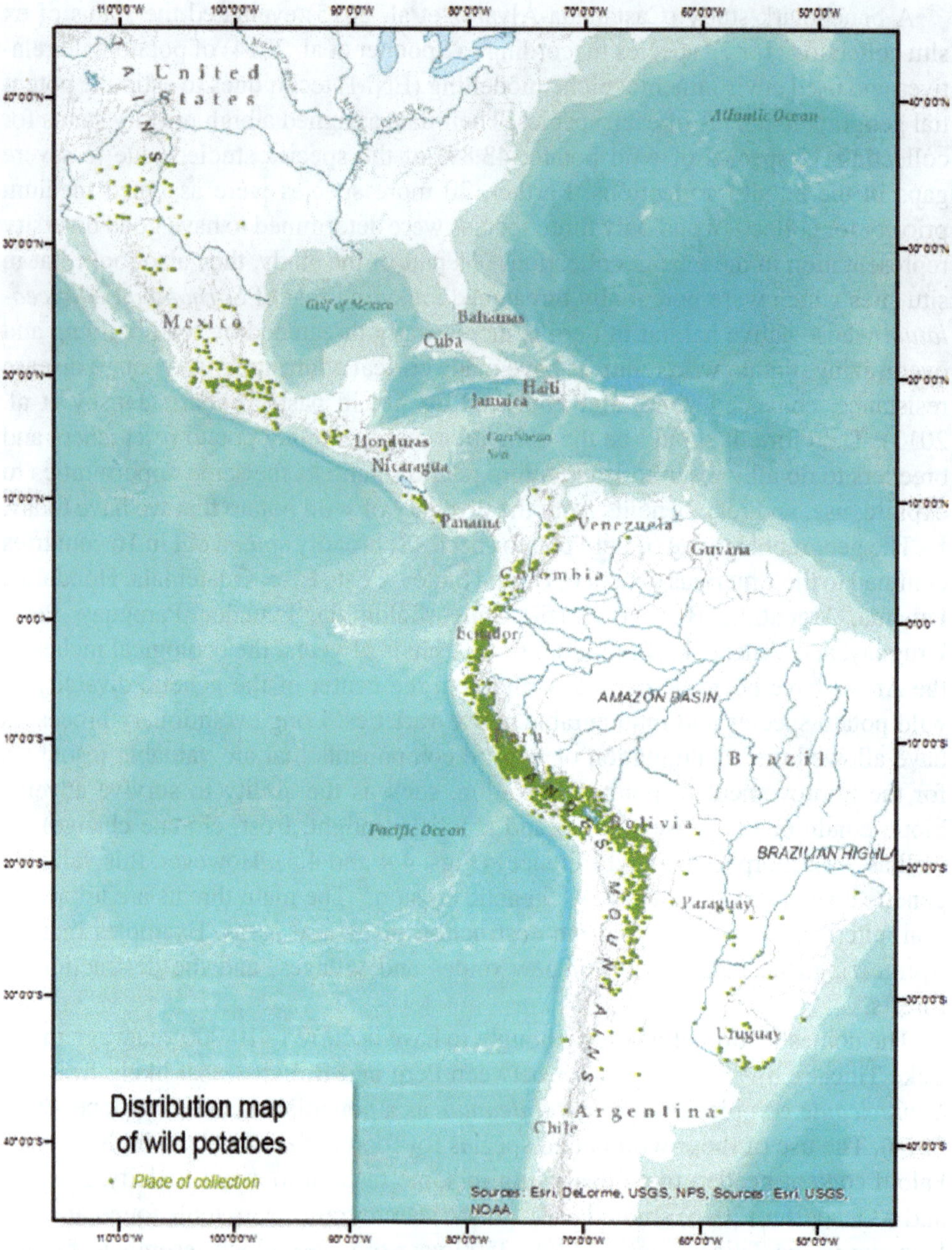

Fig. 4.3 A map of the distribution of wild potato accessions in the CIP genebank. Greens dot indicates origin of the accession based on passport data (https://cipotato.org/genebankcip/process/potato/potato-wild/)

have cultivated the tubers (Kinder et al. 2017). Domestication is believed to have given rise to increased tuber size through selection and cultivation which was associated with an increase in leaf carbon fixation and transport, reduction in glycoalkaloids, adaptation to long-day photoperiod, and reduced sexual fertility (Ryoko 2015).

Fig. 4.4 Examples of the diversity found in leaf morphology, shape, and color of various wild potato species

With domestication of early edible potato tubers, it is logical this also was the start not only of selection and cultivation but also of movement of selected potato tubers from one site to another. As early as 5000 years ago, maize, originating in Mexico, was already in S. America (Grobman et al. 2012), hence there is no reason to think that the north to south movement of food materials was not also reciprocated with a south to north movement of food such as cultivated potato. The movement of domesticated varieties most certainly also gave rise to the movement of new genetic material through the zone of native wild populations followed by intercrossing leading to increased diversity in the domesticated varieties and further selection for adaptive traits. These same adaptive traits and the diversity associated with them contributed to the adaptations which were selected for by early human populations and these same traits still exist in the wild populations today and will continue to be of critical value to breeders as long as potato improvement programs exist.

Fig. 4.5 Examples of the diversity found in fruit morphology, shape, and color in wild potato species

Thus collecting, movement, and exchange of potato and potato wild relatives has occurred for millennia, yet was mainly confined to the Americans until the Spanish conquest when potato was collected and shipped to Europe where it underwent further selection. Wild relatives were likely also collected before the last century, but reliable documentation of the collection of wild potato for genebanking is more recent. Darwin, in his famous voyage of the Beagle, collected potato from the Chilóe Islands in Chile in the 1830s which he named *S. maglia*, now known as S. *tuberosum* (Ristaino and Pfister 2016). At this time, Darwin erroneously assumed potatoes originated from these islands due to the diversity and wide spread presence. In his journals, he referenced that the crew "*stocked up on game and potatoes.*" Fuentes (2014) references collection expeditions as yearly as the 1830–1833 by A. D'Orbigny from France in Bolivia. Perhaps the best documented collections are those done by the Vavilov Institute (Supreme Council of People's Economy) in Russia starting in 1925 by S.M. Bukasov and S.V. Juzepzuk (Loskutov 1999), after whom the wild potato species *S. bukasovii* and *S. xjuzepczukii* were named, respectively. Although detailed notes from these expeditions were never published, it is known that many species of wild potato were collected in Peru, Bolivia, and Chile (Ochoa 2004), and that some of this material still is maintained at the Vavilov Institute (Prof. N. Dzyubenko, personal communication).

Collections of wild potato continued by numerous collectors from Europe, Asia, North America, and Latin American institutions in an ad hoc manner until the early 1970s (Correll 1962; Hawkes 1990; Ochoa 1962) when CIP organized a meeting of experts to form and develop an International Germplasm Bank at CIP in Lima, Peru. Three Planning Conferences were held dealing with topics ranging from exploration, collection, conservation, and evaluation of potato genetic resources ending with the recommendation of collection strategies and to make the collected material available to researchers for their use for the benefit of humanity (International Potato Center 1973–1974, 1976, 1979). In the last 50 years, CIP has organized or been involved in ~300 systematic exploration missions for potato. These included the collection of wild and native potato germplasm varieties under the direction of C.M. Ochoa and collaborating with some 245 national and international scientists including A. David, J.B. Bamberg., J.G. Hawkes, A.M. Van Harten, J.P. Hjerting, W. Hondelmann, R. Hoopes, K.A. Okada, A. Salas, D. Spooner, and J.J.C. Van Soest. These collecting missions organized from Peru, where the greatest genetic diversity of the *Solanum* section Petota (Solanaceae) wild and cultivated exists, included exploring and collecting in the 16 countries of the American continent. Numerous accessions of each of the wild potato species were collected in the unique ecological niches along the geographical distribution of each species. Of interest was the rediscovery of wild species previously described, but known only as herbarium specimens in addition to the discovery of about 20 new species (including, *S. amayanum* (Ochoa 1989a, b, c), *S. bill-hookeri* (Ochoa 1988a, b), *S. bombycinum* (Ochoa 1983a, b), *S. chilliasense* (Ochoa 1981a, b, c, d), *S. incasicum* (Ochoa 1981a, b, c, d), *S. irosinum* (Ochoa 1981a, b, c, d), *S. longiusculus* (Ochoa 1987), *S. neovavilovii* (Ochoa 1983a, b), *S. orocense* (Ochoa 1980a, b, c), *S. ortegae* (Ochoa 1998), *S. peloquinianum* (Ochoa 1980a, b, c), *S. salasianum* (Ochoa 1989a, b, c), *S. sarasarae* (Ochoa 1988a, b), *S. simplicissimum* (Ochoa 1989a, b, c), *S. sucubunense* (Ochoa 1980a, b, c), *S. tapojense* (Ochoa 1980a, b, c), *S. taulisense* (Ochoa 1981a, b, c, d)) which were characterized and described by Ochoa (Hawkes 1989, 1990; Ochoa 1962, 1990, 1999; Spooner et al. 1999, 2001). These accessions are all documented with full passport data (country and location of collection) and many are available for use globally through the CIP website (https://cipotato.org/genebank-cip/process/potato/potato-wild/).

Most collection missions only have resources to focus collections from a single site and rarely can make multiple visits to the same site. Therefore, collectors are at the mercy of annual changes in rainfall and other abiotic and biotic factors that affect flowering or tuber germination. Exceptions are the collection trips by Bamberg and del Rio (2011) and Bamberg et al. (2003) where annual expeditions have been made to monitor in situ sites of *S. jamesii* and *S. stoloniferum* (formerly *S. fendleri*) since 1992 with over 200 visits. The expeditions were initiated to recollect from sites originally sampled in 1958 and 1978 to evaluate and characterize what changes could be measured over a 40+ year period and to maximize the ex situ collection of diversity found in this species (del Rio et al. 2001). Their research has yielded insight into the collection from wild potato, and perhaps, upset some long-held dogma. For example, as expected, they found more diversity using random amplified polymorphic DNAs (RAPDs) in seed collections, compared with the collection of tubers from individual plants in *S. jamesii*, but the opposite phenomena was observed for *S. stoloniferum*, where collection of tubers had greater diversity than seeds (Bamberg et al. 2009). In another report (del Rio et al. 1997), they found using RAPD markers, there were significant differences in allele frequencies between genebank-conserved material and recollected material from the same site in all seven comparisons of *S. jamesii* (diploid outcrosser), and 12 of 16 comparisons within *S. stoloniferum* (tetraploid inbreeder). While in situ biologists would conclude that this makes sense, since the argument is that evolutionary processes continue in situ, for the authors, this meant that recollecting from the same site was not only a good idea but also is desirable to capture new alleles (Bamberg et al. 2003). Other observations included a comparison of "easy" (road side collection sites) versus "remote" (sites one needed to hike to) where both collection methods yielded valuable alleles (Bamberg et al. 2010) and that patterns of geographic structure were not valuable predictors of genetic differences in these two species inferring that wide sampling of each population was needed to capture diversity.

4.3 International Fora for Plant Genetic Resources

The Convention of Biological Diversity (https://www.cbd.int/; CBD) is the most significant international treaty for the access to genetic resources globally; yet, it has had a negative impact on the collection of potato germplasm throughout the native range of wild potato. This is due to CBD reaffirming the sovereign rights of a nation to the genetic resources falling within the country boundaries, and hence, the country of origin has exclusive rights over terms for accessing genetic resources, including cultivated and wild relatives of potato, found within its boundaries. The treaty also reaffirmed the country of origin's right to "*fair and equitable sharing of benefits arising*" from the use of their genetic resources. CBD further asserts that nations are responsible for the long-term protection and *access* to genetic resources over which they have jurisdiction. CBD also acknowledged that access to, and sharing of, genetic resources, and *relevant technologies*, are critical for meeting food and health needs of a growing world population. Thus, it is a country's responsibility to not only protect but also make available for use and share in the benefits from such use of the genetic resources it holds. Unfortunately, most countries, and the international community generally, have been ill prepared for the responsibility of regulating access and for ensuring the sharing of benefits by third-party use of genetic resources, such as germplasm related to cultivated and wild potato relatives. Mechanisms to grant permission for the collection of plant genetic resources have taken time to implement, and in the meantime the collecting of plant genetic resources ceased. The challenge has been how to make genetic resources available for use on the one hand, while on the other, to ensure equitable sharing of benefits from the commercial utilization of the genetic resources, or derivatives thereof, by multiple third-party stakeholders' years later. Further, definition of the terms under which genetic resources could be accessed and benefits shared have been defined in the Nagoya Protocol on Access and Benefit Sharing (https://www.cbd.int/abs/, *Nagoya*). All potato species, except for *S. phureja*, are covered by the International Treaty for Plant Genetic Resources for Food and Agriculture (https://www.fao.org/plant-treaty/; ITPGRFA) which provides a legal framework to distribute the genetic material for research, training, and breeding with the acceptance of a Standard Material Transfer Agreement (SMTA) by the recipient.

Presently, there is a lack of a defined, functional mechanism for granting permission for collecting germplasm from the wild in most Latin American nations. The collection of potato germplasm for ex situ conservation in the countries where the clear majority of cultivated potato landraces and wild relatives originated and exist, has been virtually nonexistent since 2000. The exception is the United States of America (USA), which is not a party to CBD, and hence germplasm from potato wild relatives have continued to be collected in the USA (Bamberg et al. 2003), although only two species [*S. stoloniferum* (previously *S. fendleri*) and *S. jamesii*] are confirmed to exist within the boundaries of the USA. In 2016, a significant change occurred when the Peruvian National Institute for Innovative Agriculture (Instituto Nacional de Innovacion Agraria, INIA) was granted a permit for the collection of potato wild relatives within Peru, and in 2017 three collection missions were carried out collecting more than 70 populations of 14+ species (putatively *S. acaule, S. acroscopicum, S. aymaraesense, S. bukasovii, S. coelestispetalum, S. marinasense, S. pillahuatense S. raphanifolium, S. sandemani, S. sparsipilum, S. tacnaense, S. urubambae, S. velardei, S. yungasense*) under the auspices of the ITPGRFA. It is hoped that this will provide a model for further collections of wild and cultivated potato within its native range.

The collection of potato in its native range should be facilitated by the inclusion of cultivated potato (except for *S. phureja*) and all wild potato species in the list of Annex 1 crops under the ITPGRFA. The ITPGRFA established a global Multilateral System (MLS) that clearly defines the terms under which access for the use of genetic resources and their derivatives would be granted for use in research, training, and breeding under a Standard Material Transfer Agreement (SMTA) (http://www.fao.org/plant-treaty/areas-of-work/the-multilateral-system/the-smta/en/), which clearly outlines benefit sharing options. Of nations where wild potato species are native, only Colombia and Mexico are not parties to ITPGRFA, and thus, it is hoped that further facilitated access to collecting, conservation, and use of cultivated and wild potato species for research, training, and breeding will commence. For Colombia and Mexico, collecting, access, and use of native potato germplasm would be subject to bilateral agreements under the Nagoya protocol framework. To our knowledge, no other collections of wild potato species for inclusion in the MLS of the ITPGRFA have been carried out since the 1990s.

4.4 Methods of Conservation of Potato Germplasm

A major challenge for ex situ genebanks is to maintain the diversity held, often thousands of different accessions, in a form where every accession is readily available and accessible for use at a moment's notice. Collections of cultivated crops are often composed of various traditional varieties, along with improved cultivars, which often do not produce large quantities of seeds/tubers making them challenging to reproduce in field conditions. Additionally, in general, wild species are more challenging for ex situ genebanks to regenerate than cultivated due to the multitude of ecological niches they have become adapted to, often requiring unique environmental conditions (i.e., photoperiods) to thrive and reproduce, along with some genotypes/species, which genetically produce limited flowers, seeds, or tubers.

With the lifespan of most research grants being less than 5 years, delays in obtaining germplasm can mean missing critical planting windows which could result in a year's delay or failure to meet deliverables for a project. Thus, availability of material from the genebanks is critical. Most crop genebanks maintain their germplasm collections as botanical seed which for the majority of plants, including potato, stores very well and can be made available with a few weeks' notice if adequate seeds exist for that accession. Genebanks will only provide seed to requestors if they have adequate seed in cold storage with good viability to ensure that the material is safeguarded and there is enough seed to keep regenerating the accession. However, unlike most crops, cultivated potato is also often maintained as clones to preserve the unique allelic combinations in the selected material which poses additional challenges for maintenance and distribution.

Where desirable allelic combinations have not been fixed by selection, as in wild potato germplasm, the collections are maintained as botanical seed or true-potato-seed (TPS). A representative number of individuals (20–50) are typically collected from a wild population and seed is regenerated and combined to form a unique genebank accession of heterogeneous seed theoretically representative of the alleles found in that population. The accession is then distributed as a population of 50–100 seed where each seed represents a distinct genotype from that population. Thus, what is distributed is a heterogeneous mix of genotypes (seed) with each genotype representing a portion of the genetic make-up of the accession or population. Different collections of the same species (individual accessions) may or may not contain the same trait(s) desired by a breeder or even have this trait at the same frequency in different seed populations distributed. For example, when assessing late blight (*Phytophthora infestans*) resistance in wild potato accessions, Perez et al. (2001) found very high uniform resistance in three accessions of *S. orophilum* (100%, 98%, and 96% resistant individuals, respectively from 48 seedlings from each accession); whereas, in three accessions of *S. wittmachii*, resistance was high in one accession but very low in another (75%, 46% and 0.02% resistant individuals, respectively from 48 seedlings from each accession).

Unlike the wild species, cultivated potato is generally maintained as clones; however, in some cases seed populations of cultivated material may be just as useful to users. One of the main advantages of maintaining cultivated potato as seed is the overall cost and staffing required to maintain a seed collection is greatly reduced compared to in vitro collections which require subculturing every year or two. Further, sterile laboratory conditions with specialized equipment are not needed to reproduce potato if maintained as seed rather than in vitro clones. Seed populations can be produced and if adequate seed is obtained, a curator would not need to regenerate the accession for several decades unless the vigor of the seed decreases or the seeds are exhausted. Additionally, seed populations would have more diversity and new allele combinations that could lead to new trait discovery.

In contrast to maintenance of cultivated material as seed, maintenance of cultivated potato as a clone, preserving the unique allelic makeup of the landrace or variety, has advantages for either direct planting or facilitating the transfer of unique allelic combinations to breeding or research programs. Maintenance of clonal potato collections can be in the field, greenhouse, or in vitro as tissue culture. If maintained in the field or greenhouse, the collection is harvested, tubers stored and annually replanted as maintaining tuber viability for longer than a year is unreliable. Field or greenhouse maintenance of the collection also carries a risk of losing accessions due to biotic (insects and diseases) or abiotic (hail or wind) challenges. Therefore, many clonal potato collections are maintained in tissue culture as in vitro plantlets, which virtually eliminates the challenges of external biotic and abiotic factors. In vitro maintenance also facilitates international distribution of the collection as tissue culture plants are maintained in a sterile environment, and once cleaned of diseases, they can be certified as disease-free material for import purposes internationally.

The cultivated potato germplasm collection at CIP originated in the early 1970s with material collected or donated from around the world. In the early 1980s, research was initiated to place the collection into in vitro culture. Few in vitro collections have been maintained for 30 years or that have been used as extensively as the CIP cultivated potato collection. CIP currently maintains the largest in vitro potato collection with 8354 potato in vitro accessions (as of October 2018), the majority of which (89.8%) are landraces (native potatoes or "papa nativas" in Spanish) originating mainly from the Andean region, with the remaining accessions being improved varieties and breeding lines.

Since the vast majority of potato germplasm in genebanks around the world was collected over 20 years ago, and some, such as the collection at the Vavilov Institute in Russia, almost 90 years ago, most potato ex situ collections would be virtually impossible to replicate or replace today if lost. For this reason, Standards 4.9 of the Genebank Standards for Plant Genetic Resources for Food and Agriculture (FAO 2014) clearly state that "*A safety duplicate sample for every original accession should be stored in a geographically distant area, under the same or better conditions than those in the original genebank.*" CIP, along with other genebanks in the CGIAR, whose collections are held in trust under the ITPGRFA, takes this a step further with a guideline for backing up the collection in two distinct locations, one nationally and a second backup internationally. In the case of TPS accessions, such as the wild relatives of potato, a sealed package of seed equivalent to one to three times what is needed for a regeneration are backed up in a remote island archipelago in Norway at the Svalbard Global Seed Vault (https://www.nordgen.org/sgsv/, SGSV). The SGSV was established 10 years ago and funded by the Norwegian government with operations coordinated by the Nordic Gene Resource Centre and the Global Crop Diversity Trust. Shipment and coordination of all seed deposits are done by the Nordic Gene Resource Centre under a MTA which clearly states that the objective of the vault is "*...to provide a safety net for the international conservation system of plant genetic resources, and to contribute to securing the maximum amount of plant genetic diversity of importance to humanity for the long-term....*" (https://www.nordgen.org/sgsv/index.php?page=welcome). Under this agreement, the material is kept by the SGSV as a *black-box* safety back-up which denotes that the depositor retains full ownership and rights to the material, with the back-up facility holding the material only as a service of storage for the depositor. As of the end of 2017, SGSV contained almost 900,000 samples of seed from different crops including seed from 136 species of wild potato, five wild potato natural hybrids and six species of cultivated potato (taxonomy according to Hawkes 1990). Under conditions at the SGSV (airtight containers, properly dried seed, stored at −18 °C), good quality TPS can survive and maintain viability for 50–100 years.

Unfortunately, there are no analogous safety backup facilities to the SGSV for clonal material. In the case of clonal collections maintained in the field, the expense is often too high to allow distinct duplicate field planting sites, and hence, these collections are all too often not backed up and diminish over time. In the case of in vitro collections, the collections can be safely backed-up by shipping duplicate copies of in vitro cultures to a distant location for safety back-up. The facility backing-up the in vitro material generally needs to have lighted controlled temperature chambers (5–10 °C) for sustained slow growth of the material. In the case of the back-up for the USDA National Potato Genebank in vitro collection, cultures were shipped to the National Laboratory for Genetic Resources Preservation in Fort Collins, CO, where they are maintained as mini-tubers at 5 °C requiring subculture and reinitiating of the mini-tubers once every 2–3 years. For other in vitro potato collections, such as the one at CIP, the entire collection is shipped periodically (semi-annually) to two sites, one in a geographically distinct location in the high Andean Plateau at a CIP field station in Huancayo, Peru and the other facility outside of Peru at the *Brazilian Agricultural Research Corporation* (EMBRAPA) labs in Brasilia, Brazil. At both sites, the material is kept in slow growth conditions at 7 °C where the cultures can survive approximately 2 years without subculturing. Again, as with the seed deposits at the SGSV, these materials are stored in a black-box arrangement.

As genebanks are the stewards of the genetic resources that they maintain, with this stewardship comes great responsibility to ensure that no known pathogens are introduced into other countries with the distribution of germplasm creating a new disease problem. Vegetatively propagated crops such as potato, sugarcane, cassava, sweetpotato, yam, banana, citrus, and strawberry can be challenging because they must be multiplied continuously when maintained in the field to keep them in optimal growth. Although field maintenance of any crop is subject to abiotic challenges, as well as, biotic challenges such as diseases and pests, and with each multiplication or movement from one field to another, there are ample opportunities for the introduction of viruses and viroids (Sastry and Zitter 2014). Diseases caused by plant viruses limit sustainable production of vegetatively propagated staple food crops and in developing countries, this problem is confounded as agronomic practices to limit disease incidence and spread is marginal if done at all. As well, the use of new clean planting material to lessen virus titers is a luxury which most small holder farmers cannot afford. Additionally, low virus titers in planting materials generally display low symptom severity which makes infected plants not easily recognized as disease containing, and subsequently problematic if used for vegetative propagation (Bosch et al. 2007) because it can further perpetuate the diseases. Often human actions and climate change can be linked to the spread and outbreak of disease in conjunction with the movement and exchange of diseased plants and the international food supply chain are drivers of new disease threats (Wilkinson et al. 2011).

Actual examples of the human spread of pathogens with potato germplasm, other than by word of mouth, are difficult to find, yet published reports do exist to verify the potential challenge. The recent introduction of East Africa cassava mosaic-like viruses from African to the South West Indian Ocean islands follows human and insect transmission of the disease (De Bruyn et al. 2012), while the spread of the disease to Oman is likely directly related to human movement of plant material in the 1960s (Khan et al. 2012). Spread of viruses across international borders can also happen by insect vectors as in the case of the expansion and subsequent pandemic of the cassava mosaic disease (CMD) in the Great Lakes region of East and Central Africa which followed tightly the outbreaks of whitefly populations (Legg et al. 2011).

Therefore, for international distribution of germplasm from genebanks, plant material is tested, treated to eliminate diseases and certified to be clean and free of pathogens of phytosanitary importance as defined by import permits prior to transporting across any international border. For virus-free production of TPS, regeneration is carried out under strict sanitary conditions in a controlled screen house with limited access and maintained free of insects and thus vectors for viruses. Arracacha virus B–oca (arracacha and oca are two Andean Root and Tubers crops cultivated and eaten throughout Peru) strain (AVB-O), potato virus T (PVT), potato spindle tuber viroid (PSTVd), and alfalfa mosaic virus (AMV) have been shown to transmit virus particles to seed when pollen from an infected plant is used for pollinations (Jones 1982; Valkonen et al. 1992; FAO/IPGRI Technical Guidelines for the Safe Movement of Germplasm n.d.)]. Therefore, at CIP, for individual TPS production mother plants are tested pre-flowering for these seed transmitted viruses (AVB-O, PVT, PSTVd, and AMV) as well as for Andean potato latent virus (APLV), potato yellowing virus (PYV), and tobacco mosaic virus (TMV). If an individual mother plant tests positive for any of the viruses listed above, it is discarded and not used as a pollen donor or as a maternal parent for TPS production.

Pathogen elimination from clonal material is much more laborious, costly, and time-consuming. This is one rationale for in vitro culture as once the plantlets are clean, the in vitro culture methodology will maintain the investment of the phytosanitary cleaning. If endogenous bacteria are present in the cultures, these are usually eliminated with antibiotics prior to the elimination of viruses. A typical and successful virus elimination scheme starts with placing the in vitro cultures into thermotherapy (32–34 °C for 1 month), followed by isolation of 0.1–0.3 mm meristems and then plantlet regeneration. The regenerated plants are tested for the presence of viruses by serological, molecular, host range testing, and grafting onto indicator plants for final assurance of virus-free status. The entire process takes 12–24 months for one round and longer if not successful on the first run. At CIP, this method is highly efficient with 90% of all potato material going through virus elimination becoming virus-free after one run. The viruses of import/export importance in clonal material include potato virus T (PVT), potato spindle tuber viroid (PSTVd), potato virus X (PVX), potato virus Y (PVY), potato leaf roll virus (PLRV), potato virus S (PVS), Andean potato mottle virus (APMV), Andean potato latent virus (APLV), potato yellowing virus (PYV), and Arracacha virus B (AVB-O).

Advances in next-generation sequencing (NGS) hold great promise for reducing the labor, time, and cost of disease screening in germplasm collections. One such technology is the detection of viruses using small RNA sequencing and assembly (sRSA) technology. This technology is being tested on potato and it is hoped to be able to reduce the time necessary for phytosanitary cleaning of clonal potato germplasm from 12–24 months to less than 6 months (Kreuze 2014). Importantly, this technology could be invaluable in the future as new viruses or diseases of import importance develop so that entire germplasm collections can be rescreened quickly and efficiently ensuring the germplasm is readily available when needed.

The maintenance of in vitro potato clonal material has its advantages, including being readily available and the maintenance of disease-free status, but it is also undisputedly one of the most expensive long-term conservation methods available. This is due to the need for sterile plant tissue culture facilities, climate-controlled growth rooms, and highly trained personnel. Maintenance of clonal material as seed *is* one alternative, yet to maintain the allelic combinations present in an accession, maintenance of the clone is the only option. Cryopreservation, freezing tissue in liquid nitrogen at −196 °C, has been used for the long-term preservation of clonal plant genetic resources. Although research in the cryopreservation of plants has been ongoing for over 50 years (Reed 2008) and high confidence exists with the technology in animal and human systems (Di Santo et al. 2012), the application of long-term preservation of plant material, despite many successes, still has its skeptics. Limited long-term studies of plant material have shown largely positive and optimistic results (Volk et al. 2008), which support the theoretical long-term cryopreservation of plant meristems for centuries. Potato was one of the early crop plants used in cryopreservation research (Bajaj 1977; Grout and Henshaw 1978; Towill 1981a, b) and there are reports of over 20 potato species having been tested for cryopreservation using several different methods (Kaczmarczyk et al. 2011; Vollmer et al. 2017). To date the most successful methods for potato cryopreservation of genebank accessions included the IPK genebank using dimethyl sulfoxide (DMSO) droplet vitrification (Kaczmarczyk et al. 2011) with over 1000 cryopreserved accessions and the CIP genebank using a Plant Vitrification Solution (PVS2) droplet vitrification (Vollmer et al. 2017) method with over 2500 accessions cryopreserved as of the end of 2017.

Although the number of accessions or samples distributed from potato ex situ collections tells us little about *how* the germplasm was used or if it was useful at all, it does provide a measure of interest in the collections. In marketing, if a product is not available and there is little information on the attributes and benefits of the product, few will buy or order it. The same is true with germplasm collections, if the germplasm is not phytosanitary cleaned, in a form which can be distributed or due to some other factor(s), such as regulatory hurdles, is unavailable, it is not accessible for use and hence will not be used. Unfortunately, the physical accessibility of potato accessions from global collections is scattered with only 11/23 (~50%) of the collections surveyed having distributed material internationally in the period from 2004 to 2006 (Global Strategy for Ex situ Conservation of Potato 2006). While this is the only statistics found on distribution from multiple global potato global genebanks and it predates the International Treaty for Plant Genetic Resources for Food and Agriculture (Carputo et al. 2013), it serves as a good example and is likely the status today that less than 50% of potato ex situ material held in genebanks globally are available today beyond the walls of the genebank or the borders of the nation in which the genebank resides. This is a reflection of the state of the art of many factors and the international regulatory environment is just one factor (reviewed above). We have also discussed the need for phytosanitary cleaning and a propagule which can be distributed as disease-free. TPS could be an asset and a valuable tool for genebanks lacking the technology for long-term maintenance in tissue culture, yet is capable of virus screening of mother plants for TPS regeneration.

Another major limitation to the use/distribution of potato germplasm is the lack of information publicly accessible about potato germplasm in the various genebanks. While Genesys (https://www.genesys-pgr.org/welcome) offers a one site shop for information on the location of 19,066 potato accessions comprising over 175 species from 21 different countries, as of the end of 2017 it does not have information on 3 (INIA, France; Vavilov, Russia; and NIAS, Japan) out of 5 of the largest potato genebanks identified in the Second Report on The State of the World's Plant Genetic Resources for Food and Agriculture (FAO 2010). This highlights the difficulty for users to find information on the accessions conserved. Even the European Union Potato Project Database (http://ecpgr.cgn.wur.nl/eupotato/) does not contain information on the French or United Kingdom potato collections, yet it has information from both CIP and USA collections. If information on the holdings of the global potato germplasm collections is not readily available, or if available but not in a universal language, it will be impossible for most users to find information on what is held in ex situ collections.

As mentioned above, one measure of the value of genetic resources conservation is whether the collection is desired by the user community. Using the *in trust* potato collection at CIP as an example of the value of conservation of diversity of ex situ collections to the breeding and research community, since 2007, CIP has received 1,399 requests for potato germplasm. A total of 48,533 samples were distributed in the last 10 years (2007–2017), following the legal framework of the ITPGRFA. This represents over 1,200 unique accessions per year or almost 25% of the *in-trust* collection distributed annually. Material was sent to 96 countries with Peru, USA, Australia, and China requesting the most potato germplasm. All material is transferred using the Standard Material Transfer Agreement (SMTA) under the ITPGRFA. It could be expected that in the genomics era, DNA distributions would steadily increase; however, there is no trend in this as physical genetic material remains in high demand suggesting use in breeding programs still requires plants for crossing. Interestingly, in a recent (2015) survey (Shirey, unpublished data), late blight resistance and yield were the most cited traits of interest among users.

In 1997, a program was initiated to return traditional cultivars and landraces to the native, indigenous farmers in Peru (repatriation), whose ancestors preserved the potato diversity. For centuries, these potato farming communities have nurtured and planted multiple, 20–40, varieties per family as an insurance policy to ensure food availability. The rationale was that in one year one group of landraces would produce well, while in other years a different group might flourish. By conserving and planting diversity, they sustained their livelihood. In recent times, many communities have lost important cultivars they had planted for generations, due to terrorism, extreme weather, climate change, and/or disease/pest pressures. The CIP repatriation program is an example of the broader user base of ex situ genebanks where the *in-trust* collection served as a reservoir specifically for the landraces the current generation of potato guardians of the Andes (Papa Arariwa) needed to revive their traditional farming systems. In some cases, it allowed them to plant again the landraces they remembered as a child. Thus, repatriation is helping to restore the diversity and productivity in the traditional Andean potato farming region (Aguilar 2016). Further, the repatriation of genebank materials has benefits beyond restoring diversity and the traditions that come with this diversity, the added benefit is the distribution of pathogen-free stocks, which can increase productivity in a single season by as much as 40% thereby contributing to food security and poverty reduction for these communities. In total, 89 communities in Peru have received over 6,000 samples representing more than 1250 accessions of native landraces or over 50% of Peruvian landraces held in the in-trust potato collection.

4.5 Characterizing Potato Diversity in Genebanks

Molecular markers, are considered fixed landmarks in a genome, and thus, can reveal crucial genetic variability (Semagn et al. 2006). Markers have been used in many crop plants to assess genetic diversity, determine population structure, establishing trait-marker associations, discover and track quantitative trait loci (QTLs), produce genetic linkage maps, assist in selection for traits, understand the influence of genotypes on phenotypes, and more, all to improve or understand crop plants.

Loss of plant species in the past centuries has triggered genetic resource conservation with a need to accurately identify each accession. Molecular tools provide an easy less laborious method to assign plant taxa, as well as, characterize certain traits (Arif et al. 2010). To address the challenge of defining the biodiversity of a crop such as potato, biochemical and molecular markers have been used, yet the wide-scale use for looking at entire ex situ collections has only recently become more feasible due to advances in next generation sequencing (NGS) which has made genotyping projects more automated and affordable for large sample sizes.

Examples of where molecular markers have been employed to gain insight in potato are included below. Markers such as isozymes, simple sequence repeats (SSRs), and rapid amplified polymorphic DNA (RAPDs) have been employed in potato for varietal identification and genetic diversity assessments (Anoumaa et al. 2017; Carputo et al. 2013; Collares et al. 2004; Ghislain et al. 2009; Hoque et al. 2013; Rocha et al. 2010; Salimi et al. 2016; Xiaoyan et al. 2016). SSR markers have also been used to support a reevaluation of the taxonomic classification and structure of the gene pool of cultivated potato into four species (Spooner et al. 2007). Amplified fragment length polymorphisms (AFLPs) were utilized to hypothesize the domestication origin for cultivated potato from the northern species of the *S. brevicaule* group (Spooner et al. 2005) which was contrary to previous reports.

Even though different types of molecular markers exist and have been employed since the 1980s, single nucleotide polymorphisms (SNPs) are increasingly used predominantly due to recent advances in genome sequencing technology, the abundance of SNPs in most crop plants, reduced labor required to collect the data, and price per data point. The affordable cost and high-throughput nature of SNP markers have made them powerful tools for genetic analysis of plant species such as potato and highly useful in breeding (Bertioli et al. 2014). Discovery of SNPs in simple genomes is relatively easy requiring collection and evaluation of sequence data; however, in complex genomes such as potato, SNP detection is more challenging due to repetitive segments of the genome and multiple ploidy levels (Mammadov et al. 2012). Genome complexity reduction methods, such as genotyping by sequencing (GBS), diversity arrays technology (DArTseq), restriction site-associated DNA sequencing (RADseq), have been developed to aid in the discovery of novel SNPs; nevertheless, it is often challenging to identify SNP markers in polyploids such as potato, cotton, canola, and wheat (Bertioli et al. 2014; Logan-Young et al. 2015; Mammadov et al. 2012) due to separating allelic versus homoeologous SNPs or determining dosage in autopolyploids, both of which increase the rate of false positives (Clevenger and Ozias-Akin 2015).

SNP arrays have been developed which allow thousands to one million genome-wide SNP markers to be assessed simultaneously in an individual assay (LaFramboise 2009), thereby reducing the cost per marker data point. The Infinium 12K V2 Potato Array contains 12,720 SNPs, including the SNPs from the original SolCAP Infinium V1 8303 Potato Array with additional markers derived from the Infinium High Confidence SNPs (69K, Hamilton et al. 2011), which were selected for improved genome coverage, candidate genes, and regions with resistance genes. Both potato SNP arrays have been used in numerous studies as a genomic tool to improve cultivated potato or gain insight on genetic attributes. The SolCAP Infinium 8303 Potato Array were selected from 69,011 high-quality SNPs derived from six commercial potato cultivars "Atlantic," "Premier Russet," "Snowden," "Bintje," "Kennebec," and "Shepody" (Hamilton et al. 2011). These SNP markers were used to measure linkage disequilibrium for genome-wide association (GWA) mapping and population structure in European diploid and tetraploid germplasm (Stich et al. 2013). Genotyping a diversity panel of 250 lines of wild species, genetic stocks, and cultivated potato revealed that changes in heterozygosity and allele dosage has not occurred in over 150 years of breeding, but clear selection for alleles in biosynthetic pathways has occurred (Hirsch et al. 2013). The SolCAP Infinium 8303 Potato Array was used to develop linkage maps (Felcher et al. 2012), genotype populations for QTL analysis (Douches et al. 2014), and assess variation in glycoalkaloid biosynthesis (Manrique-Carpintero et al. 2013, 2014). It was also used to evaluate the genetic diversity and population structure of *S. tuberosum* sbsp. *andigenum* and *S. phureja* accessions from Colombia along with identifying 23 markers associated with nine morphological traits (Berdugo-Cely et al. 2017). In a wild species study, relationships deduced from the SNP markers were generally complementary to existing taxonomic classifications for 74 *Solanum* lines representing 25 wild taxa and were also effective in resolving complex taxa boundaries among germplasm with close genetic relationships (Hardigan et al. 2015).

Sequencing efforts including expressed sequence tags (ESTs) and the reference genome for potato (Potato Genome Sequencing Consortium 2011) allowed the development of the Infinium SOLCAP array that has provided the underpinning and a valuable tool for looking more closely at potato diversity. Additionally, the ease of data collection utilizing a SNP array has allowed genotyping of large sample numbers to finally be a reality. The CIP genebank has begun a project to genotype the entire landrace in vitro potato collection utilizing the Infinium V2 12K SNP array (http://solcap.msu.edu/potato_infinium.shtml) along with low density genotyping by sequencing (GBS) of 600 wild potato accessions and over 450 cultivated potatoes. This data is forthcoming; however, it will provide unique fingerprints of a large portion of the collection for users, diversity assessment in wild and cultivated material, identification of potential duplicated material, and genetic tools for genome-wide association studies (GWAS).

An initial assessment of the Infinium V2 SNP array was made by genotyping 250 potato accessions representing 7 taxa (Ellis et al. 2018). The genome-wide SNPs on the array were well distributed across the 12 potato chromosomes ranging from 798 to 1647 SNPs per chromosome. Of the 12,720 SNPs included in the Infinium 12K V2 Potato Array, the majority yielded good-quality signal intensities that were converted into genotypes. SNPs that did not produce a signal in ≥10% of the individuals or could not be clustered were filtered out of the dataset along with SNPs noted in previous studies to be poor or questionable (http://solcap.msu.edu/potato_infinium.shtml). The three-cluster diploid calling yielded 77% (9800) of the total SNPs on the array for use in subsequent analysis. In this data set only, 2.7% of the SNPs were monomorphic with only one of the three diploid genotypic classes scored among all the accessions of the diverse taxa used in this study. Aside from these few, most of the SNPs (97.3%) were polymorphic. A previous study with the 8303-array showed similar rates with a polymorphic rate of 75% (Stich et al. 2013); yet, this study included less taxa with only 44 genotypes. As expected, the five-cluster tetraploid calling had a lower rate of monomorphic markers with only 31 of 4859 (0.6%) of the SNPs being monomorphic; however only 39.5% of the total SNPs on the array could be utilized because of the difficulty in clustering/scoring the three heterozygous classes. Heterozygosity in this data set was high in many species/genotypes and ranged from 0.1 to 81.2% (Ellis et al. 2018).

A distance matrix was calculated from the SNP data and a phylogeny was constructed along with estimates on population structure (Ellis et al. 2018). Most of the taxa did not form distinct monophyletic clades and there were mixed taxa in each clade suggesting significant gene flow, low representation of certain taxa, and/or misclassified accessions. Further, branch lengths among individuals were short suggesting low diversity between taxa and the possibility of hybrids. *S. xjuzpeczukii* (2n = 3x = 36) was the only taxa to form a distinct monophyletic clade. The accessions of *S. xjuzpeczukii* were genetically redundant. This is interesting as these accessions originated from three countries (Bolivia, Peru, and Argentina); therefore, one would expect some genetic divergence. However, given that triploids are sterile and thus represent an evolutionary dead-end they were likely spread by human migration as clones. More of the genome needs to be evaluated (i.e., reduced representation sequencing) in these triploid accessions to elucidate if genetic differences among these accessions truly exists or if they are genetic duplicates before an archival process begins of putatively redundant material.

The phylogeny further demonstrated that *S.* x*chaucha*, *S. stenotomum* subsp. *goniocalyx*, and *S. stenotomum* subsp. *stenotomum* appear to be sister taxa with ancestry to *S. phureja*. Similar suggestions have emerged in previous studies using microsatellite data (Gomez, personal communication) and are supported by the revised taxonomy of Spooner et al. (2014) where these four taxa are lumped into a single species *S. tuberosum*. The phylogenetic tree also reveals putative discrepancies in species designation of accessions which will need to be verified in the future with phenotyping data. For example, multiple *S. stenotomum* subsp. *stenotomum* (17) and *S.* x*chaucha* (9) accessions grouped with the *S. tuberosum* subsp. *andigenum* accessions. In contrast, most of the *S. stenotomum* subsp. *goniocalyx* accessions clustered together or within the *S. stenotomum* subsp. *stenotomum* group (Ellis et al. 2018).

The analysis of the SNP data on 250 accessions from the genebank provided valuable information on intra- and interspecific relationships among taxa and provided additional support for targeting the collection of phenotypic data of suspected misclassified accessions. The data suggests that some of the accessions are hybrids and that gene flow has occurred between many of the taxa. The SNP array produced unique fingerprints with a few exceptions among this diverse panel of cultivated potato. These fingerprints provide a legacy for quality management systems (QMS) of the potato collection for years into the future to assure that as accessions are handled errors do not occur, and if they do occur, fingerprints can be used to validate genetic identity. Further, these data can be used to compare accessions between genebanks to determine where unique genetic material exists on a global scale and ensure it is safely preserved for the future. Overall, the genetic analysis facilitated a better understanding of the genetic diversity, population structure, and genetic relatedness of these potato taxa.

4.6 Climate Change and Genetic Resources Collections

The increased variability of the global climate, resulting in greater extremes of temperatures punctuated by increased dry, wet, cold, warm, and generally unpredictable climatic periods, will continue to have severe impacts on crop productivity and sustainability. Estimates of the extent of temperature increases range between 2 and 6 °C (Jarvis et al. 2011). The warming climates will be accompanied by increased atmospheric CO_2, and although both factors will favor plant productivity of C_3 plants, such as potatoes, this increased productivity will be accompanied by large increases in biotic (diseases and pests) and abiotic (drought, heat, adverse and unseasonable weather) challenges that could decrease plant yield and increase crop uncertainty. Previously adapted varieties may no longer be robust in the same geographic locality due to a lack of resistance/tolerance to these biotic and abiotic factors. These factors are interrelated, stressed resistance varieties may have attenuated resistance in different environments, resistance may be needed at different developmental stages than previously required, resistance could be overwhelmed by population sizes of pests and diseases, and new pests and diseases may appear. It must also be recognized that one cannot look at a single factor such as how climate affects insect vectors and thus the spread of diseases as an effect on insects will also affect insect pollinators reducing subsequent plant populations, insect predators, and insect populations by an extended growing season. It has been estimated that with a 2 °C rise in temperature, insects will go through one to five additional life cycles/year. In general, bacteria will respond better to moisture fluctuations regardless of temperature while fungi will respond to moisture fluctuations more in cooler weather (Beed 2011).

What is needed now and in the future is genes which can be introgressed into potato to help maintain productivity with the changing climate. Crop wild relatives conserved in genebanks offer "an enormous and unimaginable potential" for the discovery of valuable and desired traits (Machida-Hirano 2015). Wild relatives of potato have been a source of disease resistance for breeding programs for over 100 years (Hawkes 1958). An example is *Phytophthora infestans* (causal agent of late blight), which is the causal agent responsible for the devastation it inflicted during the Irish potato blight in 1845–1846. It is most severe in periods of high moisture with temperatures between 7.2 and 26.8 °C. It is predicted that for each degree of warming, late blight can occur 4–7 days earlier, extending the period of infections to between 10 and 20 days. With warmer temperatures starting earlier, there is a potential threat that late blight infestation will come earlier and earlier and last longer into the season resulting in increased fungicide applications leading to a greater pressure of the evolution of resistance to the fungicides (Beed 2011). Eleven hypersensitive-type resistance genes (R genes) have been characterized from *S. demissum*, a wild relative of potato of Mexican origin, thus *S. demissum* has been used extensively as a source for late blight resistance in European and North American cultivars (Asano and Tamiya 2016; Love 1999). Other sources of resistance in wild relatives of potato include *S. bulbocastrum, S. candolleanum, S. chacoense*, *S. pinnatisectum,* and *S. stoloniferon* (Machida-Hirano 2015).

Warmer temperatures can affect resistance genes rendering them ineffective. In tomato, increased temperatures caused a complete breakdown of resistance to *Ralsonia solanacearum* (bacterial wilt), another severe disease in potato (Kuun et al. 2001; Tung et al. 1992). Thus, increased temperatures affect not only the pathogenic organisms and vectors but also the resistance genes. Therefore, multiple forms of resistance for breeding programs may be needed and genebanks could provide such multiple forms of resistance. Examples of potato wild relatives where bacterial wilt resistance has been found are *S. chacoense, S. commersonii, S. phureja*, and *S. stonotonum* (Asano and Tamiya 2016; Machida-Hirano 2015).

Increased temperatures will also favor an increase in pests including aphids, weevils, and potato tuber moth (Asano and Tamiya 2016), and it has been stated that temperature is likely the single most important factor in cold blooded insect behavior, distribution, development, survival, and reproduction (Bale et al. 2002). Pest resistance in potato wild relatives is associated with high levels of glycoalkaloids, dense hairs, and trichomes found in species such as *S. chacoense, S. polyadenium,* and *S. tarijense* (Jansky et al. 2009). Nematode populations can also be influenced by climate as observed with the Southern root-knot nematode (*Meloidogyne incognita*) in coffee which has expanded its range in recent years associated with changing climatic conditions (Beed 2011). Nematode resistance in potato wild relatives can be found in *S. hougasii* for Columbia root-knot nematode (Brown et al. 1991) and *S. vernei* and *S. acuale* for cyst nematode (Hawkes 1994).

In addition to biotic challenges, climatic changes also can greatly affect potato survival and yield simply by the stresses imposed independent of effects on biotic factors. Increases in early and late season frosts, extreme heat, delay or unseasonable rains and other severe weather events. Tolerance to these abiotic factors is as important as resistance to pests and diseases. *S. acuale*, a potato wild relative can form a carpet of tiny plants at 4500 m.a.s.l displaying extreme frost and drought tolerance (personal observation). Other frost tolerant species include *S. demissum* and *S. xjuzcepczukii*. Drought tolerance can be found in *S. chillonanum*, *S. jamesii*, and *S. okadae* (Watanabe et al. 2011) and *S. xcurtilobum* (Cabello et al. 2012).

4.7 Concluding Comments

It is fortunate that the vast majority of the diversity of cultivated potato, as well as much of that of its wild relatives, is already safely conserved in numerous genebanks around the world. These resources represent an investment to ensure potato's role in food security well into the future. These conserved potato genetic resources will provide the traits and alleles needed to continue potato's dominance on plates around the world. Increased knowledge of the molecular and biochemical pathways is critical to maximize not only the discovery but also the use of the needed attributes in breeding programs and food systems. The benefits of these ex situ collections will continue to be delivered as the hidden characteristics, some with unknown value today, continue to be uncovered. One cannot however sit back complacently because as long as there is diversity which is threatened, and not securely conserved, there will be valuable alleles, individuals, and populations that could be critical in the future.

References

Aguilar AM (2016) Informe final de siembra de papas nativas repatriadas en la comunidad capesina de Santa Cruz de Pichiu, Chana, Huari, Ancash, 51p

Anoumaa M, Yao NK, Kouam EB, Kanmegne G, Machuka E, Osama SK, Nzuki I, Kamga YB, Fonkou T, Omokolo DN (2017) Genetic diversity and core collection for potato (*Solanum tuberosum* L.) cultivars from Cameroon as revealed by SSR markers. Am J Potato Res 94:449–463

Arif IA, Bakir MA, Khan HA, Al Farhan AH, Al Homaidan AA, Bahkali AH, Al Sadoon M, Shobrak M (2010) A brief review of molecular techniquest to assess plant diversity. Int J Mol Sci 11:2079–2096

Asano K, Tamiya S (2016) Breeding of pest and disease resistant potato cultivars in Japan using classical and molecular approaches. Jpn Agric Res Q 50(1):1–6

Bajaj YPS (1977) Initiation of shoots and callus from potato-tuber sprouts and axillary buds frozen at −196 °C. Crop Improv 4:48–53

Bale JS, Masters GJ, Hodkinson ID, Awmack C, bezemer TM, Brown VK, Bitterfield J, Buse A, Coulson JC, Farrer J, Good JEG, Harrington R, Hartley S, Jones TM, Lindroth RL, Press MC, Symrnooudis I, Watt AD, Whittaker JB (2002) Herbivory in global climate change research: direct effects of rising temperatures on insect herbivores. Glob Chang Biol 8:1–16

Bamberg J, del Rio A (2011) Use of native potatoes for research and breeding. Hortscience 46(11):1444–1445

Bamberg J, del Rio A, Moreyra R (2009) Genetic consequences of clonal versus seed sampling in model populations of two wild potato species indigenous to the USA. Am J Potato Res 86:367–372

Bamberg J, del Rio A, Fernandez C, Salas A, Vega S, Zorrilla C, Roca W, Tay D (2010) Comparison of "remote" versus "easy" in situ collection locations for USA wild solanum (potato) germplasm. Am J Potato Res 87:277–284. https://doi.org/10.1007/s12230-010-9133-8

Bamberg JB, del Rio A, Huamán Z, Vega S, Salas A, Pavek J, Fernandez C, Spooner DM (2003) A decade of collecting and research on wild potatoes of the Southwest USA. Am J Potato Res 80:159–172

Beed F (2011) The impact of climate change on interdependence for microbial genetic resources for agriculture. In: Fujisaka S, Williams D, Halewood M (eds) The impact of climate change on countries interdependence on genetic resources for food and agriculture. Background Study Paper No. 48. FAO Commission on Genetic Resources for Food and Agriculture, Italy, pp 38–47

Berdugo-Cely J, Valbuena RI, Sanchez-Betancourt E, Barrero LS, Yockteng R (2017) Genetic diversity and association mapping in the Colombian central collection of *Solanum tuberosum*

L. Andigenum group using SNP markers. PLoS One 12(3):e0173039. https://doi.org/10.1371/journal.pone.0173039
Bertioli DJ, Ozias-Akins P, Chu Y, Dantas KM, Santos SP, Gouvea E, Guimaraes PM, Leal-Bertioli SC, Knapp SJ, Moretzsohn MC (2014) The use of SNP markers for linkage mapping in diploid and tetraploid peanuts. G3 (Bethesda) 4:89–96
Bosch FVD, Jeger MJ, Gilligan CA (2007) Disease control and its selection for damaging plant virus strains in vegetatively propagated staple food crops; a theoretical assessment. Proc R Soc [Biol] 247:11–18
Brown CR, Mojtahedi H, Santo GS (1991) Resistance to Columbia root-knot nematode in *Solanum* spp. and in hybrids of *S. hougasii* with tetraploid cultivated potato. Am Potato J 68:445–452
Cabello R, De Mendiburu F, Bonierbale M, Monneveux P, Roca W, Chujoy E (2012) Large-scale evaluation of potato improved varieties, genetic stocks and landraces for drought tolerance. Am J Potato Res 89:400–410
Carputo D, Alioto D, Aversano R, Garramone R, Miraglia V, Villano C, Frusciante L (2013) Genetic diversity among potato species as revealed by phenotypic resistances and SSR markers. Plant Genetic Resources 11(2): 131–139. https://doi.org/10.1017/S1479262112000500
Castañeda-Álvarez NP, de Haan S, Juárez H, Khoury CK, Achicanoy HA, Sosa CC, Bernau V, Salas A, Heider B, Simon R, Maxted N, Spooner D (2015) Ex situ conservation priorities for the wild relatives of potato (*Solanum* L. Section Petota). PLos One 10(14):e0122599
Clevenger JP, Ozias-Akin P (2015) SWEEP: a tool for filtering high quality SNPs in polyploid crops. G3 (Bethesda) 5:1797–1803
Collares EAS, Choer E, Pereira AS (2004) Characterization of potato genotypes using molecular markers. Pesqui Agropecu Bras 39:871–878
Correll DS (1962) The potato and its wild relatives. Contributions from the Texas Research Foundation 4, p 606. Texas Research Foundation, Renner, Texas
De Bruyn A, Villemot J, Lefeuvre P, Villar E, Hoareau M, Harimalala M, Abdoul-Karime AL, Abdou-Chakour C, Reynaud B, Harkins GW, Varsani A, Martin DP, Lett J-M (2012) East African cassava mosaic-like viruses from Africa to Indian Ocean islands: molecular diversity, evolutionary history and geographical dissemination of a bipartite begomovirus. BMC Evol Biol 228(12):228–246
del Rio AH, Bamberg JB, Huaman Z, Salas A, Vega SE (1997) Assessing changes in the genetic diversity of potato gene banks. 2. In situ vs ex situ. Theor Appl Genet 95:199–204
del Rio AH, Bamberg JB, Huaman Z, Salas A, Vega SE (2001) Association of ecogeographical variables and RAPD marker variation in wild potato populations of the USA. Crop Sci 41:870–878
Douches D, Hirsch CN, Manrique-Carpintero NC, Massa AN, Coombs J, Hardigan M, Bisognin D, De Jong W, Buell CR (2014) The contribution of the Solanaceae coordinated agricultural project to potato breeding. Potato Res 57:215–224
Ellis D, Chavez O, Coombs J, Soto J, Gomez R, Douches D, Panta A, Silvestre R, Anglin NL (2018) Genetic identity in genebanks: application of the SolCap 12K SNP array in fingerprinting the global in trust potato collection. Genome 61:523–537
FAO (2010) The second report on the state of the world's plant genetic resources for food and agriculture. FAO, Rome, 368p
FAO (ed) (2014) Genebank standards for plant genetic resources for food and agriculture, rev edn. FAO, Rome
FAO/IPGRI Technical Guidelines for the Safe Movement of Germplasm. https://cropgenebank.sgrp.cgiar.org/images/file/learning_space/potato_tech_guid_safe_move_germplasm.pdf
Felcher KJ, Coombs JJ, Massa AN, Hansey CN, Hamilton JP, Veilleux RE, Buell CR, Douches DS (2012) Integration of two diploid potato linkage maps with the potato genome sequence. PLoS One 12:e36347
Fuentes XC (2014) Conserving the genetic diversity of Bolivian wild potatoes. PhD thesis, Wageningen University, Wageningen, NL, 229p, ISBN 978-94-6257-168-67

Ghislain M, Nunez J, Herrera MDR, Pignataro J, Guzman F, Bonierbale M, Spooner DM (2009) Robust and highly informative microsatellite based genetic identity kit for potato. Mol Breed 23:377–388

Global Strategy for Ex situ Conservation of Potato 2006. https://www.genebanks.org/wp-content/uploads/2017/01/Potato-Strategy-2007.pdf

Grobman A, Bonavia D, Dillehay TD, Pipernod DR, Iriartef J, Holst I (2012) Preceramic maize from Paredones and Huaca Prieta, Peru. Proc Natl Acad Sci U S A 109(5):1755–1759

Grout BWW, Henshaw GG (1978) Freeze preservation of potato shoot-tip cultures. Ann Bot 42:1227–1229

Hamilton JP, Hansey CN, Whitty BR, Stoffel K, Massa AN, Deynze AV, De Jong WS, Douches DS, Buell CR (2011) Single nucleotide polymorphism discovery in elite North American potato germplasm. BMC Genomics 12:302

Hardigan MA, Bamberg J, Buell CR, Douches DS (2015) Taxonomy and genetic differentiation among wild and cultivated germplasm of *Solanum* sect Petota. Plant Genome 1:16–270

Hawkes JG (1958) Significance of wild species and primitive forms for potato breeding. Euphytica 7:257–270

Hawkes JG (1989) Nomenclatural and taxonomic notes on the infrageneric taxa of the tuber-bearing solanums (Solanaceae). Taxon 39(3):489–492

Hawkes JG (1990) The potato: evolution, biodiversity and genetic resources. Belhaven Press, London, 259p

Hawkes JG (1994) Origins of cultivated potatoes and species relationships. In: Bradshaw JE, MacKay GR (eds) Potato genetics. CAB International, Wallingford, pp 3–42

Hijmans RJ, Spooner DM (2001) Geographic distribution of wild potato species. Am J Bot 88(11):2101–2112

Hirsch CN, Hirsch CD, Felcher K, Coombs J, Zarka D, Deynze AV, Jong WD, Veilleux RE, Jansky S, Bethke P, Douches DS, Buell CR (2013) Retrospective view of North American potato (*Solanum tuberosum* L.) breeding in the 20th and 21st centuries. G3 (Bethesda) 3:1003–1013. https://doi.org/10.1534/g3.113.005595.

Hoque ME, Huq H, Moon NJ (2013) Molecular diversity analysis in potato (*Solanum tuberosum* L.) through RAPD markers. SAARC J Agric 11:95–102

International Potato Center (1973–1974) Germplasm exploration and taxonomy of potatoes. In: Program planning conferences 1973–74, Lima, Perú, pp 5–48

International Potato Center (1976) Report of the planning conferences on the exploration and maintenance of germplasm resources, Lima, Peru, Marzo 15–19, 1976, 130p

International Potato Center (1979) Report of the planning conferences on the exploration, taxonomy and maintenance of potato germplasm III. CIP, Lima, Peru, 193p

Jansky SH, Simon R, Spooner DM (2009) A test of taxonomic predictability: resistance to Colorado potato beetle in wild relatives of potato. J Econ Entomol 102:442–431

Jansky SH, Dempewolf H, Camadro EL, Simon R, Zimnoch-Guzowska E, Bisognin DA, Bonierbale M (2013) A case for crop wild relative preservation and use in potato. Crop Sci 53:746–754

Jarvis A, Ramirez J, Hansen J, Leibing C (2011) Crop and forage genetic resources: international interdependence in the face of climate change. In: Fujisaka S, Williams D, Halewood M (eds) The impact of climate change on countries interdependence on genetic resources for food and agriculture. Background Study Paper No. 48. FAO Commission on Genetic Resources for Food and Agriculture, Italy, pp 7–83

Jones RAC (1982) Tests for transmission of four potato viruses through potato true seed. Ann Appl Biol 100:315–320

Kaczmarczyk A, Rokka V-M, Keller ERJ (2011) Potato shoot tip cryopreservation. A review. Potato Res 54:45–79. https://doi.org/10.1007/s11540-010-9169-7

Khan AJ, Akhtar S, Al-Matrushi AM, Briddon RW (2012) Introduction of East African cassava mosaic Zanzibar virus to Oman harks back to "Zanzibar, the capital of Oman". Virus Genes 46(1):195–198. https://doi.org/10.1007/s11262-012-0838-2

Kinder DH, Adams KR, Wilson HJ (2017) *Solanum Jamesii*: evidence for cultivation of wild potato tubers by ancestral Puebloan groups. J Ethnobiol 37(2):218–240
Kreuze J (2014) siRNA deep sequencing and assembly: piecing together viral infections. In: Gullino ML, Bonants PJM (eds) Detection and diagnostics of plant pathogens, plant pathology in the 21st century, vol 5. Springer, Dordrecht, 200p. https://doi.org/10.1007/978-94-017-9020-8_2
Kuun GK, Okole B, Bornmann L (2001) Protection of phenypropanoid metabolism by prior heat treatment in *Lycopersicon esculentum* exposed to *Ralstonia solanacearum*. Plant Physiol Biochem 39:871–880
LaFramboise T (2009) Single nucleotide polymorphism arrays: a decade of biological, computational and technological advances. Nucleic Acids Res 37:4181–4193
Legg JP, Jeremiah SC, Obiero HM, Maruthi MN, Ndyetabula I, Okao-Okuja G, Bouwmeester H, Bigirimana S, Tata-Hangy W, Gashaka G, Mkamilo G, Alicai T, Kumar LP (2011) Comparing the regional epidemiology of the cassava mosaic and cassava brown streak virus pandemics in Africa. Virus Res 159(2):161–170
Logan-Young CJ, Yu JZ, Verma SK, Percy RG, Pepper AE (2015) SNP discovery in complex allotetraploid genomes (*Gossypium spp.*, Malvaceae) using genotyping by sequencing. Appl Plant Sci 3:1400077. https://doi.org/10.1007/s12229-014-9146-y
Loskutov IG (1999) Vavilov and his institute. A history of the world collection of plant genetic resources in Russia. International Plant Genetic Resources Institute, Rome. ISBN 92-9043-412-0
Louderback LA, Pavlik BM (2017) Starch granule evidence for the earliest potato use in North America. Proc Natl Acad Sci U S A 114(29):7606–7610
Love SL (1999) Founding clones, major contributing ancestors, and exotic progenitors of prominent North American potato cultivars. Am J Potato Res 76:263–272
Machida-Hirano R (2015) Diversity in potato genetic resources. Breed Sci 65:26–40
Mammadov J, Aggarwal R, Buyyarapu R, Kumpatla S (2012) SNP markers and their impact on plant breeding. Int J Plant Genomics 2012:728398. https://doi.org/10.1155/2012/728398
Manrique-Carpintero NC, Tokuhisa JG, Ginzberg I, Holliday JA, Veilleux RE (2013) Sequence diversity in coding regions of candidate genes in the glycoalkaloid biosynthetic pathway of wild potato species. G3 (Bethesda) 3:1467–1479
Manrique-Carpintero NC, Tokuhisa JG, Ginzberg I, Veilleux RE (2014) Allelic variation in genes contributing to glycoalkaloid biosynthesis in a diploid interspecific population of potato. Theor Appl Genet 127:391–405
Ochoa CM (1962) Los Solanun Tuberiferos silvestres del Peru Secc. Tubernarium, sub-secc. Hyperbasarthrum), Lima, Peru
Ochoa CM (1980a) New taxa of Solanum from Peru and Bolivia. Phytologia 46(4):223–225
Ochoa CM (1980b) New tuber-bearing Solanum from Colombia. Phytologia 46(7):495–497. ISSN 0031-9430
Ochoa CM (1980c) Solanum peloquinianum: a new wild potato species from the Peruvian Andes. Am Potato J 57(1):33–35
Ochoa CM (1981a) [Solanum chilliasense, a new tuber bearing species of the Series Piurana]. Solanum chilliasense, nueva especie tuberifera de la serie Piurana. Lorentzia (Argentina) 4:9–11
Ochoa CM (1981b) [Solanum taulisense, a new Peruvian tuber bearing species]. Solanum taulisense, nueva especie tuberifera peruana. Lorentzia (Argentina). 4:13–15
Ochoa CM (1981c) Solanum irosinum, new Peruvian tuber bearing species resistant to Phytophthora infestans. Am Potato J 58(3):131–133. ISSN 0003-0589
Ochoa CM (1981d) Two new tuber-bearing Solanum from South America. Phytologia 48:229–232
Ochoa CM (1983a) Solanum bombycinum, a new tuber-bearing tetraploid species from Bolivia. Am Potato J 60(11):849–852. ISSN 0003-0589
Ochoa CM (1983b) Solanum neovavilovii: a new wild potato species from Bolivia. Am Potato J 60(12):919–923. ISSN 0003-0589
Ochoa CM (1987) Solanum longiusculus (Sect. petota), nova specie peruviana. Phytologia 63(5):329–330. ISSN 0031-9430

Ochoa CM (1988a) [Solanum sarasarae (Sect. Petota), a new species from Perú]. Solanum sarasarae (Sect. Petota), nova specie peruviana. Phytologia 64(4):245–246

Ochoa CM (1988b) Solanum bill-hookeri: new wild potato species from Peru. Am Potato J 65(12):737–740

Ochoa CM (1989a) [Solanum Ser. Simplicissima, new serie Tuberifera of Sect. Petota (Solanaceae)]. Solanum Ser. Simplicissima, nueva serie Tuberifera de la Sect. Petota (Solanaceae). Revista de la Academia Colombiana de Ciencias Exactas, Físicas y Naturales 17(65):321–323. ISSN 0370-3908

Ochoa CM (1989b) Solanum amayanum: a new wild Peruvian potato species. Am Potato J 66(1):1–4. ISSN 0003-0589

Ochoa CM (1989c) Solanum salasianum: new wild tuber-bearing species from Peru. Am Potato J 66(4):235–238. ISSN 0003-0589

Ochoa CM (1990) The potatoes of South America: Bolivia. Cambridge University Press, Cambridge, 512p. ISBN 0-521-38024-3

Ochoa CM (1998) Solanum ortegae: a new Peruvian species from Sect. petota. Phytologia 85(4):271–272. ISSN 0031-9430

Ochoa CM (1999) Las papas de Sudamérica: Perú (parte 1). Allen Press, Kansas, 1036p. ISBN 92-9060-197-3

Ochoa CM (2004) The potatoes of South America: Peru The wild species, part 1. International Potato Center, Peru, 1036p. ISBN 9290601981

Perez W, Salas A, Raymundo R, Huaman Z, Nelson R, Bonierbale M (2001) Evaluation of wild potato species for resistance to Late Blight. In: Scientist and farmer, partners in research for the 21st century, Program Report International Potato Center, Lima, Peru, 480p

Potato Genome Sequencing Consortium (2011) Genome sequence and analysis of the tuber crop potato. Nature 475:189–195

Reed BM (2008) Cryopreservation—practical considerations. In: Reed BM (ed) Plant cryopreservation a practical guide. Springer Science, New York, 513p

Ristaino JB, Pfister DH (2016) "What a painfully interesting Subject": Charles Darwin's studies of potato late blight. BioScience 66(12):1035–1045

Rocha EA, Paiva LV, de Carvalho HH, Guimaraes CT (2010) Molecular characterization and genetic diversity of potato cultivars using SSR and RAPD markers. Crop Breed Appl Biotechnol 10:204–210

Rumold CU, Aldenderfer MS (2016) Late archaic–early formative period microbotanical evidencefor potato at Jiskairumoko in the Titicaca Basin of southern Peru. Proc Natl Acad Sci U S A 13(48):13672–13677

Ryoko M-H (2015) Diversity in potato genetic resources. Breed Sci 65:26–40

Salimi H, bahar M, Mirlohi A, Talebi M (2016) Assessment of the genetic diversity among potato cultivars from different geographical áreas using the genomic and EST microsatellites. Iran J Biotechnol 14:270–277

Di Santo M, Tarozzi N, Nadalini N, Borini A. (2012) Human sperm cryopreservation: Update on techniques, effect on DNA integrity, and implications for ART. Adv Urol 2012:854837. https://doi.org/10.1155/2012/854837.

Sastry KS, Zitter TA (2014) Plant virus and viroid diseases in the tropics, volume 2: epidemiology and management. Springer, Dordrecht, pp 149–480. https://doi.org/10.1007/978-94-007-7820-7_2

Semagn K, Bjornstad A, Ndjiondjop MN (2006) An overview of molecular marker methods for plants. Afr J Biotechnol 5:2540–2568

Spooner DM, Salas Lopez A, Huaman Z, Hijmans RJ (1999) Wild potato collecting expedition in southern Peru (Department of Apurimac, Arequipa, Cusco, Moquegua, Puno, Tacna) in 1998: taxonomy and new genetic resources. Am J Potato Res 76(3):103–119. ISSN 1099-209X

Spooner DM, Salas AR, Huaman Z, Torres Maita RV, Hoekstra R, Schuler K, Hijmans RJ (2001) Taxonomy and new collections of wild potato species in central and southern Perú. Am J Potato Res 78(3):197–208. ISSN 1099-209X

Spooner DM, McLean K, Ramsay G, Waugh R, Bryan GJ (2005) A single domestication for potato based on multilocus amplified fragment length polymorphism genotyping. Proc Natl Acad Sci U S A 102:14694–14699

Spooner DM, Nunez J, Trujillo G, Herrera MDR, Guzman F, Ghislain M (2007) Extensive simple sequence repeat genotyping of potato landraces supports a major reevaluation of their gene pool structure and classification. Proc Natl Acad Sci U S A 104:19398–19403

Spooner DM, Ghislain M, Simon R, Jansky SH, Gavrilenko T (2014) Systematics, diversity, genetics, and evolution of wild and cultivated potatoes. Bot Rev 80:283–383. https://doi.org/10.1007/s12229-014-9146-y.

Stich B, Urbany C, Hoffmann P, Gebhardt C (2013) Population structure and linkage disequilibrium in diploid and tetraploid potato revealed by genome wide high density genotyping using the SolCap SNP array. Plant Breed 132:718–724

Towill LE (1981a) *Solanum etuberosum*—a model for studying the cryobiology of shoot-tips in the tuberbearing Solanum species. Plant Sci Lett 20:315–324

Towill LE (1981b) Survival at low temperatures of shoot-tips from cultivars of *Solanum tuberosum* group Tuberosum. CryoLett 2:373–382

Tung PX, Hermsen JG, Zaag P, Vander S, Schmiedche P (1992) Effects of heat tolerance on expression of resistance to *Pseudomonas solanacearum* in potato. Potato Res 35:321–328

Valkonen JPT, Pehu E, Watanabe K (1992) Symptom expression and seed transmission of alfalfa mosaic virus and potato yellowing virus (SB-22) in *Solanum brevidens* and *Setuberosum.* Potato Res 35:403–410

Volk GM, Wadell J, Bonnart R, Towill L, Ellis D, Lauffman M (2008) High viability of dormant *Malus* buds after 10 years of storage in liquid nitrogen vapour. CryoLetters 29:89–94

Vollmer R, Villagaray R, Egúsquiza V, Cardenas JE, Castro M, Chavez O, Anglin NA, Ellis D (2017) A large-scale viability assessment of the potato cryobank at the International Potato Center (CIP). In Vitro Cell Dev Biol Plant 53(4):309–317. https://doi.org/10.1007/s11627-017-9846-1

Watanabe KN, Kikuchi A, Shimazaki T, Asahima M (2011) Salt and drought stress tolerance in transgenic potatoes and wild species. Potato Res 54:319–324

Wilkinson K, Grant WP, Green LE, Hunter S, Jeger MJ, Lowe P, Medley GF, Mills P, Philipson J, Poppy GM, Waage J (2011) Infectious diseases of animals and plants: an interdisciplinary approach. Philos Trans R Soc B Biol Sci 366(1573):1933–1942

Xiaoyan S, Chunzhi Z, Ying L, Shuangshuang F, Qing Y, Sanwen H (2016) SSR analysis of genetic diversity among 192 diploid potato cultivars. Hortic Plant J 2:163–171

5

Potato Genetics and Genomics: Gene Editing and Genomic Selection

Marc Ghislain and David S. Douches

Abstract During the last decade, genomics research has generated new insights into potato genetics and made possible new strategies for varietal improvement. The most commonly grown and eaten potato is an autotetraploid, highly heterozygote crop suffering from rapid inbreeding depression. The genetic improvement of the potato presents numerous challenges using conventional tetraploid breeding techniques. However, novel breeding technologies are now available to increase precision and gains for varietal improvement. The public availability of the first potato genome sequence has created new ways to identify the genetic determinants of key traits of the potato as well as ways to use this knowledge for speeding up variety development. Genomic selection applied to tetraploid breeding promises to increase prediction of progeny performance by a more efficient selection of parents. Diploid hybrid breeding is finally making its way two decades after discovering a suppressor gene of the self-incompatibility locus of diploid potatoes. Direct gene transfer into existing varieties of major genes for key traits has been successful but biotech potato development has been constrained by public perception and issues related to the regulation of the technology. Although genome or gene editing is still in its primary stage in potato, it has already been successful in modifying gene expression in a controlled way, and it might face a lower regulatory burden and easier adoption than biotech, transgenic potatoes. Concluding on an optimistic note, we have many reasons, and evidence is starting to mount, that potato crop improvement is finally benefiting from decades of investment in molecular genetics and that the future hold the promises of faster releases of more robust varieties to pest, disease, and climatic extremes, as well as nutritionally enhanced varieties to feed an ever-growing world population.

M. Ghislain (✉)
International Potato Center, Nairobi, Kenya
e-mail: m.ghislain@cgiar.org

D. S. Douches
Michigan State University, East Lansing, MI, USA
e-mail: douchesd@msu.edu

5.1 At the Crossroad of Potato Improvement

5.1.1 The Numerous Challenges of Tetraploid Potato Breeding

Potato breeding has a long and successful history of improved variety released after breeding from mostly advanced breeding clones and landraces of tetraploid nature from essentially two genepools, the short-day adapted upland Andigenum and the long-day adapted lowland Chilotanum (Bonierbale et al., this volume; Spooner et al. 2007). The latter group has given rise to the modern cultivar well-adapted to the Northern hemisphere referred to as the Tuberosum group (Gavrilenko et al. 2013). In the US and Canada, variety replacement has been particularly disappointing in potato since turnover is in the range of several decades, unlike grain crops (Walker 1994). In developing countries, there has been marked change in variety adoption as evidenced by a study on CIP-related varieties which grew from nothing to beyond one million hectares in 35 years (Thiele et al. 2008). A more recent study by Gatto et al. (2018) provides further insight on the impact of CIP's breeding efforts, and documents that in China, the world's largest potato producer, there are over one million hectares planted with varieties that trace back to CIP pedigrees. Furthermore, CIP's genetic footprint in China reaches above 35% of all varieties currently in use, be it through the registration of CIP advanced breeding lines as varieties or through using CIP's elite breeding lines as parents in crosses initiated by Chinese breeding programs. The main bottleneck explaining the difference in varietal turnover is market-driven mostly by processing industry in the US and Canada, unwilling to adjust their manufacturing processes to the cooking and frying attributes of the new varieties. Regardless, potato breeding has shown the ability to deliver varieties with market-demanded processing qualities and new traits for higher resilience against climatic extremes, pest and diseases threats, and with enhanced nutritional qualities.

Phenotypic recurrent selection has been the method of choice to select for improved potato lines starting usually with about 100,000 seedlings from 200 to 300 crosses followed by clonal selection over many years (Bradshaw 2009, 2017). A recent review of the history of conventional potato breeding revealed many examples of important varieties been released after 30 years or more of crossing and clonal selection when an optimal timeline should be 13–14 years (Bradshaw 2009; Jansky and Spooner 2017). The main reason for such long cycle for variety development is the quantitative nature of most important traits, the rapid inbreeding depression, and the low intensity of selection in early generations. The propagation through tubers adds on delays due to low multiplication rate and ease of contamination with pathogens which delay bulking enough quality seed tubers for multilocation field selection.

The genetic base of potato varieties grown in large commercial area is relatively narrow compared to the accessible gene pool for conventional breeding of the potato. This is likely due to the narrow genetic base of the original tetraploid sources from Andigenum and Tuberosum which were used as starting material for breeding.

Ellis et al. (this volume) provide a detailed current status of the gene pool and the germplasm of potato. Many wild potato species can be crossed with cultivated potatoes directly or via another wild species used as a bridge (Plaisted and Hoopes 1989). However, only a fraction of useful genes from wild species have been introgressed successfully into modern potato varieties. About 40% of the wild species carry interesting genetic traits value for pests, diseases and abiotic stresses (Sood et al. 2017). Major genes for disease resistance from wild relatives of the potato were introgressed into breeding lines for late blight (*S. demissum, S. bulbocastanum*), viruses (*S. stoloniferum*), and nematodes (*S. spegazzini, S. vernei*) resistance (Bradshaw and Ramsay 2005; Bradshaw 2009; Finkers-Tomczak et al. 2011).

The real contribution of wild relatives to modern potato varieties is likely underestimated due to uncertainty in pedigree information and quantitative nature of many important traits of the potato. *S. acaule,* which has been a source of disease resistance and abiotic stress tolerance but has been more used as a bridge between wild species and the cultivated potato (Watanabe et al. 1992). From the first cross between *S. bulbocastanum* bearing late blight-resistance genes and the bridging wild species *S. acaule*, 46 years of crossing and selection with cultivated potato, first diploid *S. Phureja* and then tetraploid *S. tuberosum*, were necessary to release the late blight-resistant varieties "Bionica" and "Toluca" (Haverkort et al. 2009). Introgression of wild species genomes into the cultivated groups has been facilitated by unreduced (*2n*) gametes in diploid potatoes and was shown to be highest in group Tuberosum because of intense breeding effort using a dozen of wild species (Plaisted and Hoopes 1989; Hardigan et al. 2017). Wild species carry genes for wild characteristics which are introduced as genetic drag with the disease-resistance genes lead to the notion that there could be a tradeoff between disease resistance and yield (Ning et al. 2017). This might have contributed to the limited use of wild species in potato breeding.

Assuming allelic combination has to consider one positive allele from a wild species and three neutral alleles from cultivated potato, the introgression of only ten positive alleles from wild species is only one in a million genotypes [$(1/4)^n$ where $n = 10$ genes]. This number becomes quickly without practical reach considering epistatic effects from the cultivated potato alleles, and that each cross redistributes the 20 or so quantitative trait loci which are priority traits of modern varieties (Bradshaw 2017). Potato being an auto-tetraploid clonally propagated crop has also accumulated rare mutations and epigenetic changes in alleles otherwise identical. This complicates further the straightforward exploitation of emerging molecular breeding approaches (Visser et al. 2014).

Increased selection intensity before clonal selection has been proposed by progeny tests and full-sib family selection (Bradshaw et al. 1995, 2000). Marker assisted selection can also screen at early stage major genes and Quantitative Trait Loci (QTL) with large effects (Gebhardt 2013; Sharma et al. 2014). In recent years, markers flanking major genes and QTL were developed for resistance to viruses (Mihovilovich et al. 2014; del Rosario et al. 2018), tuber starch and yield (Schönhals et al. 2016), and other important traits (Ramakrishnan et al. 2015). If applied at the early clonal generation stage and multiplexed, marker-assisted selection can be cost-effective (Slater et al. 2013). Estimated breeding value for traits which can be inferred from pedigree information was also proposed to accelerate the intensity of section and result in shorter breeding cycle (Slater et al. 2014a, b). Genomic selection is also proposed to improve combining unknown QTL at an early stage in the breeding process (Slater et al. 2016). New high density and high throughput polymorphic marker systems have been developed for potato (Hamilton et al. 2011; Uitdewilligen et al. 2013; Vos et al. 2015). Using a panel of 83 cultivars of mostly European origin, a high frequency of relatively rare variants and/or haplotypes, with 61% of the variants having a minor allele frequency below 5%, was found which can be explained by the limited number of meiosis separating these cultivars (Uitdewilligen et al. 2013). Recent estimates of linkage disequilibrium in modern potato cultivar populations confirmed the relatively limited number of meiosis separating modern cultivars and therefore limited power of Genome Wide Assisted Studies (GWAS) for allele/gene discovery (Vos et al. 2017; Sharma et al. 2018).

Hence, ways to improve conventional potato breeding exist and are under development but the fundamental inherent limitations of the narrow genetic base of advanced tetraploid potato germplasm used by breeders, the rapid inbreeding depression, and the low multiplication rate of seed tubers call for new methods and tools to improve the potato which will be complementary to conventional tetraploid breeding for some and an alternative for others.

5.1.2 New Potato Breeding Technologies

Potato genetic improvement has taken shortcuts many times to circumvent the limited genepool accessible by crossing and the tedious phenotypic recurrent selection.

Mutagenesis has long been used to improve yield, quality, biotic and abiotic stress resistance, and tolerance of many crops (Maluszynski et al. 1995). According to this review, more than 1,700 mutant varieties involving 154 plant species have been officially released. However, the tedious process of segregating out the rare positive mutation from the negative ones represent a bottleneck for potato crop improvement. Nevertheless, a novel form of doing mutagenesis is making a surprising come-back for potato crop improvement as mentioned below.

Somatic hybridization has been used in potato to bypass sexual incompatibilities between cultivated potato and wild species for about 40 years (Tiwari et al. 2018). These authors reported successful fusion products obtained from 23 *Solanum* species that were characterized for multiple traits. Numerous studies were generated from somatic hybrids to understand the genetic architecture of important traits including isolating important genes. However, no variety has apparently been released from breeding somatic hybrids with cultivated potato likely due to the limitation of tetraploid potato breeding to efficiently remove undesirable alleles.

Although direct gene transfer through transgenics has a much shorter history in potato crop improvement than conventional breeding approaches, it has been highly successful, though for a limited number of traits for which natural genetic variation was not readily available. Virus resistance was the first trait successfully engineered in potato in the late 1980s, and a commercial cultivar was first reported with the combined resistance to PVX and PVY (Lawson et al. 1990). Soon after, an insect resistance was engineered and led to the production of new commercial cultivars (Perlak et al. 1993). This first generation of biotech potatoes were commercialized under the name of NewLeaf™ from 1995 through 2001 in the United States and Canada, but potato processors and retailers realized soon that the NewLeaf potatoes were going to increase their costs without a share of their benefits which precipitated their decline (Thornton 2003). By 2004, none of them were commercialized anymore. For the next decade, no new biotech potato was released. Recently, a series of potato biotech varieties have been released with reduced bruising and browning first and then with late blight resistance, low acrylamide potential, reduced black spot, and lowered reducing sugars while others are near to be released (see below). The opportunities for engineering new traits that bring benefits to the producer and the consumer are numerous and applicable to potato (Halterman et al. 2016). However, the public acceptance of biotech crops remains volatile and unpredictable, and the current lack of science-based regulatory frameworks in many potato producing countries constrain the scope of genetic engineering to products with sizable benefits to both producers and consumers unachievable by other means.

Genome (gene) editing is the most recent and significant genetic engineering technique targeting specific DNA sequences in the crops' genome (Scheben et al. 2017; Yin et al. 2017). Targeted mutagenesis of specific genes for knock-out, deletion, or allelic changes are now possible with a final product free of foreign DNA. Potato has already been shown to be amenable to genome editing and even to develop novel useful products (Butler et al. 2015, 2016; Clasen et al. 2016; Nicolia et al. 2015; Wang et al. 2015; Andersson et al. 2017; Ma et al. 2017). The two editing tools, TALEN and CRISPR/Cas9, must access the genome without integration of foreign DNA since it cannot be eliminated by crossing without losing most of the qualities of the original commercial variety. Transient expression by PEG-mediated protoplast transfection or Agrobacterium-mediated leaf infiltration generated the intended mutation, but the absence of a selectable agent and somatic variation of plants regenerated from protoplasts can make these strategies labor-intensive. Delivery of the editing reagent may be mediated by virus vectors, but their spread and elimination pose additional difficulties. Hence, genome editing technology offers great opportunities in potato but is still at its first stage of development.

True hybrid potato is a new potato breeding strategy which is increasingly been regarded as the game-changing solution to many of the pitfalls of conventional tetraploid breeding. The sparking step goes two decades back with the discovery of a self-incompatibility inhibitor gene in the wild species *S. chacoense* (Hosaka and Hanneman 1998a, b). The *S* locus inhibitor gene (*Sli*) was introgressed into diploid cultivated potato and shown to confer self-compatibility

(Phumichai et al. 2005). Soon after, few breeding programs started to introgress the *Sli* gene into their diploid potato lines and obtained S3 diploid lines with 80% homozygosity and good agronomic performance including yield (Lindhout et al. 2011). In parallel, an inbred line of *S. chacoense* (M6) was developed to produce recombinant inbred line populations (Jansky et al. 2014). As stated in the title of an opinion paper by a large community of US potato geneticists and breeders, it is proposed to "reinvent the potato as a diploid inbred-based line crop" (Jansky et al. 2016). New sources of self-compatibility system are needed to circumvent the use of the wild species *S. chacoense*. Within the diploid cultivated potato germplasm, self-compatible landraces exist, though rare, but can be used to develop inbred lines from distinct gene pools such as the Stenotomum group and Phureja group (Haynes and Guedes 2018). Recently, an even more promising new system has been developed by knocking out the self-incompati-bility gene *S-RNase* using the CRISPR–Cas9 gene editing system (Ye et al. 2018).

In addition to obtaining quick genetic gain by fixing major genes for disease resistance or other important traits and exploiting heterosis by hybridization, the hybrid variety propagation is via true seeds. The use of botanical seeds has long been known to be an extremely interesting alternative to tuber seeds because of it low weight, lower content of pathogens, good storability, option for beneficial coating, and high multiplication rate. Previous work by CIP and other potato research organizations on the concept of True Potato Seed (TPS) aimed to complement traditional seed systems by the use of botanical seed as a mean to propagate potato. However, its actual adoption by farmers has been much less than originally expected. Breeding for good parental clones from tetraploid breeding lines led to the development of several varieties but adoption remained conditioned to reduced or scarcity of seed tuber supply at affordable prices (Almekinders et al. 2009). Recently, a TPS variety, Oliver F1, was devel-oped by the Dutch breeding company Bejo Zaden B.V. (http://www.bejo.com/magazine/bejo-introduces-its-first-true-potato-seed-variety) and is now under deployment in some African countries where quality seed availability is rare. Many years of breeding to develop superior parental inbred lines with disease-resistance genes adapted to the various agroecologies and markets are needed but the potential benefits that could be derived from true hybrid potato seeds are immense. The next decades will tell us whether reinventing the potato as hybrid varieties from diploid inbred parental lines will be adopted by small-holder farmers in developing countries who are the likely first adopters of this new technology.

5.2 The Genome of the Potato

5.2.1 Cultivated, Wild Potato Genome Sequences Towards a Pan-Genome

The identification of the first cultivated potato genome sequence is and will remain a turning point in the history of potato science. Prior to its discovery, genetic markers were associated to genetic determinants of traits breeders and geneticists had been working on. A fraction of the genes was known while transcriptomes were describing their expression in tissues, at different times, and under various environmental situations. Candidate genes were tested for association with these quantitative trait loci but for the most, genes underlying QTL remained unknown. The potato genome sequence brought together all this genetic knowledge into a physical perspective for the first time. Eighty-six percent of the 844 Mb genome was assembled into 12 pseudomolecules where 39,031 protein-coding genes were predicted (The Potato Genome Sequencing Consortium 2011). The potato whose genome was sequenced is a homozygote diploid plant obtained after chromosome doubling of a monoploid derived by anther culture of a heterozygous diploid potato from the *Solanum tuberosum* Group Phureja (Paz and Veilleux 1999). This cultivar groups are diploid short-day adapted cultivars producing tubers lacking dormancy. They occur throughout the eastern slope of the Andes from western Venezuela to central Bolivia at elevation between 2000 and 3400 masl (Ochoa 1990). Other genome sequence from cultivated potato, in particular from the Group Andigenum and Tuberosum including modern cultivars are still missing and expected to reveal insight into the domestication/wild species contributions to the various groups of cultivars. The difficulties lie in the presence of four genome sequences derived by auto-ploidization and multiple introgression of chromosome segments from wild species (Rodríguez et al. 2010; Spooner et al. 2014). This makes assembly and phasing particularly difficult and only recently claims of successful assembly of all four genome sequences from modern potato cultivars were made (*NRGene* at http://www.nrgene.com). The comparison of 99 Mb of genome from a potato of the group Tuberosum with the DM potato genome sequence revealed collinearity and high sequence identity (The Potato Genome Sequencing Consortium 2011). A year after the release of the potato genome sequence, the tomato genome sequence was published together with its closest wild relative and compared to the potato genome sequence (The Tomato Genome Consortium 2012). As known from previous cytological and comparative genetic mapping and synteny studies, the tomato genome presents very similar chromosomal organization but nine large and several smaller inversions. The euchromatic, gene rich regions diverge by 8.7%, whereas the intergenic and repeat-rich heterochromatic regions diverge by 30%. The potato genome sequence was greatly improved by ordering and reordering 93% of the previously assembled genome into 12 pseudomolecules representing the 12 chromosomes of the potato (Sharma et al. 2013). This genome sequence continues to be the sole publicly available genome sequence of a cultivated potato. It is accessible through a friendly

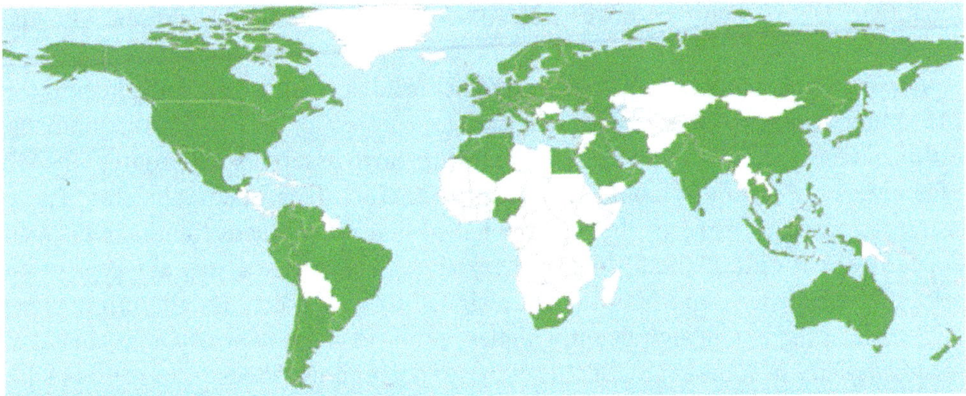

Fig. 5.1 Geographic distribution of SpudDB users. Filled in countries were the source of ten or more unique visits to SpudDB (http://solanaceae.plantbiology.msu.edu/) from October 2017 to October 2018 (Courtesy from John Hamilton, Robin Buell Michigan State University)

web genome browser hosted and maintained by the Buell Lab at Michigan State University in United States and includes annotation datasets, phenotypic and genotypic data from a diversity panel of 250 potato clones (Hirsch et al. 2013). This genomic resource is actively used by potato scientists worldwide (Fig. 5.1).

The initial efforts of the potato sequencing consortium were on resolving the two genome sequences of the dihaploid clone *S. tuberosum* Group Tuberosum RH89-039-16 (RH), but these could not be fully assembled in spite of the availability of the DM sequence. Higher level of heterogeneity was found among the two RH genomes than between RH and DM genomes (The Potato Genome Sequencing Consortium 2011). About 5% of the RH genome sequence (free of repetitive sequences) were aligned with the DM genome sequence and found to be mainly collinear with 97.5% sequence identity, whereas the two RH genome sequences presented 96.5% sequence identity. However, when larger RH genome sequences were obtained, loss of collinearity was frequently observed for the euchromatic region and the three highly diverged pericentric heterochromatin haplotypes of the chromosome 5 (de Boer et al. 2015). These findings stress the importance of sequencing other cultivated potato genome and perform de novo assembly.

Being a diploid and polyploid crop with frequent inbreeding depression and wild species introgression, genome sequence diversity is expected and has contributed to the difficulties of assembling more genome sequences from the cultivated potato.

The genome sequence of a wild species, *Solanum commersonii* was assembled using the potato genome sequence as reference (Aversano et al. 2015). This species has interesting sources of resistance to important diseases of the potato and is known for its freezing resistance and cold acclimation. The species has been used recently in breeding potato for resistance to bacterial wilt in potato and increased levels of resistance were observed (Carputo et al. 2009; Boschi et al. 2017). Flow cytometry estimated a total genome size of 830 Mb. The genome appears to be slightly smaller mainly due to differences in the intragenic regions, to have lower amount of repetitive DNA, and to have 126 cold-related genes not present in the *S. tuberosum* genome.

Recently, the genome sequence of another wild species was resolved using the M6 inbred clone of *Solanum chacoense* (Leisner et al. 2018). Flow cytometry estimated the genome to be 882 Mb. Using a de novo assembly procedure, 508 Mb of the genome assembly could be used to construct 12 pseudomolecules. These were compared to those of the first published genome sequence and shown with concordance for all of them. Interestingly, the genotype used was a 7-generation selfed *S. chacoense* plant but retained residual heterozygosity on all chromosome with three of them with significantly higher proportion. It is too early to assume that heterozygosity in some region is due either to deleterious alleles or to regions with reduced recombination. Genome annotation for gene-models revealed the presence of 37,740 genes. The *S. chacoense* genome sequence is a new resource for identifying important genes of key traits in population derived from M6.

The pan-genome of the cultivated potato covering traditional landraces (diploids to pentaploids), and modern potato cultivars of the Andigenum and Chilotanum gene pools represents today a huge endeavor due to its extraordinary diversity. When available, it would be a powerful resource for breeders to understand the genome structure of the cultivated potato between the core genome with genes present in all cultivars and the dispensable genome made of genes present only in some cultivar groups. The concept is not restricted to modern cultivars but can include wild relatives, or higher taxonomic level (Vernikos et al. 2015). Clearly, more genomes of wild species are also needed to be assembled to improve our understanding of the interspecific genome variation. Ten years after the beginning of sequencing the potato genome, it is worth noting that only one cultivated potato genome is publicly available unlike maize or rice. This highlights the complexity of resolving uneven heterozygosity of the two or four genomes present in wild and cultivated potato. New sequencing technologies and genome assembly software are about to deliver the genomes sequences from heterozygous potatoes. This is highly desirable due to the diversity of species that have contributed to the potato.

5.2.2 The Genome Plasticity of the Cultivated Potato

Comparative analysis of genome sequences in a small panel of closely related potatoes revealed extensive genome plasticity in potato (Hardigan et al. 2016). This study used a panel of doubled monoploid potatoes derived from *S. tuberosum* Group Phureja landraces with limited introgression from Group Stenotomum, Group Tuberosum, and *Solanum chacoense*. Large regions of the potato genome bearing stress-related gene families are duplicated or deleted revealing a possible evolutionary adaptation response to environmental stresses. Copy number variation (CNV) assessed with a minimum 100-bp size revealed that about 30% of the genes are duplicated or deleted in this panel of 12 closely related potatoes. The duplicated regions varied from 500 bp to 575 kb, with total CNV calls per individual varying from 2,978 to 10,532 located preferentially in intergenic sequences in pericentromeric region of the chromosomes. This genome plasticity concerns >7000 genes referred to as dispensable genes. A remarkably high level of genome heterogeneity is found in diploid potato, which is retained through clonal propagation.

Genome heterogeneity is responsible for differential gene expression observed among the genes of tetraploid cultivars (Pham et al. 2017). Genome-wide study of genomic variation and transcription in a panel of six North American tetraploid cultivars revealed the importance of preferential allele expression often associated with evolutionarily conserved genes. Additive allele expression genes in leaves and tubers were only slightly more abundant than preferred-allele expression genes. This can be due to the differential presence of regulatory sequences (promoters, enhancers) but also to structural differences (chromatin structure, epigenetic control). Copy number was frequent; about 40% of the genes from each cultivar were in variable copy number. Again here, copy number variation seemed to be more recent and concerning genes involved in response to biotic and abiotic stresses.

Resequencing of the genomes of a representative sample of cultivated potatoes revealed about 2622 genes under domestication selection, with only 14–16% shared by the North American modern potato cultivars and the Andigenum landraces (Hardigan et al. 2017). This relatively small original gene set suggests a relatively short original common domestication of cultivated potato which diverged into two geographically distinct and long-day adapted cultivar groups by the contribution of wild species. An equally plausible interpretation is two independent domestication events from distinct wild species. This hypothesis has been debated since the early days of potato taxonomy at the beginning of the twentieth century opposing the Russian and the English taxonomist schools advocating respectively multiple and single origin of the cultivated potato (reviewed in Spooner et al. 2014). The absence of extant wild species closely related to the ancestor species of the Southern domestication is the weakest support to this hypothesis (Spooner et al. 2012). The Hardigan study revealed the role of specific wild Solanum species in the evolution of the long-day adapted *S. tuberosum* cultivar group and adaptation to upland and lowland distinguishing the Andigenum and Chilotanum groups. However, both cultivated groups presented a significant contribution from the domestication progenitor *Solanum candolleanum* suggesting the differential contribution from wild species occurred after the domestication from the *S. candollearum* progenitor. Considering variants from the regions of introgression of wild species DNA, the nuclear phylogeny resolved the Chilotanum group and modern cultivars as deriving from the Andigenum group. This study brings closer to closure of a century-old controversy on the independent domestication event leading to the Chilotanum group of cultivars.

5.2.3 New Genomic Tools for Potato Improvement

Potato genomic resources are gradually expanding since the availability of the first potato genome sequence from an Andean potato landrace of the Phureja group (Hirsch et al. 2014). Partial genome sequences are available from dihaploid from modern cultivars and fully resolved haplotypes from tetraploid potato cultivars have been recently achieved. Transcriptomes corresponding to these genome sequences and similar ones have been produced under many important developmental and stress conditions.

The exploitation of potato genomic resources in modern cultivar development is mostly exemplified by the use of the Single Nucleotide Polymorphisms (SNP) arrays developed by the potato community (Douches et al. 2014). Several generations of SNP arrays were generated building on the original Infinium 8303 SNP array (Felcher et al. 2012). As listed by Hirsch et al. (2014), the SolCAP array was used to understand variation for glycoalkaloid biosynthesis in wild and cultivated potato, genotype several diversity panels for a retrospective view of North American potato breeding, for a taxonomic alignment, and for genetic structure of European potato cultivars. Since then, it has been used for genetic mapping in populations derived from a diploid inbred parent (Endelman and Jansky 2016; Peterson et al. 2016), genetic mapping of agronomic traits (Manrique-Carpintero et al. 2015), combined with other SNPs to extend its use to European potato breeding germplasm (Vos et al. 2015), assess linkage decay and testing GWAS models (Sharma et al. 2018), and test genetic identity of accessions in genebanks (Ellis et al. 2018).

5.3 From Genomes to the Genes of the Potato

5.3.1 Gene Discovery Facilitated by the Genome Sequence

Genomics-derived strategies for gene discovery have emerged with the availability of high density markers, decrease in sequencing costs, and the increasing power of bioinformatics.

GWAS has the potential to associate markers with regions, genes, underlying the phenotypic variation of trait of interest, and therefore to increase the effectiveness of potato breeding efforts. Unlike marker association studies based on biparental populations, GWAS is not constrained by the performance of one single genotype as the sole source of an allele of interest, and instead it exploits the power of large populations to identify marker-trait associations. A recent review of GWAS in potato highlighted the importance of understanding the structure (kinship) of the population under study (Sharma et al. 2018). Potato populations made of varieties and breeding lines have been studied to establish Linkage Disequilibrium (LD) between adjacent makers. This is an important parameter of the population under study because the shorter it is the higher is the significance of the association. LD decay in earlier studies were found to present large variation (1–10 cM until equilibrium) depending on the population, the locus, and the type of markers (reviewed in Spooner et al. 2014). The first whole-genome scan of LD decay on a large European potato cultivar population estimated LD decay to 5 cM (D'Hoop et al. 2010), concluding that association studies can be performed at moderate marker densities. Since the advent of SNP arrays, new GWAS have been conducted and revealed the power of this mapping approach over the biparental mapping (Stich et al. 2013). An extended SNP array of the 8303 SolCAP (SolSTW) was used to genotype 569 potato cultivars with 20k SNP markers (Vos et al. 2017). Although this study used a different estimator of LD decay than previous studies, it was found to be in the range of 1.5 Mb for old potato cultivars and 0.6 Mb for those of the second half of twentieth century, values which are compatible with the known limited number of meiosis (5–10) having taken place in the development of these European cultivar populations (Gebhardt et al. 2004; van Berloo et al. 2007). The most recent study using the SolCAP SNP array on a large European cultivar population of 351 tetraploid potatoes estimated LD decay in different regions (short and long arms, and pericentromeric heterochromatin) of each chromosome (Sharma et al. 2018). Again here, their estimates were in the range of 2.73 Mb for euchromatin and 3.27 Mb for whole chromosomal regions. Hence, most studies of LD decay report a modest decay of LD in European potato cultivar populations ranging from 0.6 to 20 Mb depending on the region and chromosome. Interestingly, smaller values of 0.3 Mb in chromosome 4 to 8 Mb in chromosome 8 were estimated for a population of 652 Andigenum cultivars (Berdugo-Cely et al. 2017). The lower distance for LD decay in these native cultivars is expected though a much lower distance could have been anticipated for a population from cultivar domesticated between 8,000 BC and 11,500 BC based on fossil evidence from the dry coast of Peru and south-central Chile (Spooner et al. 2014). It does appear that GWAS in potato can be successful at a modest marker density conferred by current SNP arrays in particular for traits with large QTL effects. However, GWAS alone will not be sufficient to associate markers directly to a specific gene contributing to the trait of interest in potato cultivar populations due to the limited number of meiotic recombination.

Annotation of the potato genome revealed the large family of plant resistance (R) genes discovered by motif sharing (nucleotide-binding site and leucine-rich repeat domain, NB-LRR) with an estimated number per haploid genome of 438 (Jupe et al. 2012). By rescreening the potato genome for NB-LRR target sequences, a total of 755 *R* gene homolog were identified (Jupe et al. 2013). This *R* gene enrichment and sequencing (RenSeq) method was applied to identify markers co-segregating with *R* genes for LB resistance and rapidly clone them (Jupe et al. 2013; Witek et al. 2016; Chen et al. 2018). A derived application of this genome-wide gene discovery is the diagnostic resistance gene enrichment sequencing (dRenSeq) which identifies full *R* genes and their homologs in breeding materials (Armstrong et al. 2018). Combined strategies to identify or clone, multiple resistance genes for diseases such as late blight, viruses, and nematodes will speed up the development of new cultivar with stacked resistance genes.

5.3.2 Progress Toward Next Generation of Potato Varieties

The exploitation of the pan-genome of the potato for varietal improvement will increase as more genomes are sequenced and traits phenotyped more accurately in broad germplasm. However, genetic gain will continue to be low in tetraploid breeding though faster and more predictable by the application of genomic selection. New breeding technologies and diploid hybrid breeding can generate unachievable genetic gains by tetraploid breeding.

Direct gene transfer in potato has been successful in generating disease resistance varieties since the early days of genetic engineering (Halterman et al. 2016). Existing widely grown varieties were genetically upgraded by addition of transgenes conferring resistance to pest and diseases, improved processing qualities, and consumer preferences (Table 5.1).

These transgenes produced new pest and pathogen toxins, silenced incoming viruses or endo-genes, or new enzymes for metabolite engineering. After a short life, the first generation of biotech potatoes were withdrawn as reported above. However, a renewed interest of the industry lead to the release of new biotech potatoes in the US (Waltz 2015). A long awaited released came about the same time in Argentina with a PVY virus-resistant potato variety (Bravo-Almonacid and Segretin 2016). With the exception of the latter, all biotech potato released so far were developed by the private sector. Efforts towards future release of late blight-resistant varieties have increased in the last years (Table 5.1). A 10-year research project in The Netherlands developed transgenic and cisgenic potatoes from four varieties with single and multiple *R* genes (Haverkort et al. 2016). *R* gene stacking was shown to confer high levels of resistance in the filed over several seasons (Zhu et al. 2012; Haesaert et al. 2015). One of these was even fully tested for regulatory approval (Storck et al. 2012). This biotech variety, Fortuna, was unfortunately withdrawn from regulatory approval because of the unfavorable European environment. In the US, a 5-year project aimed at the release in Indonesia and Bangladesh of late blight-resistant local varieties with three *R* genes (https://www.canr.msu.edu/biotechpp/index). These three-*R*-gene biotech potatoes are also the focus of a project aiming at release in sub-Saharan African countries potato varieties with an extremely high level of genetic tolerance to late blight, the most devastating disease in potato, caused by *Phytophtora infestans*, unrivalled by the genetic tolerance achieved to date through conventional breeding(Ghislain et al. 2018). Biotech potatoes have been field tested under natural infection for five seasons and have not shown any lesions caused by *P. infestans* (Fig. 5.2).

The latter two projects are benefiting from the release in the US of the Innate potato with late blight resistance for which the regulatory dossier is publicly available (Clark et al. 2014). One of the important costs in regulatory dossier development is the toxicity assessment of the new proteins for which the 3*R* gene technology can build a weight of evidence instead of costly purification, stability, and gavage testing (Habig et al. 2018). Therefore, when both projects estimated their regulatory costs, these were found to be reasonable unlike those reported by

Table 5.1 Traits of biotech potato from potatoes approved for food and cultivation (source ISAAA GM crop database)

Trait(s)	Trade name	Developer	First approved for food	First approved for cultivation
Colorado Potato Beetle resistance[a]	New Leaf™ Russet Burbank potato	Monsanto Co.	CAN USA (1995); AUS JPN NZL (2001); PHL (2003); KOR (2004)	USA (1994); CAN (1995)
	Atlantic NewLeaf™ potato	Monsanto Co.	CAN MEX USA (1996); AUS NZL (2001)	USA (1995); CAN (1997)
	Superior NewLeaf™ potato	Monsanto Co.	CAN (1995); USA (1996); MEX (1996); AUS JPN NZL (2001); PHL (2003); KOR (2004)	USA (1995)
Colorado Potato Beetle and PVY resistance[a]	New Leaf™ Y Russet Burbank potato	Monsanto Co.	USA (1998); CAN (1999); AUS JPN MEX NZL (2001); PHL KOR (2004)	USA (1997); CAN (1999)
	Shepody NewLeaf™ Y potato	Monsanto Co.	USA (1998); CAN (1999); AUS MEX NZL (2001); JPN PHL (2003); KOR (2004)	USA (1997); CAN (2001)
	Hi-Lite NewLeaf™ Y potato	Monsanto Co.	USA (1998)	
Modified starch (high amylose)[a]	Amflora™	BASF	EU (2010)	EU (2010)
	Starch Potato	BASF	USA (2014)	
Low asparagine (acrylamide), low black spot bruise	Innate® Cultivate	JR Simplot Co.	USA (2014); CAN (2016); AUS JPN MEX MYZ NZL (2017)	USA (2014); CAN (2016)
	Innate® Generate	JR Simplot Co.	USA (2014); CAN (2016); AUS MEX NZL (2017)	USA (2014); CAN (2016)
	Innate® Accelerate	JR Simplot Co.	USA (2014); CAN (2016); AUS MEX NZL (2017)	USA (2014); CAN (2016)
Low asparagine (acrylamide), low black spot bruise, late blight resistance	n/a (Russet Burbank)	JR Simplot Co.	USA (2015); AUS CAN NZL (2017)	USA (2015); CAN (2017)
	Innate® Acclimate	JR Simplot Co.	USA (2016); AUS CAN NZL (2017)	USA (2015); CAN (2017)
	Innate® Hibernate	JR Simplot Co.	USA (2016); AUS CAN NZL (2017)	USA (2015); CAN (2017)
PVY resistance	n/a (Spunta)	Technoplant Argentina	ARG (2018)	ARG (2018)

Countries are represented by three letter codes
[a]Refers to products phased out of the market

Fig. 5.2 Confined field trial conducted in Uganda with transgenic potato with three *R* genes (dark green plots), developed as described by Ghislain et al. (2018), and nontransgenic potato varieties (severely damaged plots)

larger players for commodities like maize (Kalaitzandonakes et al. 2006; Schiek et al. 2016). However, the adoption of biotech potatoes remains challenging due to negative perception by a large part of the public unfamiliar with the challenges and potential solutions to improve agriculture production. The long-standing opposition to industrialization of agriculture, the concerns about multinational corporate dominance, the lack of trust in risk assessment of regulatory agencies, the growing conflict of interest of the organic industry, and the fear of unknown manipulations of our food, have delayed the approval and adoption of biotech crops. The release of biotech potatoes addressing a major long-lasting threat on its production which calls back bad memories to Europeans and North Americans, may well result in a perception change provided strong public education is developed (Hallerman and Grabau 2016).

Gene editing in potato has already passed the stage of proof-of-concept as reviewed above. There are yet no potato products on the market, but gene-edited varieties will soon be released with traits governed by known existing genes whose regulation and allele structure determine the trait value (Table 5.2).

It is important to realize that gene editing is a complement to transgenesis, not replacement, because it is limited to the existing endogenous genes of the potato. Disruptive news came up recently when the European Court of Justice passed a judgment that genome edited crops should be regulated using the same regulatory framework as the transgenic crops (Callaway 2018). This decision is reminiscent of a previous one in 2012 when the European Food Safety Authority concluded that cisgenic crops should be regulated as transgenic crops (EFSA 2012). This European

Table 5.2 Traits targeted by genome editing in potato and opportunities for improving pest and disease resistance as well as nutritional qualities of the potato

Trait	Target gene	Expected impact	References
Heat tolerance (high yield under higher temperature)	Heat-shock cognate 70 (HSc70)	Enhanced yields of potato grown under lowland tropics	Trapero-Mozos et al. (2018)
Virus resistance (PVY potyvirus resistance)	Eukaryotic translation initiation factor 4E (eIF4E)	Reduce yield loss due to PVY and enhance tuber seed quality	Arcibal et al. (2016)
Reduced accumulation of reducing sugars	Vacuolar invertase (VInv)	Improved qualities of processed potato	Clasen et al. (2016)
Reduced acrylamide in processed products	Vacuolar invertase gene VInv and the asparagine synthetase genes StAS1 and StAS2	Reduction of acrylamide formation under extreme cooking temperature	Zhu et al. (2016)
Decreased accumulation of glycoalkaloids	Sterol side chain reductase 2 (St-SSR2)	Release of advanced breeding potato lines with elevated SGA	Sawai et al. (2014)
Inbreeding tolerance	S-RNase alleles (Sp3 and Sp4)	Generation of self-compatible diploid potato for developing hybrid potato varieties	Ye et al. (2018)
LB resistance	Ethylene response factor StERF3; 6 susceptibility genes; DND1 gene	Reduction of production losses and reduced costs of production	Tian et al. (2015), Sun et al. (2016a, b)
VitA biofortification	beta carotene hydroxylase *b-ch* gene	Enrichment in beta carotene in potato (precursor of VitA)	Van Eck et al. (2007)

decision will impact agricultural biotechnology innovation negatively not only in Europe but also in developing countries.

Hybrid breeding in potato has already been tested by farmers in developing countries and has received great excitement by the potato crop improvement actors in spite of the initial skepticism (Lindhout et al. 2017). The first yield assessment of hybrid varieties was conducted in two locations; the Netherlands and the Democratic Republic of Congo (de de Vries et al. 2016). In the latter, the best hybrid variety yielded three to four times the national average in sub-Saharan Africa (SSA) countries, whereas it yielded only half of the yield of conventional varieties in the Netherlands. A confounding factor is the type of seeds and health status that will need to be factored out for more precise yield comparison between hybrid and conventional potato. Nevertheless, the possibility of combining complementary traits from the parents, obtaining heterosis from hybridization of inbred parents, avoiding pathogen load of seed tubers, and facilitating transport of true seeds leaves no doubt that hybrid varieties will attract a lot of interest in the developing world.

5.4 Concluding Remarks

Despite early optimism, and unlike in other crops, the vast insight gained from its genes and genome has not been steadily translated into substantial genetic progress in potato, through either molecular breeding or transgenic approaches. Among the issues behind the above, the genetic complexity of tetraploid potatoes, issues related with public acceptance of transgenic crops, and a critical mass smaller than in other crops stand out as the most salient ones. Regardless, we remain confident that recent scientific developments, such as an increased focus on developing hybrid varieties at the 2X level, are one of the main factors that will change the above-described trend, since a main advantage of dealing with 2X instead of 4X genetics is a much more straightforward application of molecular approaches, as demonstrated already by the routinary use of such technologies in other Solanaceous crops such as potato, and to a lesser extent, pepper. In addition, early reports on the use of genomic selection in potato have demonstrated its ability to circumvent many of the pitfalls observed when QTL were used to attempt increasing the effectiveness of potato breeding efforts. The continuous reduction of DNA sequencing will enable collecting sequencing data on a larger scale than before, further facilitating both the identification of genomic regions associated with traits of economic importance, and a better understanding of quantitative traits in potato. Regarding the use of gene editing approaches, although they provide a much more targeted ability to modify the potato's genome, the full realization of its potential to facilitate the development of varieties carrying genetic alleles not hitherto found in the germplasm available will, by and large, depend on how the public acceptance of genetic modification evolves, both in developed and developing countries.

References

Almekinders CJM, Chujoy E, Thiele G (2009) The use of true potato seed as pro-poor technology: the efforts of an International Agricultural Research Institute to innovating potato production. Potato Res 52(4):275–293. https://doi.org/10.1007/s11540-009-9142-5

Andersson M, Turesson H, Nicolia A, Fält AS, Samuelsson M, Hofvander P (2017) Efficient targeted multiallelic mutagenesis in tetraploid potato (*Solanum tuberosum*) by transient CRISPR-Cas9 expression in protoplasts. Plant Cell Rep 36(1):117–128. https://doi.org/10.1007/s00299-016-2062-3

Arcibal E, Gold KM, Flaherty S, Jiang J, Jahn M, Rakotondrafara AM (2016) A mutant eIF4E confers resistance to potato virus Y strains and is inherited in a dominant manner in the potato varieties Atlantic and Russet Norkotah. Am J Potato Res 93(1):64–71. https://doi.org/10.1007/s12230-015-9489-x

Armstrong MR, Vossen J, Lim TY, Hutten RCB, Xu J, Strachan SM, Harrower B, Champouret N, Gilroy EM, Hein I (2018) Tracking disease resistance deployment in potato breeding by enrichment sequencing. Plant Biotechnol J 17(2):540–549. https://doi.org/10.1111/pbi.12997

Aversano R, Contaldi F, Ercolano MR, Grosso V, Iorizzo M, Tatino F, Delledonne M (2015) The *Solanum commersonii* genome sequence provides insights into adaptation to stress conditions and genome evolution of wild potato relatives. Plant Cell 27:954–968. https://doi.org/10.1105/tpc.114.135954

Berdugo-Cely J, Valbuena RI, Sanchez-Betancourt E, Barrero LS, Yockteng R (2017) Genetic diversity and association mapping in the Colombian Central Collection of *Solanum tuberosum* L. *Andigenum* group using SNPs markers. Plos One 12(3):e0173039. https://doi.org/10.1371/journal.pone.0173039

Boschi F, Schvartzman C, Murchio S, Ferreira V, Siri MI, Galván GA, Dalla-Rizza M (2017) Enhanced bacterial wilt resistance in potato through expression of *Arabidopsis* EFR and introgression of quantitative resistance from *Solanum commersonii*. Front Plant Sci 8:1642. https://doi.org/10.3389/fpls.2017.01642

Bradshaw JE (2009) Potato breeding at the Scottish Plant Breeding Station and the Scottish Crop Research Institute: 1920–2008. Potato Res 52:141–172. https://doi.org/10.1007/s11540-009-9126-5

Bradshaw JE (2017) Review and analysis of limitations in ways to improve conventional potato breeding. Potato Res 60:171–193. https://doi.org/10.1007/s11540-017-9346-z

Bradshaw JE, Ramsay G (2005) Utilisation of the commonwealth potato collection in potato breeding. Euphytica 146(1–2):9–19. https://doi.org/10.1007/s10681-005-3881-4

Bradshaw JE, Stewart HE, Wastie RL, Dale MFB, Phillips MS (1995) Use of seedling progeny tests for genetical studies as part of a potato (*Solanum tuberosum* subsp tuberosum) breeding programme. Theor Appl Genet 90:899–905

Bradshaw JE, Todd D, Wilson RN (2000) Use of tuber progeny tests for genetical studies as part of a potato (*Solanum tuberosum* subsp tuberosum) breeding programme. Theor Appl Genet 100:772–781

Bravo-Almonacid FF, Segretin MF (2016) Status of transgenic crops in Argentina. In: Collinge DB (ed) Biotechnology for plant disease control. Wiley Blackwell, New York, pp 275–283

Butler NM, Atkins PA, Voytas DF, Douches DS (2015) Generation and inheritance of targeted mutations in potato (*Solanum tuberosum* L.) using the CRISPR/Cas system. PloS One 10(12):e0144591. https://doi.org/10.1371/journal.pone.0144591

Butler NM, Baltes NJ, Voytas DF, Douches DS (2016) Geminivirus-mediated genome editing in potato (*Solanum tuberosum* L.) using sequence-specific nucleases. Front Plant Sci 7:1045. https://doi.org/10.3389/fpls.2016.01045

Callaway E (2018) CRISPR plants now subject to tough GM laws in European Union. Nature 560:16. https://doi.org/10.1038/d41586-018-05814-6

Carputo D, Aversano R, Barone A, Di Matteo A, Iorizzo M, Sigillo L, Zoina A, Frusciante L (2009) Resistance to *Ralstonia solanacearum* of sexual hybrids between *Solanum commersonii* and *S. tuberosum*. Am J Potato Res 86:196–202. https://doi.org/10.1007/s12230-009-9072-4

Chen X, Lewandowska D, Armstrong MR, Baker K, Lim T-Y, Bayer M, Harrower B, McLean K, Jupe F, Witek K, Lees AK, Jones JD, Bryan GJ, Hein I (2018) Identification and rapid mapping of a gene conferring broad-spectrum late blight resistance in the diploid potato species *Solanum verrucosum* through DNA capture technologies. Theor Appl Genet 131(6):1287–1297. https://doi.org/10.1007/s00122-018-3078-6)

Clark P, Habig J, Ye J, Collinge S (2014) Petition for determination of nonregulated status for Innate™ potatoes with late blight resistance, low acrylamide potential, reduced black spot, and lowered reducing sugars: Russet Burbank event W8. JR Simplot Company Petition JRS01 (USDA Petition 14-093-01p)

Clasen BM, Stoddard TJ, Luo S, Demorest ZL, Li J, Cedrone F, Coffman A (2016) Improving cold storage and processing traits in potato through targeted gene knockout. Plant Biotechnol J 14(1):169–176. https://doi.org/10.1111/pbi.12370

de Boer JM, Datema E, Tang XM, Borm TJA, Bakker EH, van Eck HJ, van Ham RCHJ, de Jong H, Visser RGF, Bachem CWB (2015) Homologues of potato chromosome 5 show variable collinearity in the euchromatin, but dramatic absence of sequence similarity in the pericentromeric heterochromatin. BMC Genomics 16(1):374. https://doi.org/10.1186/s12864-015-1578-1

D'Hoop BB, Paulo MJ, Kowitwanich K, Sengers M, Visser RGF, van Eck HJ, van Eeuwijk FA (2010) Population structure and linkage disequilibrium unravelled in tetraploid potato. Theor Appl Genet 121(6):1151–1170. https://doi.org/10.1007/s00122-010-1379-5

Douches D, Hirsch CN, Manrique-Carpintero NC, Massa AN, Coombs J, Hardigan M, Bisognin D, De Jong W, Buell CR (2014) The contribution of the Solanaceae Coordinated Agricultural Project to potato breeding. Potato Res 57:215–224. https://doi.org/10.1007/s11540-014-9267-z

EFSA Panel on Genetically Modified Organisms (GMO) (2012) Scientific opinion addressing the safety assessment of plants developed through cisgenesis and intragenesis. EFSA J 10(2):2561[33p]. https://doi.org/10.2903/jefsa20122561

Ellis D, Chavez O, Coombs JJ, Soto JV, Gomez R, Douches DS, Anglin NL (2018) Genetic identity in genebanks: application of the SolCAP 12K SNP array in fingerprinting and diversity analysis in the global in trust potato collection. Genome 61(7):523–537. https://doi.org/10.1139/gen-2017-0201

Endelman JB, Jansky SH (2016) Genetic mapping with an inbred line-derived F2 population in potato. Theor Appl Genet 129(5):935–943. https://doi.org/10.1007/s00122-016-2673-7

Felcher KJ, Coombs JJ, Massa AN, Hansey CN, Hamilton JP, Veilleux RE, Buell CR, Douches DS (2012) Integration of two diploid potato linkage maps with the potato genome sequence. PloS One 7(4):e36347. https://doi.org/10.1371/journal.pone.0036347

Finkers-Tomczak A, Bakker E, de Boer J, van der Vossen E, Achenbach U, Golas T, Suryaningrat S, Smant G, Bakker J, Goverse A (2011) Comparative sequence analysis of the potato cyst nematode resistance locus *H1* reveals a major lack of co-linearity between three haplotypes in potato (*Solanum tuberosum* ssp.). Theor Appl Genet 122(3):595–608. https://doi.org/10.1007/s00122-010-1472-9

Gatto M, Hareau G, Pradel W, Suárez V, Qin J (2018) Release and adoption of improved potato varieties in Southeast and South Asia. International Potato Center (CIP), Lima, Peru, 45p. ISBN 978-92-9060-501-0. Social Sciences Working Paper 2018-2, 42p

Gavrilenko T, Antonova O, Shuvalova A, Krylova E, Alpatyeva N, Spooner DM, Novikova L (2013) Genetic diversity and origin of cultivated potatoes based on plastid microsatellite polymorphism. Genet Resour Crop Evol 60(7):1997–2015. https://doi.org/10.1007/s10722-013-9968-1

Gebhardt C (2013) Bridging the gap between genome analysis and precision breeding in potato. Trends Genet 29:248–256. https://doi.org/10.1016/j.tig.2012.11.006

Gebhardt C, Ballvora A, Walkemeier B, Oberhagemannand P, Schuler K (2004) Assessing genetic potential in germplasm collections of crop plants by marker-trait association: a case study for potatoes with quantitative variation of resistance to late blight and maturity type. Mol Breeding 13(1):93–102

Ghislain M, Byarugaba A, Magembe E, Njoroge A, Rivera C, Roman ML, Tovar JC, Gamboa S, Forbes G, Kreuze J, Barekye A, Kiggundu A (2018) Stacking three late blight resistance genes from wild species directly into African highland potato varieties confers complete field resistance to local blight races. Plant Biotech J 17(6):1119–1129. https://doi.org/10.1111/pbi.13042

Habig JW, Rowland A, Pence MG, Zhong CX (2018) Food safety evaluation for R-proteins introduced by biotechnology: a case study of VNT1 in late blight protected potatoes. Regul Toxicol Pharmacol 95:66–74. https://doi.org/10.1016/j.yrtph.2018.03.008

Haesaert G, Vossen JH, Custers R, De Loose M, Haverkort A, Heremans B, Hutten R, Kessel G, Landschoot S, Van Droogenbroeck B, Visser RGF, Gheysen G (2015) Transformation of the potato variety Desiree with single or multiple resistance genes increases resistance to late blight under field conditions. Crop Prot 77:163–175. https://doi.org/10.1016/j.cropro.2015.07.018

Hallerman E, Grabau E (2016) Crop biotechnology: a pivotal moment for global acceptance. Food Energy Secur 5(1):3–17. https://doi.org/10.1002/fes3.76

Halterman D, Guenthner J, Collinge S, Butler N, Douches D (2016) Biotech potatoes in the 21st century: 20 years since the first biotech potato. Am J Potato Res 93(1):1–20. https://doi.org/10.1007/s12230-015-9485-1

Hamilton JP, Hansey CN, Whitty BR, Buell CR (2011) Single nucleotide polymorphism discovery in elite North American potato germplasm. BMC Genomics 12(1):302. https://doi.org/10.1186/1471-2164-12-302

Hardigan MA, Crisovan E, Hamilton JP, Kim J, Laimbeer P, Leisner CP, Manrique-Carpintero NC, Newton L, Pham GM, Vaillancourt B, Yang X, Zeng Z, Douches DS, Jiang J, Veilleux RE, Buell CR (2016) Genome reduction uncovers a large dispensable genome and adaptive role for copy number variation in asexually propagated *Solanum tuberosum*. Plant Cell 28:388–405. https://doi.org/10.1105/tpc.15.00538

Hardigan MA, Laimbeer FPE, Newton L, Crisovan E, Hamilton JP, Vaillancourt B, Wiegert-Riningera K, Wooda JC, Douches DS, Farréa EM, Veilleux RE, Buell CR (2017) Genome diversity of tuber-bearing Solanum uncovers complex evolutionary history and targets of domestication in the cultivated potato. Proc Nat Acad Sci 114(46):E9999–E10008. https://doi.org/10.1073/pnas.1714380114

Haverkort AJ, Boonekamp PM, Hutten R, Jacobsen E, Lotz LAP, Kessel GJT, Visser RGF (2016) Durable late blight resistance in potato through dynamic varieties obtained by cisgenesis: scientific and societal advances in the DuR*Ph* project. Potato Res 59(1):35–66. https://doi.org/10.1007/s11540-015-9312-6

Haverkort AJ, Struik PC, Visser RGF, Jacobsen E (2009) Applied biotechnology to combat late blight in potato caused by *Phytophthora infestans*. Potato Res 52:249–264. https://doi.org/10.1007/s11540-009-9136-3

Haynes KG, Guedes ML (2018) Self-compatibility in a diploid hybrid population of *Solanum phureja–S. stenotomum*. Am J Potato Res 95(6):729–734. https://doi.org/10.1007/s12230-018-9680-y

Hirsch CD, Hamilton JP, Childs KL, Cepela J, Crisovan E, Vaillancourt B, Hirsch CN, Habermann M, Neal B, and Buell CR (2014) Spud DB: A resource for mining sequences genotypes and phenotypes to accelerate potato breeding. Plant Genome 7(1):1–12. https://doi.org/10.3835/plantgenome2013.12.0042

Hirsch CN, Hirsch CD, Felcher K, Coombs J, Zarka D, Van Deynze A, De Jong W, Veilleux RE, Jansky S, Bethke P, Douches DS, Buell CR (2013) Retrospective view of North American potato (*Solanum tuberosum* L.) breeding in the 20th and 21st centuries. G3: Genes, Genomes, Genetics 3(6):1003–1013. https://doi.org/10.1534/g3.113.005595

Hosaka K, Hanneman RE (1998a) Genetics of self-compatibility in a self-incompatible wild diploid potato species *Solanum chacoense*: I detection of an *S* locus inhibitor (*Sli*) gene. Euphytica 99:191–197

Hosaka K, Hanneman RE (1998b) Genetics of self-compatibility in a self-incompatible wild diploid potato species *Solanum chacoense*: 2 localization of an *S* locus inhibitor (*Sli*) gene on the potato genome using DNA markers. Euphytica 103:265–271

Jansky SH, Charkowski AO, Douches DS, Gusmini G, Richael C, Paul C, Spooner DM, Novy RG, De Jong H, De Jong WS, Bamberg JB, Thompson L, Bizimungu B, Holm DG, Brown CR, Haynes KG, Vidyasagar R, Veilleux RE, Miller JC, Bradeen JM, Jiang JM (2016) Reinventing potato as a diploid inbred line-based crop. Crop Sci 11:1–11. https://doi.org/10.2135/cropsci2015.12.0740

Jansky SH, Chungand Y, Kittipadukal P (2014) M6: a diploid potato inbred line for use in breeding and genetics research. J Plant Regist 8:195–199. https://doi.org/10.3198/jpr2013.05.0024crg

Jansky SH, Spooner DM (2017) The evolution of potato breeding. Plant Breed Rev 41:169–211

Jupe F, Pritchard L, Etherington GJ, MacKenzie K, Cock PJA, Wright F, Sharma SK, Bryan GJ, Jones JDG, Hein I (2012) Identification and localisation of the NB-LRR gene family within the potato genome. BMC Genomics 13(1):75. https://doi.org/10.1186/1471-2164-13-75

Jupe F, Witek K, Verweij W, Śliwka J, Pritchard L, Etherington GJ, Maclean D, Cock PJ, Leggett RM, Bryan GJ, Cardle L, Hein I, Jones JDG (2013) Resistance gene enrichment sequencing (RenSeq) enables reannotation of the NB-LRR gene family from sequenced plant genomes and rapid mapping of resistance loci in segregating populations. Plant J 76(3):530–544. https://doi.org/10.1111/tpj.12307

Kalaitzandonakes N, Alston JM, Bradford KJ (2006) Compliance costs for regulatory approval of new biotech crops. In: Just RE, Alston JM, Zilberman D (eds) Regulating agricultural biotechnology: economics and policy. Natural resource management and policy, vol 30. Springer, Boston, pp 37–57. https://doi.org/10.1007/978-0-387-36953-2_3

Lawson C, Kaniewski W, Haley L, Rozman R, Newell C, Sanders P, Tumer NE (1990) Engineering resistance to mixed virus infection in a commercial potato cultivar: resistance to potato virus X and potato virus Y in transgenic Russet Burbank. Nat Biotechol 8(2):127

Leisner CP, Hamilton JP, Crisovan E, Manrique-Carpintero NC, Marand AP, Newton L, Pham GM, Jiang J, Douches DS, Jansky SH, Buell CR (2018) Genome sequence of M6, a diploid inbred clone of the high-glycoalkaloid-producing tuber-bearing potato species *Solanum chacoense* reveals residual heterozygosity. Plant J 94(3):562–570. https://doi.org/10.1111/tpj.13857

Lindhout P, Meijer D, Schotte T, Hutten RCB, Visser RGF, Eck HJ (2011) Towards F1 hybrid seed potato breeding. Potato Res 54:301–312. https://doi.org/10.1007/s11540-011-9196-z

Lindhout P, De Vries M, Ter Maat M, Ying S, Viquez-Zamora M, Van Heusden S (2017) Hybrid potato breeding for improved varieties. In: Achieving sustainable cultivation of potatoes, vol. 1. Breeding, nutritional and sensory quality. Burleigh Dodds Science Publishing, Cambridge, p 24. https://doi.org/10.19103/AS.2016.0016.04

Ma J, Xiang H, Donnelly DJ, Meng FR, Xu H, Durnford D, Li XQ (2017) Genome editing in potato plants by agrobacterium-mediated transient expression of transcription activator-like effector nucleases. Plant Biotechnol Rep 11(5):249–258. https://doi.org/10.1007/s11816-017-0448-5

Maluszynski M, Ahloowalia BS, Sigurbjörnsson B (1995) Application of *in vivo* and *in vitro* mutation techniques for crop improvement. Euphytica 85(1–3):303–315

Manrique-Carpintero NC, Coombs JJ, Cui Y, Veilleux RE, Buell CR, Douches DS (2015) Genetic map and QTL analysis of agronomic traits in a diploid potato population using single nucleotide polymorphism markers. Crop Sci 55(6):2566–2579. https://doi.org/10.2135/cropsci2014.10.0745

Mihovilovich E, Aponte M, Lindqvist-Kreuze H, Bonierbale M (2014) An RGA-derived SCAR marker linked to PLRV resistance from *Solanum tuberosum* ssp. andigena. Plant Mol Biol Rep 32(1):117–128

Nicolia A, Proux-Wéra E, Åhman I, Onkokesung N, Andersson M, Andreasson E, Zhu LH (2015) Targeted gene mutation in tetraploid potato through transient TALEN expression in protoplasts. J Biotechnol 204:17–24. https://doi.org/10.1016/j.jbiotec.2015.03.021

Ning Y, Liu W, Wang GL (2017) Balancing immunity and yield in crop plants. Trends Plant Sci 22(12):1069–1079. https://doi.org/10.1016/j.tplants.2017.09.010

Ochoa CM (1990) The potatoes of South America: Bolivia. Cambridge University Press, Cambridge

Paz MM, Veilleux RE (1999) Influence of culture medium and *in vitro* conditions on shoot regeneration in *Solanum phureja* monoploids and fertility of regenerated doubled monoploids. Plant Breed 118:53–57

Perlak FJ, Stone TB, Muskopf YM, Petersen LJ, Parker GB, McPherson SA, Fischhoff DA (1993) Genetically improved potatoes: protection from damage by Colorado potato beetles. Plant Mol Biol 22(2):313–321

Peterson BA, Holt SH, Laimbeer FPE, Doulis AG, Coombs J, Douches DS, Hardigan MA, Buell CR, Veilleux RE (2016) Self-fertility in a cultivated diploid potato population examined with the Infinium 8303 potato single-nucleotide polymorphism Array. Plant Genome 9(3):1–13. https://doi.org/10.3835/plantgenome2016.01.0003

Pham GM, Newton L, Wiegert-Rininger K, Vaillancourt B, Douches DS, Buell CR (2017) Extensive genome heterogeneity leads to preferential allele expression and copy number-dependent expression in cultivated potato. Plant J 92(4):624–637. https://doi.org/10.1111/tpj.13706

Phumichai C, Mori M, Kobayashi A, Kamijimaand O, Hosaka K (2005) Toward the development of highly homozygous diploid potato lines using the self-compatibility controlling *Sli* gene. Genome 48:977–984. https://doi.org/10.1139/G05-066

Plaisted RL, Hoopes RW (1989) The past record and future prospects for the use of exotic potato germplasm. Am Potato J 66:603–627

Ramakrishnan AP, Ritland CE, Sevillano RHB, Riseman A (2015) Review of potato molecular markers to enhance trait selection. Am J Potato Res 92(4):455–472. https://doi.org/10.1007/s12230-015-9455-7

Rodríguez F, Ghislain M, Clausen AM, Jansky SH, Spooner DM (2010) Hybrid origins of cultivated potatoes. Theor Appl Genet 121(6):1187–1198. https://doi.org/10.1007/s00122-010-1422-6

del Rosario Herrera M, Vidalon LJ, Montenegro JD, Riccio C, Guzman F, Bartolini I, Ghislain M (2018) Molecular and genetic characterization of the *Ryadg* locus on chromosome XI from Andigena potatoes conferring extreme resistance to potato virus Y. Theor Appl Genet 131(9):1925–1938. doi: https://doi.org/10.1007/s00122-018-3123-5

Sawai S, Ohyama K, Yasumoto S, Seki H, Sakuma T, Yamamoto T, Takebayashi Y, Kojima M, Sakakibara H, Aoki T, Muranaka T, Kazuki Saito K, Umemotog N (2014) Sterol side chain reductase 2 is a key enzyme in the biosynthesis of cholesterol, the common precursor of toxic steroidal glycoalkaloids in potato. Plant Cell 26:3763–3774. https://doi.org/10.1105/tpc.114.130096

Scheben A, Wolter F, Batley J, Puchta H, Edwards D (2017) Towards CRISPR/Cas crops–bringing together genomics and genome editing. New Phytologist 216(3):682–698. https://doi.org/10.1111/nph.14702

Schiek B, Hareau G, Baguma Y, Medakker A, Douches DS, Shotkoski F, Ghislain M (2016) Demystification of GM crop costs: releasing late blight resistant potato varieties as public goods in developing countries. Int J Biotechnol 14:112–131

Schönhals EM, Ortega F, Barandalla L, Aragones A, De Galarreta JR, Liao JC, Sanetomo R, Walkemeier B, Tacke E, Ritter E, Gebhardt C (2016) Identification and reproducibility of diagnostic DNA markers for tuber starch and yield optimization in a novel association mapping population of potato (*Solanum tuberosum* L.). Theor Appl Genet 129(4):767–785. https://doi.org/10.1007/s00122-016-2665-7

Sharma R, Bhardwaj V, Dalamu D, Kaushik SK, Singh BP, Sharma SK, Umamaheshwari R, Baswaraj R, Kumar V, Gebhardt C (2014) Identification of elite potato genotypes possessing multiple disease resistance genes through molecular approaches. Sci Hortic 179:204–211. https://doi.org/10.1016/j.scienta.2014.09.018

Sharma SK, Bolser D, de Boer J, Sønderkær M, Amoros W et al (2013) Construction of reference chromosome-scale pseudomolecules for potato: integrating the potato genome with genetic and physical maps. G3 (Bethesda) 3(11):2031–2047. https://doi.org/10.1534/g3.113.007153

Sharma SK, MacKenzie K, McLean K, Dale F, Daniels S, Bryan GJ (2018) Linkage disequilibrium and evaluation of genome-wide association mapping models in tetraploid potato. G3 (Bethesda) 8(10):3185–3202. https://doi.org/10.1534/g3.118.200377

Slater AT, Cogan NOI, Forster JW (2013) Cost analysis of the application of marker-assisted selection in potato breeding. Mol Breeding 32(2):299–310. https://doi.org/10.1007/s11032-013-9871-7

Slater AT, Cogan NOI, Hayes BJ, Schultz L, Dale MFB, Bryan GJ, Forster JW (2014a) Improving breeding efficiency in potato using molecular and quantitative genetics. Theor Appl Genet 127:2279–2292. https://doi.org/10.1007/s00122-014-2386-8

Slater AT, Wilson GM, Cogan NOI, Forster JW, Hayes BJ (2014b) Improving the analysis of low heritability complex traits for enhanced genetic gain in potato. Theor Appl Genet 127:809–820. https://doi.org/10.1007/s00122-013-2258-7

Slater AT, Cogan NOI, Forster JW, Hayes BJ, Daetwyler HD (2016) Improving genetic gain with genomic selection in autotetraploid potato. Plant Genome 9(3):1–15. https://doi.org/10.3835/plantgenome2016.02.0021

Sood S, Bhardwaj V, Pandey SK, Chakrabarti SK (2017) History of potato breeding: improvement diversification and diversity. In: Chakrabarti SK, Xie C, Tiwari JK (eds) The potato genome. Springer, Cham, pp 31–72

Spooner DM, Núñez J, Trujillo G, del Rosario Herrera M, Guzmán F, Ghislain M (2007) Extensive simple sequence repeat genotyping of potato landraces supports a major reevaluation of their gene pool structure and classification. Proc Nat Acad Sci 104(49):19398–19403. https://doi.org/10.1073/pnas.0709796104

Spooner DM, Jansky SH, Clausen A, del Rosario Herrera M, Ghislain M (2012) The enigma of *Solanum maglia* in the origin of the Chilean cultivated potato *Solanum tuberosum* Chilotanum Group. Econ Bot 66(1):12–21

Spooner DM, Ghislain M, Simon R, Jansky SH, Gavrilenko T (2014) Systematics, diversity, genetics, and evolution of wild and cultivated potatoes. Bot Rev 80:283–383. https://doi.org/10.1007/s12229-014-9146-y

Stich B, Urbany C, Hoffmann P, Gebhardt C (2013) Population structure and linkage disequilibrium in diploid and tetraploid potato revealed by genome-wide high-density genotyping using the SolCAP SNP array. Plant Breeding 132(6):718–724. https://doi.org/10.1111/pbr.12102

Storck T, Böhme T, Schultheiss H (2012) Fortuna et al. Status and perspectives of GM approaches to fight late blight. In: Thirteenth EuroBlight workshop St. Petersburg (Russia), 9–12 Oct 2011. PPO-Special Report 15:45–48.

Sun K, Wolters AMA, Loonen AE, Huibers RP, van der Vlugt R, Goverse A, Bai Y (2016a) Down-regulation of *Arabidopsis DND1* orthologs in potato and tomato leads to broad-spectrum resistance to late blight and powdery mildew. Transgenic Res 25(2):123–138. https://doi.org/10.1007/s11248-015-9921-5

Sun K, Wolters AMA, Vossen JH, Rouwet ME, Loonen AE, Jacobsen E, Bai Y (2016b) Silencing of six susceptibility genes results in potato late blight resistance. Transgenic Res 25(5):731–742. https://doi.org/10.1007/s11248-016-9964-2

The Potato Genome Sequencing Consortium (2011) Genome sequence and analysis of the tuber crop potato. Nature 475:189–195. https://doi.org/10.1038/nature10158

The Tomato Genome Consortium (2012) The tomato genome sequence provides insights into fleshy fruit evolution. Nature 485:635–641. https://doi.org/10.1038/nature11119

Thiele G, Hareau G, Suarez V, Chujoy E, Bonierbale M, Maldonado L (2008) Varietal change in potatoes in developing countries and the contribution of the International Potato Center: 1972–2007. Social Sciences Working Paper (2008-6)

Thornton, M., 2003. The rise and fall of Newleaf potatoes. Biotechnology: Science and society at a crossroads. Nat. Agric. Biotech. Council Rep. 15:235–243.

Tian Z, He Q, Wang H, Liu Y, Zhang Y, Shao F, Xie C (2015) The potato ERF transcription factor StERF3 negatively regulates resistance to *Phytophthora infestans* and salt tolerance in potato. Plant Cell Physiol 56(5):992–1005. https://doi.org/10.1093/pcp/pcv025

Tiwari JK, Devi S, Ali N, Luthra SK, Kumar V, Bhardwaj V, Chakrabarti SK (2018) Progress in somatic hybridization research in potato during the past 40 years. Plant Cell Tiss Org 132:225–238. https://doi.org/10.1007/s11240-017-1327-z

Trapero-Mozos A, Morris WL, Ducreux LJ, McLean K, Stephens J, Torrance L, Taylor MA (2018) Engineering heat tolerance in potato by temperature-dependent expression of a specific allele of *HEAT-SHOCK COGNATE 70*. Plant Biotechnol J 16(1):197–207. https://doi.org/10.1111/pbi.12760

Uitdewilligen JG, Wolters AML, D'hoop BB, Borm TJA, Visser RGF, van Eck HJ (2013) A next generation sequencing method for genotyping-by-sequencing of highly heterozygous autotetraploid potato. PLoS One 8(5):e62355. https://doi.org/10.1371/journalpone0062355

van Berloo R, Hutten RCB, van Eck HJ, Visser RGF (2007) An online potato pedigree database resource. Potato Res 50(1):45–57. https://doi.org/10.1007/s11540-007-9028-3

Van Eck J, Conlin B, Garvin DF, Mason H, Navarre DA, Brown CR (2007) Enhancing beta-carotene content in potato by RNAi-mediated silencing of the beta-carotene hydroxylase gene. Am J Potato Res 84(4):331

Vernikos G, Medini D, Riley DR, Tettelin H (2015) Ten years of pan-genome analyses. Curr Opin Microbiol 23:148–154. https://doi.org/10.1016/j.mib.2014.11.016

Visser RGF, Bachem CWB, Borm T, de Boer J, Van Eck HJ, Finkers R, van der Linden G, Maliepaard CA, Uitdewilligen JGAML, Voorrips R, Vos P, Wolters AMA (2014) Possibilities and challenges of the potato genome sequence. Potato Res 3–4:327–330. https://doi.org/10.1007/s11540-015-9282-8

Vos PG, Uitdewilligen JGAML, Voorrips RE, Visser RGF, van Eck HJ (2015) Development and analysis of a 20K SNP array for potato (*Solanum tuberosum*): an insight into the breeding history. Theor Appl Genet 128:2387–2401. https://doi.org/10.1007/s00122-015-2593-y

Vos PG, Paulo MJ, Voorrips RE, Visser RGF, van Eck HJ, van Eeuwijk FA (2017) Evaluation of LD decay and various LD-decay estimators in simulated and SNP-array data of tetraploid potato. Theor Appl Genet 130(1):123–135. https://doi.org/10.1007/s00122-016-2798-8

de Vries M, ter Maat M, Lindhout P (2016) The potential of hybrid potato for East-Africa. Open Agric 1(1):151–156. https://doi.org/10.1515/opag-2016-0020

Walker TS (1994) Patterns and implications of varietal change in potatoes. Working paper 1994-3 Social Sciences Department, International Potato Center (CIP) Lima, Peru

Waltz E (2015) USDA approves next-generation GM potato. Nat Biotechnol 33:12–13

Wang S, Zhang S, Wang W, Xiong X, Meng F, Cui X (2015) Efficient targeted mutagenesis in potato by the CRISPR/Cas9 system. Plant Cell Rep 34(9):1473–1476. https://doi.org/10.1007/s00299-015-1816-7

Watanabe K, Arbizu C, Schmiediche PE (1992) Potato germplasm enhancement with disomic tetraploid *Solanum acaule* I. Efficiency of introgression. Genome 35(1):53–57. https://doi.org/10.1139/g92-009

Witek K, Jupe F, Witek AI, Baker D, Clark MD, Jones JD (2016) Accelerated cloning of a potato late blight–resistance gene using RenSeq and SMRT sequencing. Nat Biotechnol 34:656–660. https://doi.org/10.1038/nbt.3540

Ye M, Peng Z, Tang D, Yang Z, Li D, Xu Y, Huang S (2018) Generation of self-compatible diploid potato by knockout of *S-RNase*. Nat Plants 4:651–654. https://doi.org/10.1038/s41477-018-0218-6

Yin K, Gao C, Qiu JL (2017) Progress and prospects in plant genome editing. Nat Plants 3:17107. https://doi.org/10.1038/nplants.2017.107

Zhu S, Li Y, Vossen JH, Visser RG, Jacobsen E (2012) Functional stacking of three resistance genes against *Phytophthora infestans* in potato. Transgenic Res 21:89–99. https://doi.org/10.1007/s11248-011-9510-1

Zhu X, Gong H, He Q, Zeng Z, Busse JS, Jin W, Bethke PC, Jiang J (2016) Silencing of vacuolar invertase and asparagine synthetase genes and its impact on acrylamide formation of fried potato products. Plant Biotechnol J 14(2):709–718. https://doi.org/10.1111/pbi.12421

6

Breeding of Potatoes: Population Improvement and Variety Development

Meredith W. Bonierbale, Walter R. Amoros, Elisa Salas, and Walter de Jong

Abstract The breeding of crop plants is a highly effective means of increasing agricultural productivity in a sustainable and environmentally safe way. Prebreeding and population improvement not only capture essential genetic resources and move desired traits along variety development pipelines but also help assure the creation of broad and dynamic gene pools to meet future, unanticipated needs. To efficiently meet multiple breeding objectives requires both interdisciplinary collaboration and a grasp of a wide range of scientific knowledge and expertise. This chapter addresses a range of topics that define and govern potato breeding, drawing from the experiences of both international and regional potato breeding programs, to orient readers to the interlinked components of population improvement and variety development. Using a case study approach to discuss breeding objectives together with respective implications for breeding needs, methods, and awareness-raising approaches for impact, we detail some key research and achievements contributing to current state of the art. Major populations under improvement at the International Potato Center along with breeding objectives and trait levels selected are described in terms of the agroecologies or uses they address in developing country national programs; these are contrasted with a discussion of the Cornell University program that is oriented to the northeastern US. A sample stage gate process, accelerated multi-trait selection schemes, heritability and heterosis exploitation, genomic selection, data management, and end user consultations are introduced in the contexts of these two programs. The topic of this chapter is supported and augmented with further details on subjects closely related to potato breeding, provided in chapters contributed to

M. W. Bonierbale
International Potato Center, Lima, Peru

Duquesa Business Centre, Manilva, Malaga, Spain

W. R. Amoros · E. Salas (✉)
International Potato Center, Lima, Peru
e-mail: CIP-Breeding1@cgiar.org; e.salas@cgiar.org

W. de Jong
Cornell University, Ithaca, NY, USA
e-mail: wsd2@cornell.edu

this volume by Ortiz and Mihovilovich, Ghislain and Douches, Burgos et al., and Ellis et al. The authors hope that the content serves to orient researchers and managers in countries with different degrees of development to plan and succeed in impactful potato improvement programs.

6.1 Implications of Genetics, Genepools, and Biology for Potato Breeding

6.1.1 Key Features

Commercial potato of world importance is a heterozygous, autotetraploid, clonal crop ($2n = 4x = 48$). Modern varieties are the products of extensive breeding between different cultivar groups and wild species. Potato varieties grown outside of South America since the end of the sixteenth century, as well as landrace (indigenous) cultivars grown in lowland Chile and in the high Andes are referred to as *Solanum tuberosum*, within which several groups are recognized (Spooner et al. 2014).

The genetic resources available for potato improvement comprise a polyploid series ($2n = 2x = 24$ to $2n = 6x = 72$) with genetic features that facilitate gene transfer across ploidy levels. Farmers' landrace varieties dominate potato production in the Andean center of origin and diversity; as cultivars they offer tremendous trait diversity in readily useable form. Landrace cultivars, improved varieties and wild potato species in *Solanum* section *Petota* comprise GPs (gene pools) 1 and 2, according to the gene pool concept of Harlan and de Wet (1971) and are relatively straightforward to use in breeding. The wild tuber-bearing potatoes (section *Petota*) include some 200 species (see Spooner et al. 2014 for a review of taxonomic treatments). These typically produce only small tubers, which often contain high levels of toxic glycoalkaloids, and many require short days for tuberization. Domestication of wild potatoes for use as food likely involved selection for increased tuber size and reduced glycoalkaloid content, probably simultaneously in multiple locations in Andean and coastal regions of South America (Ugent et al. 1982, 1987).

Unreduced gametes, self-incompatibility, and inter-specific reproductive barriers played key roles in polyploidization and the maintenance of species boundaries during the evolution and domestication of potato, and knowledge of their genetics is useful in germplasm enhancement. Diploid potatoes are out-crossing due to a system of gametophytic self-incompatibility (Pushkarnath 1942; De Nettancourt 1977), which prevents inbreeding and thereby promotes intraspecific genetic variation. Tetraploid potato, on the other hand, is self-compatible. The breakdown of the gametophytic self-incompatibility system that operates in diploids is a common phenomenon in angiosperm polyploids (Frankel and Galun 1977; Levin 1983), but the molecular mechanism is not known (Comai 2005). Selfing results in severe inbreeding depression in most potatoes. Self-incompatibility at the diploid level, and inbreeding depression in both diploids and tetraploids, make it difficult to eliminate unfavorable alleles or drive favorable alleles to fixation. Several genes governing reproductive isolation, crossability, and ploidy are further discussed in the chapter contributed to this volume by Ortiz and Mihovilovic.

Potato varieties are maintained by clonal (vegetative, asexual) reproduction. Potato tubers are modified stems and comprise the vegetative "seed" used to propagate a variety. In potatoes with an even number of chromosomes (2x, 4x, 6x) it is possible to produce sexual (botanical) seed which provides for the generation of new genotypes. Once variation has been created in the form of sexual seed, any seedling has the potential to become a new variety via clonal propagation. Although most potatoes produce a large number of botanical seeds per fruit (~200) as well as many fruits per plant, clonal propagation results in a low propagation coefficient (five to tenfold per generation). Clonal breeding facilitates intentional and unintentional exposure of candidate varieties to pathogens. This helps breeders eliminate undesirable genotypes, but also necessitates steps to reduce exposure to detrimental viruses that are transmitted through vegetative (tuber) seed. In common with other root and tuber crops the limited ability to phenotype and potato for desirable morphological or developmental features is challenged by the underground location of the harvested product.

6.1.2 Genome Constitution and Variation

Maximum heterozygosity has been considered essential for performance of tetraploid potato, with inbreeding leading to reduced vigor and yield, flower bud abortion, lack of flower bud formation, and sterility (De Jong and Rowe 1971; Mendiburu and Peloquin 1977). However, empirical research with hybrid families has suggested that poor performance may be due to the expression of recessive alleles (De Jong and Rowe 1971) or that the presence of certain alleles may be more important for high yield (Bonierbale et al. 1993). Genomic studies have revealed that tri-allelic and tetra-allelic single nucleotide polymorphisms (SNPs) are rare in potato cultivars (Hirsch et al. 2013; 2014), though these might be expected in outstanding clones if yield or vigor were associated with maximum heterozygosity. Nevertheless, when the products of individual genes are amplified by PCR (polymerase chain reaction) it is not uncommon to find three alleles at a locus.

Xu et al. (2011) proposed heterozygosity as the key feature enabling the frequent occurrence of gene presence/absence variants and other potentially deleterious mutations in the genome of the heterozygous diploid potato clone "RH." Next-generation (short read length) sequencing of 807 genes from 83 potato cultivars revealed a tremendous amount of genetic variation in potato. On average, there is one variant (SNP or indel) every 24 base pair (bp) in exons, and one variant every 15 bp in introns (Uitdewilligen et al. 2013). The average minor allele frequency of a variant is low, though, at 0.14, and 61% of variants have minor allele frequencies less than 0.05 (Uitdewilligen et al. 2013). Given the sequence variation, and keeping in mind that tetraploid potato can contain up to four alleles at each locus, it is easy to understand why so much phenotypic variation results every time two heterozygous potato clones are crossed. Considerable effects due to dominance and epistasis are possible, and experience with such out-crossing polyploids has shown that the genetic variation due to dominance and epistatic effects is large compared to that seen in diploid crops (Gruneberg et al. 2009). This suggests that heterosis largely determines the performance of out-crossing, clonally propagated crops, although its basis in potato has not been fully elucidated.

6.1.3 The Cost of Increasing Genetic Variation

Most plant breeding training programs place heavy emphasis on the need to increase genetic variation. There is value in this, of course, as wild species contain many useful traits, genes/alleles not present in modern cultivars. But there is also a considerable cost that is not appreciated by those unfamiliar with potato breeding. Wild potato species contain countless alleles that are undesirable for potato production, and when a cultivated potato is crossed with a wild accession, oftentimes the offspring is quite poorly adapted. Many further generations of crossing are needed to eliminate the undesirable alleles, while keeping the desired ones. When embarking on such a venture it is important to realize upfront that the process typically takes decades. It may be possible to use molecular markers to speed the process up, by selecting against the donor genome (and for the gene of interest) after each cycle of crossing, but the process will still be much slower than adding a gene to cultivated potato by *Agrobacterium*-mediated transformation, or by editing an existing cultivated allele to a desired wild species allele with CRISPR-Cas9 (https://www.yourgenome.org/facts/what-is-crispr-cas9), once the respective target genes are known. It is worth noting that the "adaptation gap" between wild species germplasm and cultivated germplasm is growing over time, as breeders continually work to improve cultivated germplasm by increasing the frequency of desirable alleles and decreasing the frequency of undesirable ones. Despite the challenges, support for germplasm enhancement programs that strategically bridge this gap and enable the continuous influx of valuable genes from crop wild relatives through improved populations and into varieties is critical to meeting the world's growing need for food in the face of climate change.

6.1.4 Genetic Enhancement

Potato breeding strategies frequently include research to efficiently access traits from beyond the variety-ready germplasm base. Such trait research or pre-breeding is expected to result in new materials, methods, tools, knowledge, and approaches to support the breeding process rather than in finished products or varieties. Genetic and biochemical research often provides insights and tools that enable gains in traits toward new breeding objectives or improved program results. Pre-breeding is conducted in parallel with mainstream breeding activities, such as by developing support populations that are upgraded for trait levels or improved for agronomic traits so that new types of diversity can be introduced from un-adapted to adapted germplasm without impeding advance toward established breeding goals.

Before embarking on breeding for new traits or using uncharacterized germplasm, it is important to assess and consider positive and negative correlations among traits. Trait correlation influences the success of cross combinations and can determine breeding progress when multiple traits are concerned. Negative trait associations can be critical in achieving breeding progress and influence the choice of parents as well as the selection approach that will be most successful. Embarking on medium- to long-term population development does not result in new varieties after a single recombination and selection cycle, but should result in better parents that will help meet the medium- to long-term objectives of breeding programs.

Support populations are useful for enhancing diversity from un-adapted germplasm to avoid introducing undesirable features into advanced breeding populations. In the case of potato, wild and landrace relatives often carry undesired agro-morphological traits like deep eyes, small tuber size and late maturity, a requirement for short days, or high glycoalkaloid content, in addition to untapped resistance to biotic and abiotic factors and nutritional traits. Those undesired traits may be eliminated from hybrid populations by backcrossing to improved types, or source populations may be enhanced for agronomic traits before desired traits are transferred to improved genepools. The large majority of potato's genetic resources are diploid, and breeding at this level results in faster genetic gains than breeding at the tetraploid level. Thus, particularly for multigenic traits, pre-breeding in diploid source germplasm before incorporating new traits into tetraploid breeding populations can be very effective.

Introgressing novel traits from distant wild species such as those in the tertiary gene pool (GP3) may require the use of bridge species to circumvent interspecific reproductive barriers, and unreduced gametes to transfer traits across ploidy levels. When interspecific crossing is possible, backcross schemes are modified from those used for inbred crops, such that a different genotype of the recipient germplasm (adapted type) is used in each cycle of crossing with the trait donor or selected hybrid to avoid inbreeding depression. This is illustrated by Gaiero et al. (2017) who introgressed partial resistance to bacterial wilt, caused by *Ralstonia solanacearum*, from the sexually incompatible GP3 species *S. commersonii* by bridge crosses with 2x *S. tuberosum* Phureja Group and successive backcrosses with different *S. tuberosum* Group Tuberosum genotypes. The resulting advanced backcross progenies are now being used by several breeding programs due to the acute need for resistance to bacterial wilt.

6.1.5 *Case Study 1: Genetic Enhancement and Incorporation of Iron Content from Diploid into Tetraploid Cultivated Potatoes*

With support from HarvestPlus (www.harvestplus.org) advocating the breeding of staple crops for micronutrient density, CIP has sought to increase the content of iron and zinc in potato through an inter-ploid breeding strategy. Recurrent selection in a base population of landrace potatoes took advantage of greater response to selection at the diploid level, and resulted in a population of diploid potato reaching 35 ppm iron and 30 ppm zinc from a baseline of 20 and 16 ppm, respectively. Elevated iron and zinc contents achieved in the source germplasm were incorporated into advanced, tetraploid populations via unreduced gametes. The resulting tetraploid potato population had iron and zinc concentrations twofold higher than baseline levels, which approaches the breeding targets for human populations with deficiencies of these minerals and high potato intake (Bonierbale et al. 2007; Section 6.3.4). Further work remains to be done to identify clones adapted to target agro-ecologies that maintain as favorable a package of traits as possible, including new levels of iron or zinc concentrations in resilient, consumer-accepted table potatoes.

Analyzing biofortified populations under recurrent selection in diploid potato germplasm, Paget et al. (2014) found moderate to high and positive correlation between iron (Fe) and zinc (Zn) contents from cycle I ($r^2 = 0.45$) and from cycle II ($r^2 = 0.72$), indicating that evaluation and selection for one of these traits will result in concomitant increase in the other. Negative genetic correlations were found between dry matter and Fe, Zn, Ca, and vitamin C contents (genetic correlation close to zero for vitamin C in Cycle 1) when analyzed on a dry-weight basis. In contrast, the same genetic correlation estimates were positive (but small) when analyzed on a fresh-weight basis (Table 6.1).

The genetic correlations in this example are strong enough that under multi-trait selection the breeding population is improved simultaneously for iron and zinc concentrations (Fig. 6.1a, b), whereas the average tuber weight of the population decreases, and the number of tubers per plant increases (Fig. 6.1c, d). A reduction

Table 6.1 Additive genetic correlations (are in bold) and confident intervals (are in brackets) for Fe, Zn, Ca, vitamin C, and tuber dry matter content from a multivariate analysis of Cycle1 data (Model 4) estimated on a dry and fresh-weight basis

Trait	Iron	Zinc	Calcium	Vitamin C
Cycle I (dry weight basis)				
Zinc	**0.45** [0.32 0.64]			
Calcium	**0.04** [−0.23 0.34]	**0.12** [−0.15 0.39]		
Vitamin C	**−0.01** [−0.18 0.29]	**0.10** [−0.15 0.30]	**0.05** [−0.27 0.33]	
Dry matter	**−0.23** [−0.42 −0.06]	**−0.24** [−0.41 −0.07]	**−0.19** [−0.36 0.07]	**−0.06** [−0.28 0.10]
Cycle II (dry weight basis)				
Zinc	**0.72** [0.42 0.88]			
Calcium	**0.35** [−0.04 0.61]	**0.57** [0.18 0.76]		
Dry matter	**−0.34** [−0.61 0.08]	**−0.38** [−0.66 0.10]	**−0.14** [−0.49 0.20]	
Cycle II (fresh weight basis)				
Zinc	**0.61** [0.33 0.84]			
Calcium	**0.07** [−0.32 0.52]	**0.45** [−0.02 0.77]		
Dry matter	**0.18** [−0.13 0.36]	**0.14** [−0.13 0.38]	**0.05** [−0.23 0.27]	

From: Paget et al. (2014)

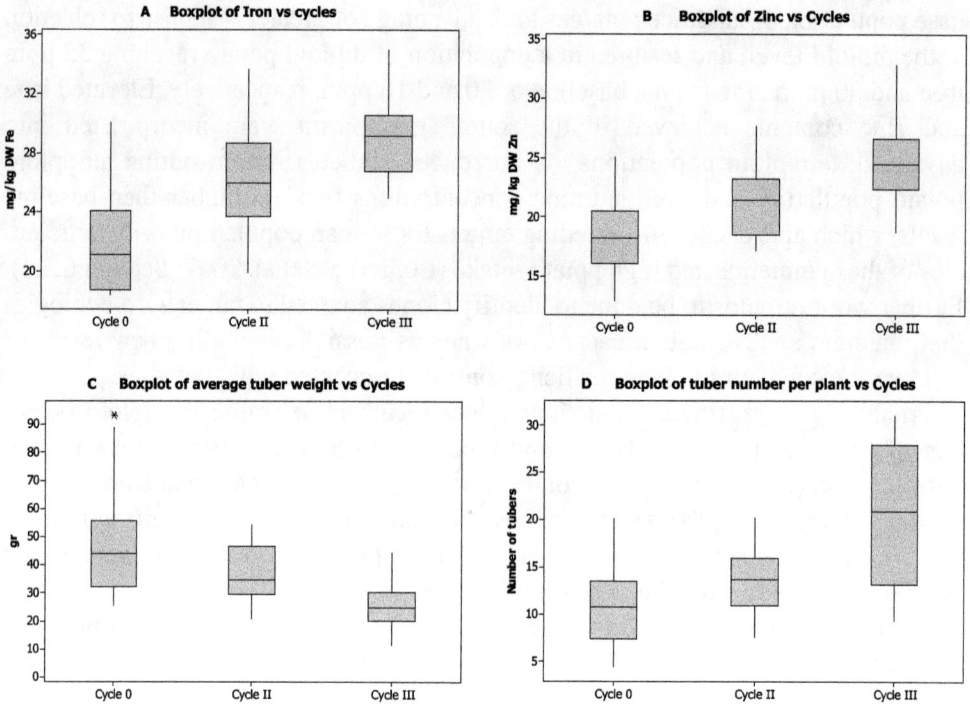

Fig. 6.1 Box plots of ranges of trait variation through three cycles of recurrent selection in diploid potato germplasm. (**a**) Iron concentration; (**b**) zinc concentration; (**c**) average tuber weight (reflecting size); (**d**) Tuber number per plant

in tuber size was encountered during population improvement at the diploid level even though it was considered in selection. This loss was mitigated, however, by returning to the tetraploid level via 4x–2x crosses by which the gains in the mineral contents realized at the 2x level were transferred to advanced populations.

6.2 Principles of Potato Breeding Methods and Approaches

The principal advantage of breeding clonally propagated crops is that each clonal variety is fixed and simple to maintain. Genetic purity is less of an issue in vegetatively—than in sexually—propagated crops. One substantial disadvantage of vegetative propagation though is that diseases are easily transmitted across clonal generations during propagation; another is that potato planting material is bulky and perishable by nature and the production of healthy material is expensive.

The single most challenging aspect of potato breeding is the identification of superior individuals that combine as many high priority traits, and as few weaknesses, as possible from a given cohort of F1 progeny in a reasonable time frame. Additional important challenges include the improvement of support populations, and the selection or construction of parents. The parents used for crossing are highly heterozygous—exceptions are inbreeding lines generated by self-fertilization or doubled-monoploid production. Each potato seed that results from a heterozygous × heterozygous cross differs, at many loci, from any other seed from the same cross. The resulting heterozygous genotypes are subject to selection after being fixed (stabilized in genetic terms) by clonal propagation.

Polyploidy, heterozygosity, and heterosis make the identification of good parents particularly challenging. At present, the performance of a parent can only be determined after the fact, that is, by looking at its progeny. Advances in genomic selection may make it possible to identify good parents in advance. Heterozygous tetraploid potato genotypes harbor great allelic diversity and interactions that are responsible for their performance as clones. As parental clones are not inbred, a genotype can never be reconstituted after sexual crossing or self-pollination. A practical consequence of this for potato breeding is that many, many, traits segregate in the progeny of any parent or parental combination. Directional breeding results in incremental changes in gene frequencies among progenies.

In the course of potato breeding an "F1 clone hybrid" is generally crossed with another "F1 clone hybrid," and the progeny is heterogeneous and has an extreme large segregation variance. A good parent generates a large genetic variation around a high family mean for a given trait. Once heterogeneous and heterozygous progenies are generated by crossing two potato clones, selection of clones within a given pool of genetic variation for variety development is conceptually, if not technically, straightforward. All the genetic advantages of clonally propagated crops can be used for variety development, and the genotype that will finally be released is among the progeny immediately after the initial crossing.

Inter-group crosses are important in population improvement of clonally propagated crops. Gruneberg et al. (2009) have suggested that this aspect of clone breeding is often neglected and may be the reason for low breeding progress in many clonally propagated crops when compared to the improvement of sexually reproducing ones. Complementary germplasm groups have not been identified in potato but strategies to maximize heterozygosity have been proposed to increase yield in tetraploid potatoes (e.g., Chase 1963). However, it is unlikely that the direct relationship of maximum heterozygosity and yield will extend to crosses involving un adapted germplasm (Bonierbale et al. 1993). Thresholds for heterosis were suggested upon finding of increased yield and vigor in two-way hybrids (cultivated × wild species) with respect to crosses within the cultivated genepool, but no additional increments in hybrids involving cultivated potato and two wild species (3-way hybrids) (Sanford and Hanneman 1982). Thus, it is unlikely that heterotic groups could be established in potato germplasm on the basis of genetic distance measures or maximum heterozygosity alone.

6.2.1 *Population Improvement*

Medium- to long-term genetic gain can be achieved over sequential cycles of crossing and selection. Recurrent selection is defined as reselection generation after generation, with inter mating of selected plants to produce the recombinant population for the next cycle of selection. The goal of recurrent selection is to improve the mean performance of a population of plants; a secondary goal, but nevertheless also important, is to maintain as much genetic variability as possible. Open recurrent selection is a method for improving the mean performance of a population while maintaining and increasing genetic variability by periodically introducing new sources of traits under selection. In genetic enhancement a few cycles of recombination facilitate the breakdown of linkage blocks so that desired traits from unadapted germplasm can be carried forward and undesirable ones left behind, minimizing the effects of linkage drag thus shortening the time required to introgress or incorporate new diversity into advanced populations. Depending on the stage of a population under improvement, selections may be considered for use as varieties, or as parents that will contribute to further gains toward complex breeding objectives. Figure 6.2

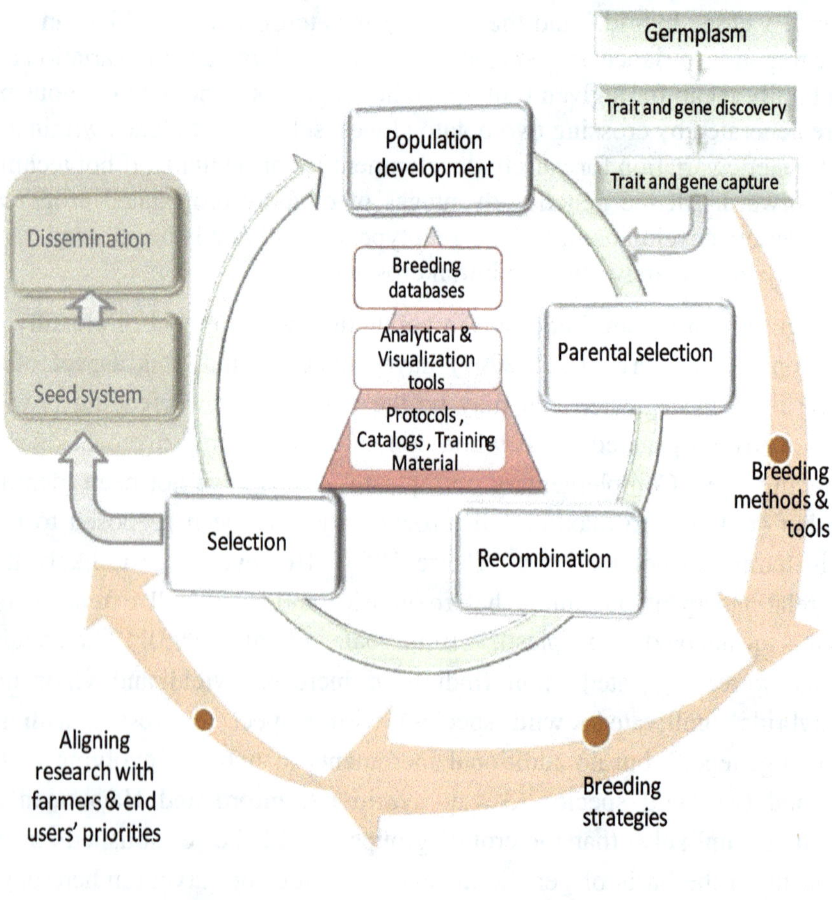

Fig. 6.2 Schematic representation of components of a potato breeding program. (Courtesy of Elisa Mihovilovich)

depicts simultaneous and integrated practices of recurrent selection, genetic enhancement, and variety selection in a potato breeding program.

In polyploid potato, more than one allele per locus can be transferred to the next generation in gametes, and thus, in contrast to diploids, the genetic variation due to dominance influences the response to selection in population improvement as long as the population is not in equilibrium. In tetraploid potato, a population is usually not in equilibrium after recombining parental material in controlled crossings, and 1/3 of the dominance variance is exploitable for selection progress when selection takes place on the female and male side [for further details see Wricke and Weber (1986) and Gallais (2004). The exploitation of the dominance variance in population improvement, in combination with faster genetic gains in diploids than in tetraploids, and the inheritance of 2n gametes as described by Ortiz (1998) provide great diversity and flexibility for potato improvement.

6.2.2 Crossing Parents

The choice of parents is an important step in any breeding program, especially for medium- and long-term breeding progress. The identification and number of crosses to perform at a given breeding stage is a factor of the knowledge available on potential parents (i.e., trait constitution and breeding value), the breeding objective, and facilities and resources available for crossing and selection. As mentioned above, the polyploid and highly heterozygous nature of potato mean that dominance and epistatic effects contribute considerably to clone performance, and for this reason, little is known about the value of a parent or specific cross combination until it has been tested.

Parental value can be assessed strategically through the conduct of progeny tests in appropriate breeding designs, the evaluation of pedigrees, or empirically through observation of selection ratios. Even when hundreds of crosses are made, it is often observed that the best clones trace back to very few crosses. Thus, it is desirable to predict which genotypes will be the best parents, since if this is known, efforts can be concentrated on the generation and the evaluation of the most promising combinations. Regardless of how much information is available to support the choice of parents, a practical approach is to sow around 200 seeds of every cross to be evaluated, and then sow more seed or emphasize the parents of those combinations that performed well in future years.

Parental value is a function of the genetic constitution of a trait donor. Complex traits are often comprised of several components, and defining those that contribute reliably to influence trait values is a prerequisite to successful identification of parents. Accurate evaluation of progeny is the most effective means to identify superior progenitors of inherited characteristics. Sprague and Tatum (1941) introduced the concepts of general combining ability (GCA) and specific combining ability (SCA) to distinguish between the average performance of parents in crosses (GCA) and the deviation of individual crosses from the average of crosses (SCA).

Studies of combining ability make it possible to identify parents (those with good GCA) that will perform well in most crosses, and also facilitate the development of superior hybrids through the use of parents with high SCA. Genetic analyses to measure combining ability further allow breeders to estimate genetic effects and parameters such as trait variance, covariance, correlations among traits, heritability, and the relative importance of additive and dominance variation. GCA represents mainly additive and additive × additive types of genetic variance. Thus, when a potato clone is selected as having good GCA, it means that the algebraic sum of the additive and additive × additive epistasis effects it passed on to its offspring produced a favorable result in excess of the average of all the offspring tested. GCA has a conceptual implication that each line being evaluated is tested against a large random sample of some specified population (Plaisted et al. 1962). In actual tests for GCA, however, the testers (usually males) often represent only a limited sample of a population.

6.2.3 Mating Designs

The term "mating design" refers to the mating of parents in a systematic plan of crosses to determine genetic parameters and/or parental value. These procedures are particularly useful for the identification of plants that will be the most effective parents in pre-breeding or recurrent selection programs. However, systematic matings can be difficult to achieve because of sterility and incompatibility encountered during crossing. Mating designs have been classified according to the number of factors to be analyzed, the parents, and modalities of combination. Some mating designs are used more extensively than others, but each has its advantages and disadvantages depending on the reference population under consideration and the information desired. Following are descriptions of some of the most important designs for identifying superior progenitors to help assure genetic gains in a potato breeding program. Figure 6.3a–c, illustrates some of these mating designs.

Diallel Analysis is the best way to determine the combining ability of parents. It consists of the analysis of a set of crosses produced involving "*n*" lines in all possible combinations, a so-called diallel cross. The diallel mating design has been used extensively in potato germplasm enhancement and can be very useful if properly analyzed and interpreted. The analysis of diallels provides information on GCA and SCA of parents and their crosses and makes it possible to determine if reciprocal crosses give equivalent results.

The most commonly used *methods of diallel designs* are those proposed by:

- Griffing (1956), by which general and specific combining abilities are estimated;
- Gardner and Heberhart (1966), by which the variety and heterosis are evaluated; and
- Hayman (1954), in which information regarding additive and dominance effects for a characteristic and the genetic values of the parents is used.

Fig. 6.3 Schematic representation of different mating designs in which parent lines are indicated with numbers, and the cross combinations between them are represented with checkmarks in shaded cells. (**a**) Partial dialel, (**b**) Line × tester, and (**c**) Design II

A

Partial diallel									
		Males							
		1	2	3	4	5	6	7	8
Females	1				√	√	√		
	2					√	√	√	
	3						√	√	√
	4							√	√
	5								√
	6								
	7								
	8								

B

Line x Tester				
		Testers		
		1	2	3
Lines	1	√	√	√
	2	√	√	√
	3	√	√	√
	4	√	√	√
	5	√	√	√
	6	√	√	√
	7	√	√	√

C

Design II					
		Males			
		5	6	7	8
Females	1	√	√	√	√
	2	√	√	√	√
	3	√	√	√	√
	4	√	√	√	√

One of the most used is *Griffing's Method 2*, which estimates GCA and SCA) relating mainly to additive and non-additive gene effects (dominance and epistasis).

6.2.3.1 Partial Diallel

Analyses involving "*n*" parents in all possible combinations become unmanageable as the number of lines (*n*) increases. On the other hand, if only a small number of parents are tested, the estimates of combining ability tend to have a large sampling error. These difficulties led to the development of sampling crosses produced by large numbers of parents, without affecting the efficiency of the diallel technique. In a normal diallel, each line is involved in ($n - 1$) crosses. Kempthorne and Curnow (1961) presented the concept of the partial diallel design in which only a random sample of crosses, say of size's, is analyzed where "*s*" is less than $n - 1$.

6.2.3.2 Line × Tester Design

The Line × Tester Design also provides information about the GCA and SCA of parents, and is also helpful in estimating various types of gene effects. The *crossing plan* of this design is as follows:

"*l*" lines are crossed to "*t*" testers so that $l \times t$ full-sib progenies are produced;

These progenies, with or without parents (i.e., testers and lines), are then tested in a replicated trial using a suitable experimental design, say randomized block design.

If there are three testers and seven lines, there are 7 × 3 = 21 crosses. For evaluation, the 21 crosses along with 10 parents, for a total of 31 entries, might be tested in a randomized complete block design with four replications. For this case, uniform planting material must be produced to enable inclusion of parental clones and progeny in the same trial.

6.2.3.3 Design II

In mating Design II (or Factorial Design), described by Comstock and Robinson (1948), the genetic information is similar to that obtained with Diallel Analysis.

Different sets of parents are used as males and females. If a set of eight parents is included in the design II, 16 crosses will be obtained. This design is advantageous when not all clones to be tested are male or female fertile.

6.2.4 Breeding Values

The estimated breeding value (EBV) of an individual can be calculated on the basis of pedigrees and performance in the course of a breeding program, and does not rely on the conduct of mating designs. Using appropriate statistical analysis, breeding value predicts how useful each individual would be as a parent; it expresses the ability of a parent to pass on superior trait levels to its offspring and is used for ranking breeding performance of an individual as a parent relative to the population average. The use of information on the individual and all relatives greatly increases the accuracy of selection in a breeding program (Lynch and Walsh 1998).

The calculation of breeding value goes beyond the typical estimation of genotypic or parental values with models based on fixed effects, by enabling the estimation of random effects of a mixed model. Mixed linear models are able to model different covariance structures and thus provide an improved representation of the underlying random and error components of variance (Oakey et al. 2007). The application of mixed models to estimate breeding values uses pedigree information to model and exploit genetic correlation among relatives and applies flexible variance–covariance structures for genotype-by-environment interaction to accurately predict performance (Piepho et al. 2008). The improved accuracy afforded by the use of all data in a breeding program with mixed models allows the analysis of repeated measures, unbalanced design experiments, spatial data, and multi-environment trials.

Breeding value has been successfully applied in several crops (sugarcane, eucalyptus, soybean, maize, and even potato). Slater et al. (2014) illustrate selection in potato based on breeding values. They conclude that using best linear unbiased prediction (BLUP) and pedigree to estimate breeding values can result in increased genetic gains for low heritability traits in auto tetraploid potato. Theory and applicability of breeding values in quantitative trait improvement are illustrated in Bernardo (2002) Relevant statistical packages that fit linear mixed models to large data sets using the Residual Maximum Likelihood (REML), approach, such as the Asreml-R reference manual (Butler et al. 2009) provide powerful software for the use of breeding values in plant breeding.

In animal breeding, EBVs are the basis for marketing breeding parents and they provide breeders with critical information for selection decisions. In applying EBVs, it is important to achieve a balance between the different groups of traits and to place emphasis on those traits that are important to the objective population, markets, and environment. It is not feasible, nor always desirable to seek high EBVs for all traits in a single progenitor. In fact, a comprehensive range of EBVs has the advantage that it is possible to avoid extremes in particular traits and select for animals/genotypes with balanced overall performance. The method is particularly useful in non-inbred populations and potato breeders should pay careful attention to this analytical approach.

6.2.5 *Early Versus Late Generation Selection*

In early generations it is possible to select for highly heritable traits for which accurate assessment of a genotype can be carried out on one or a few plants, but for more genetically complex traits, and for traits where interplant competition is an issue, it is necessary to evaluate clones in multi-plant plots planted as blocks under homogenous field conditions. The aim is an unbiased comparison of genotypes within blocks. The number of plants per plot and the number of replications or blocks depends on the breeding stage. The low propagation coefficient of potato (about ten depending on the propagation method used) limits the amount of planting material available at each stage of selection and is one reason that potato breeding is relatively slow (at least 8–10 years and usually more, from crossing until variety release). Large numbers of genotypes can be assessed for simple traits in early stages when small amounts of seeds are available; while more complex traits are assessed in later stages when larger quantities of seeds are needed for replicated trials, though less genetic diversity is represented.

Multistage selection can be managed in subsequent steps from early to later stages, or with indices that use a weight assigned to each trait. In practice, some characters are selected sequentially especially where there is clearly a lowest acceptable value (tuber size, shape, and color as well as pest and disease resistance), while others are selected simultaneously by aggregating characters into an index (often an intuitively formed index such as score values of overall performance.

6.2.6 *Case Study 2: The Use of a Selection Index in Potato Breeding*

An early-generation (seedling stage) selection method was applied at CIP for the identification of families and individuals that tuberize well under long days and high temperatures. Five groups of families generated by intercrossing CIP's best advanced parents from two populations and long day-adapted varieties underwent a greenhouse test during summer, applying two photoperiods: natural short days of 12 h and simulated long days using lamps to extend the photoperiod to 16 h. Four plant morphology prototypes were identified according to patterns of above- and below-ground growth including branching, stolon, and tuber formation (Fig. 6.4). Individual plants of each family were evaluated at harvest taking into account parameters including breeder's preference, a tuberization score based on the four patterns, tuber uniformity and physiological disorders including sprouting tubers, knobbiness, chain tubers, and cracking.

A selection index (SI) was built assigning a weight to each of these parameters, three for breeders' preference (BP), two for tuberization score (TS) and tuber uniformity (TU), and one for physiological disorders: sprouting (Sp), knobbiness (Kn), chained tubers (Ch), and cracks (Cr)

$$SI_n = BP_n(3) + TS_n(2) + TU_n(2) + Sp_n + Kn_n + Ch_n + Cr_n$$

Families with a high frequency of progenies exceeding an estimated selection index of 54 were selected. Seven of these selected families and three unselected were used in validation studies under field conditions.

Fig. 6.4 Plant morphology and growth patterns observed among families in early generation screening for adaptation to warm, long day environments. (**a**) Typical morphology under 12:12 h light:dark regime; (**b**) Typical morphology under 16:8 h light:dark regime; (**c**) Above-ground morphology of representative plants with growth patterns I, II, III, and IV; Below-ground morphology of representative plants with growth patterns I, II, III, and IV

6.2.7 *Stability and Adaptation*

Breeding programs typically breed for several locations rather than just one. Hence field evaluations that underpin selection must simulate a range of environments. For this reason, within limits of the propagation coefficient, clones are tested in plots within homogenous blocks at several locations over several years. The objective of multilocation trials is to assess promising genotypes with respect to narrow and broad adaptation. In practical potato breeding, "broad adaptation," the ability to perform well in a range of environments, appears essential for a clone to have any meaningful commercial success.

The wide range of quality preferences and the numerous pests and diseases addressed by potato breeding programs dictates the exposures required for selection. Decentralized selection is generally required by a program with multiple target environments and is best designed to address sets of traits that can logistically be assessed together in target environments or sites that represent them well. Along the course of selection, a combination of on-station and on-farm trials are usually performed.

National programs requesting potato breeding materials from CIP are provided with germplasm from different selection stages depending on their capacity to evaluate clones and the relative suitability of advanced germplasm to their production targets. Advanced clones and parents are available as in vitro plantlets. The genotypes provided meet multiple selection criteria for the given target environment, as assessed in similar testing sites. Upon receipt they are propagated and subject to multilocational trials before use as parents in national programs or testing as varieties.

6.2.8 *Case Study 3: Breeding and Variety Development in Bangladesh*

Breeding for stable yields and multiple disease resistance at CIP takes advantage of germplasm collections, broad-based advanced populations, and environmental diversity in Peru. One such population is the advanced lowland tropics virus resistant population, which is adapted to dry arid regions where virus pressure is high. With support from GIZ/BMZ and follow-up support from USAID, CIP provided 35 potato breeding lines of this population as in vitro plants to the Tuber Crop Research Centre of the Bangladesh Agricultural Research Institute in 2009. After local evaluation over a period of 7 years, a salt tolerant variety (BARI Alu-72) and a heat tolerant variety (BARI Alu-73) adapted to Bangladesh were released.

When distant environments can only be partially simulated in CIP's breeding program, samples of true seed families (TSF) from good parents can be distributed. TSF are available for national programs with the capacity to carry out early generation selection as well as variety identification trials. One advantage in working with TSF is the considerable reduction in time-to-release compared to that required when receiving advanced clones. Conducting selection with true seed in target environments can save 6 years or more, i.e., time that would otherwise be required for CIP breeders to conduct selection from the same TSF at their trial sites and recommend elite clones as in vitro plants.

When national programs test CIP germplasm as TSF, valuable information is returned to CIP regarding family performance (i.e., frequency of selected clones in each family), and this serves to refine the concept of adaptation and best bet materials for that location. National programs that invest in identifying parents based on local performance, and then utilize these parents for crossing and recurrent selection, benefit from knowing which breeding materials will serve those best and are likely to experience significantly faster genetic gains over the medium to long term.

6.2.9 *Case Study 4: Collaborative Breeding in Vietnam*

Advanced germplasm from CIP's lowland tropic virus resistant (LTVR) and late blight heat tolerant (LBHT) breeding populations developed in Peru were dispatched to Vietnam in the form of in vitro plants from 2005 to 2010. The materials were used in two ways: (i) for direct variety testing, and (ii) in cross-breeding to improve adaptation to local conditions.

Results of the collaboration demonstrated both the direct utility of clones from Peru in Vietnam—through the identification of two outstanding clones recommended for variety testing, and the advantage of local breeding, yielding, after three rapid cycles of recurrent selection, a new generation of elite clones with higher yields than the elite clones provided from Peru. The yield increase from local breeding and selection was realized in a considerably less time (9 years) than would have been required for the development, introduction, and testing of a new cycle of elite clones from CIP (at least 12 years) (Fig. 6.5a, b).

There are several advantages of testing early generation potato progenies at more than one location. Information from contrasting environments can be combined if the breeder tests a clone at two or more locations; those few clones that are broadly adapted can be identified early in the selection process. In addition, if one location experiences low disease pressure, daughter tubers from that site can be used to provide clean planting material for the next generation of selection. The healthy seed plot provides for an assessment of performance in the absence of major production constraints, while various pressures can be applied at additional sites. Alternatively, a protected environment (e.g., quarantine screen house) can be used to maintain a healthy copy of each selected genotype, but unlike the maintenance of a healthy

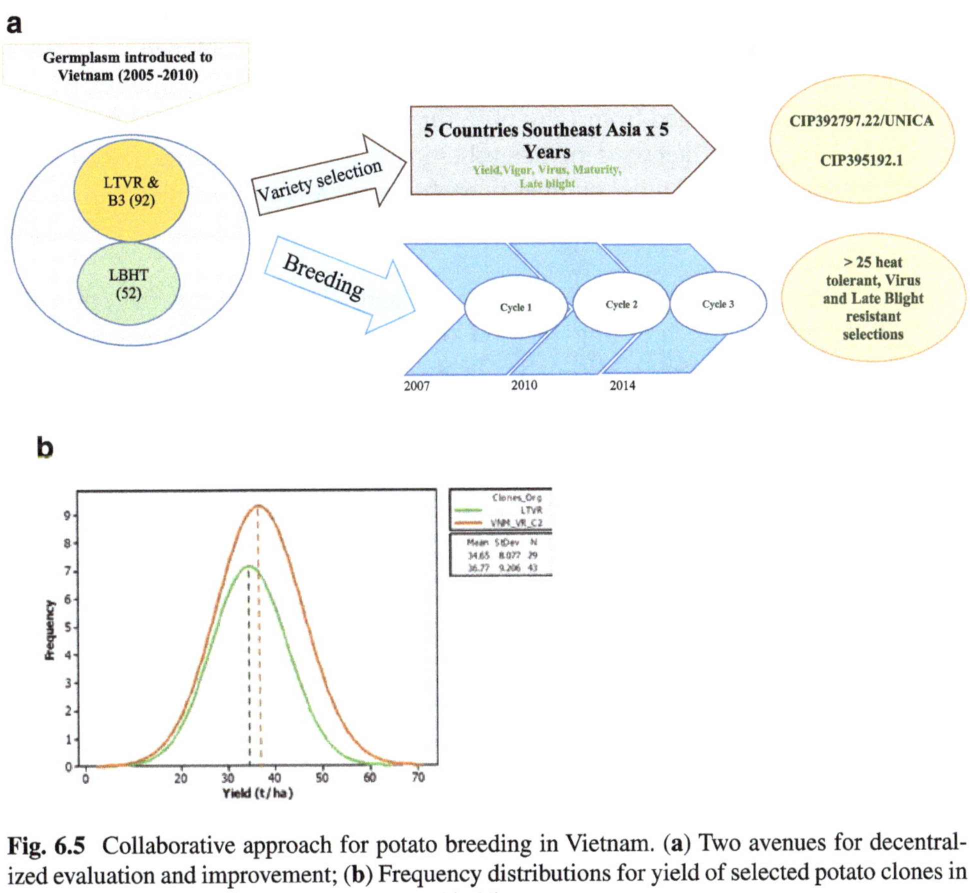

Fig. 6.5 Collaborative approach for potato breeding in Vietnam. (**a**) Two avenues for decentralized evaluation and improvement; (**b**) Frequency distributions for yield of selected potato clones in three cycles of recurrent selection conducted in Vietnam

seed plot, this typically does not allow for significant production of seed to support evaluation along the breeding cycle.

Rapid and real-time exchange of information on selection percentage between breeders and selectors helps inform breeders on the future crossing plans most likely to meet the requestor's needs. This simple data can confirm the breeders' concept of GCA and suitability of test environments with respect to the targeted ones. Breeders can use the information returned from selectors regarding percentage of selection in controlled cross families to exploit heterosis through cross combinations. Selection of parents and cross prediction can probably be improved by systematic collection of data on the percent of clones selected in families by the assignment of parents with good GCA or mutual SCA to complementary gene pools. This would benefit both CIP's and the national programs' breeding efforts, if it resulted in increased frequency of high-yielding clones with needed traits, or potentially, in yield jumps.

6.2.10 Importance and Relation of GCA and Heterosis in Potato

Heterosis was examined in a new tetraploid potato hybrid population obtained by crossing members of two advanced breeding populations developed at CIP for adaptation to the highlands (B3) and subtropical lowlands (LTVR). Significant positive heterotic values ranged from 18 to 60% for mid-parent heterosis and from 16 to 162% for best parent heterosis or heterobeltiosis. Positive heterotic values were found for tuber yield and tuber number in each environment. Significant GxE interaction on heterosis showed a differential effect of the environment on the magnitude of heterosis expression.

The expression of heterobeltiosis within B3 and LTVR confirms the effectiveness of the population breeding method on maintaining a broad genetic base (Mendoza and Sawyer 1985). Mid-parent heterosis can be the result of the combination of good levels of GCA of parents and some level of SCA of the cross. A reciprocal recurrent selection scheme (RRS) has been initiated to benefit from intercrossing between best parents of two complementary populations, B3 and LTVR (RRS), also known as recurrent reciprocal half sib selection, is a form of recurrent selection used to improve both GCA and SCA of a population for a character or characters.

6.3 Potato Breeding Procedures Overview

6.3.1 Hybridization

Heterozygous parents are recombined in controlled biparental combinations or by use of bulk pollen to create new variation in the form of sexual seed. The crossing block is established after selecting the parents to be used as females and as males, and the "planting on a brick method" or modifications of it can be practiced to encourage flowering and fruit set over a period of about 10 weeks. In the "brick method," potato tubers are placed on bricks partially buried in the soil. The tubers sprout and stems are allowed to grow. After some time the soil around the base of the full-size plant is washed away with a spray of water, which exposes emerging stolons and small tubers, but does not disrupt the root system that has penetrated in and around the brick. Thereafter, stolons and tuber that emerge from the base of the plant are removed. Potato has a complete flower with five anthers and a stigma that becomes receptive as pollen is shed from open flowers. Emasculation is practiced to prevent selfing when genetic studies will be performed, when parental value will be assessed in progeny tests and for best control of pedigrees (record-keeping).

Crossing at CIP is enhanced by flowering induction. Under long day conditions most potato clones flower to some extent. Typically, *S. tuberosum Phureja* Group flowers under both long and short-day photoperiods. Clones of Group *tuberosum*, however, usually will not flower under short days. In addition, there are many clones, especially those of early maturity, that rarely flower under any conditions or that flower sparsely and over a very short time period. For these reasons it is often desirable to induce flowering. This can be done by either of two methods: "grafting" or "planting on a brick." In the grafting method shoots of potato are grafted onto the stems of tomato plants. In both methods the idea is to prevent potato from developing beyond the vegetative flowering phase to the tuberization-senescense phase. In addition to these mechanical methods, other traditional flowering-enhancing practices may be followed such as long-day lighting, sprays of 40 ppm gibberellic acid at 4-day intervals, and heavy nitrogen fertilization. Temperature and humidity are also important. Ideal temperatures are 20 °C day and 16 °C night. Humidity should be 80% or higher. Procedures and crossing techniques for potato breeding as practiced at CIP are documented at https://research.cip.cgiar.org/potatoknowledge/proceduretechniques.php. A modification of "planting on a brick" using peat pots is shown. The open source software CIPCROSS (see Section 6.3.3) can be used to document all aspects of the crossing block, including pedigrees, storage, and inventory using bar coding for accuracy and efficiency.

6.3.2 *Selection Schemes*

Selection of potato varieties from botanical seed is conducted in several steps. Figure 6.6 shows how the crossing of two parents might be followed by five sequential steps in time (one selection step conducted with seedling plants and four subsequent selection steps conducted with cloned plants from tubers of the seedling genotypes). This scheme is an oversimplification since many crosses or families are developed simultaneously and many seeds of each family are subject to evaluation and selection at the same time, while here only three seedlings in each of 11 families are shown.

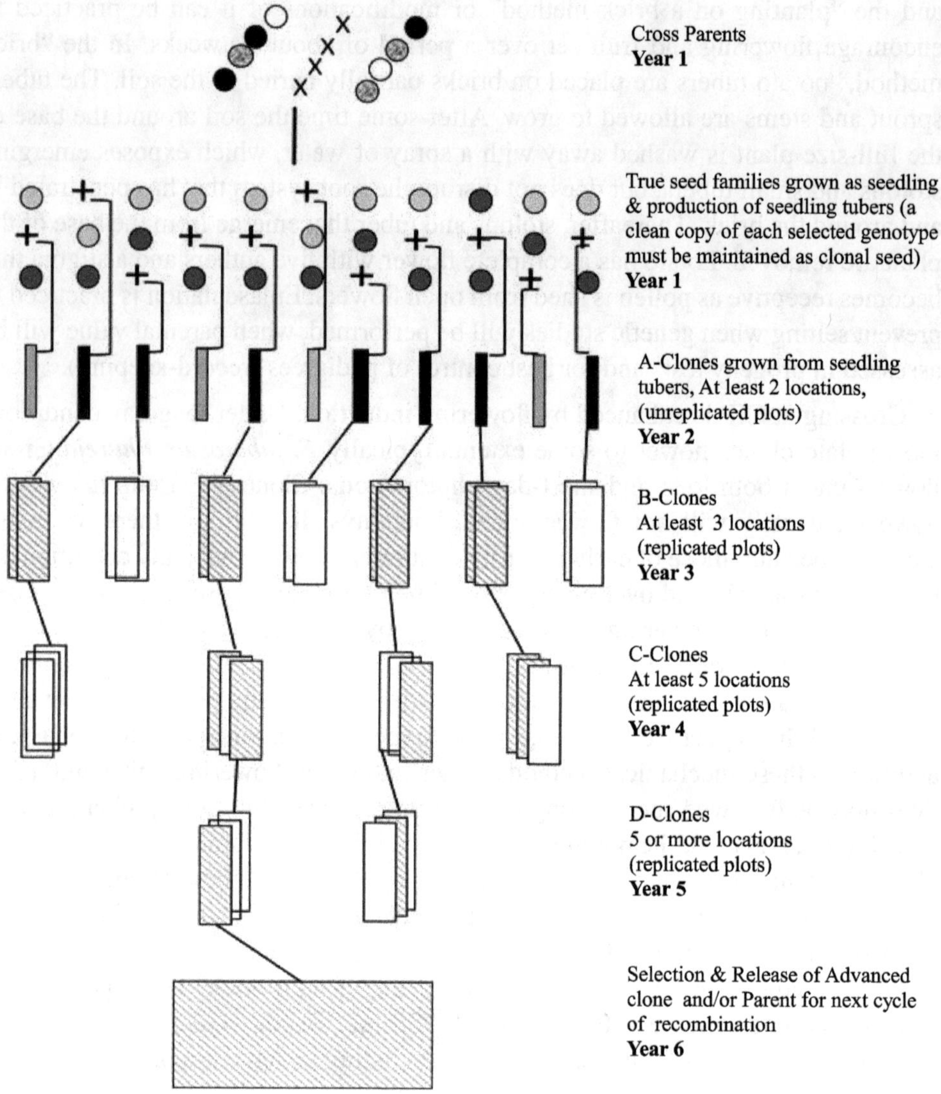

Fig. 6.6 Multilocation Clonal selection Scheme. (Modified from Gruneberg et al. 2009)

In principle, if the true seed plant could be cloned in large quantities and the population could be assessed with adequate accuracy, it is theoretically possible to select the "best" genotype in the first year, since maximum variation is present at the start of each selection cycle and no genetic changes are realized during it. In practice though, the individual genotypes selected in early stages of each clonal selection cycle are subject to propagation via tubers to permit adequate plot size and replication for later-stage assessments.

The overall selection scheme should consider the vulnerability of clonal planting material to disease. At least some tubers of every clone should be planted in a disease-free location, or time and resources should be allocated for the eventual elimination of pathogens from successful clones before they can be distributed among geographic regions.

Many breeding programs use only one location at the early selection stage, preferring to evaluate in favorable conditions such as on station, while also producing healthy seed for use in later selection stages. This location should be as free from virus pressure as possible to help keep seed clean over the years. For example, the New York breeding program makes all evaluations for the first 2 clonal years in a single location, going to a second location (for an unreplicated evaluation of yield) in the third clonal generation, and multiple locations (for replicated yield trials) starting in the fourth clonal year. Typical numbers of genotypes per cycle in this program are: 20,000 seedlings in year 1, 18,000 four hill plots in year 2, 1500 twenty-hill plots in year 3, and 250 hundred-hill plots in year 4.

Nevertheless, the first clonal generation can be evaluated at more than one location if sufficient clonal seed can be produced. For simultaneous evaluation at up to three locations in the first clonal generation, true seed families (TSF) are first converted to tuber families (TF) so that at least four copies of each genotype from each family are produced. Methods for evaluating TF include (1) planting three to five tubers of each genotype in a single row at a single location or (2) equivalent samples of each tuber family are evaluated in two or more locations, using one tuber per genotype and location. Multilocational testing of first clonal generation tuber families can be done with, or without the identification and labeling of individual genotypes. Detailed procedures for the production of tuber families followed at CIP are illustrated at: https://research.cip.cgiar.org/potatoknowledge/tuberfamilies.php

6.3.3 *Data Management and Analytical Tools*

Standardized information on the performance of progenies and selected clones across environments is necessary in order for breeders to efficiently make decisions about selection and variety release. The methods of data generation and processing that are utilized in plant breeding have radically changed in recent years. With the advancement of new high throughput technologies, data have grown in terms of quantity as well as complexity. However, the significance of the information that is hidden in newly generated experimental data can often be deciphered only by linking it to other data, collected previously and/or by others. Collaboration that makes it possible to connect disparate data sources and analyze them in meaningful ways with other researchers requires robust but practical data management solutions. CIP has developed an on line Global Trial Data Management System: https://research.cip.cgiar.org/gtdms/. Three key components are CIPCROSS, HIDAP, and Field Book Registry.

CIPCROSS is a botanical seed inventory tracking system for clonal crop breeding that enables tracking of breeding materials from crossing blocks and botanical seed inventories through to seed distribution. CIPCROSS is open source software, available online, and comprises two main tools: (a) The Crossing Management

System (CIPCROSS Tool v1.1 for Pocket PC) used the main development platforms Visual Basic 2008 for Windows Mobile OS and Microsoft SQL Server Compact 3.5 for Windows Mobile software. This tool facilitates the barcode labeling of parents and crosses, storing the information in a database and generating automated reports and (b) The Botanical Seed Inventory System (CIPSIS Tool v1.1) was designed as an easy-to-use web application. It uses PHP, Yii Booster, Bootstrap, CSS, Java Script, HTML and MySQL database programming languages. This tool facilitates the recording and accessing of information on the location of botanical seed in the storage facility, documentation of germination tests, updates on seed stocks, information about crosses and breeder's name. The system helps breeders develop, integrate, and organize their information in a database, avoid typing mistakes and saving work time (24 person–hours are saved per 14,250 labels). The Roots & Tubers Base centralized data repository facilitates access to the information with queries and filters for advanced searches (Fig. 6.7—Users can download these tools through the web page of "The Global Trial data Management System from CIP" https://research.cip.cgiar.org/gtdms/)

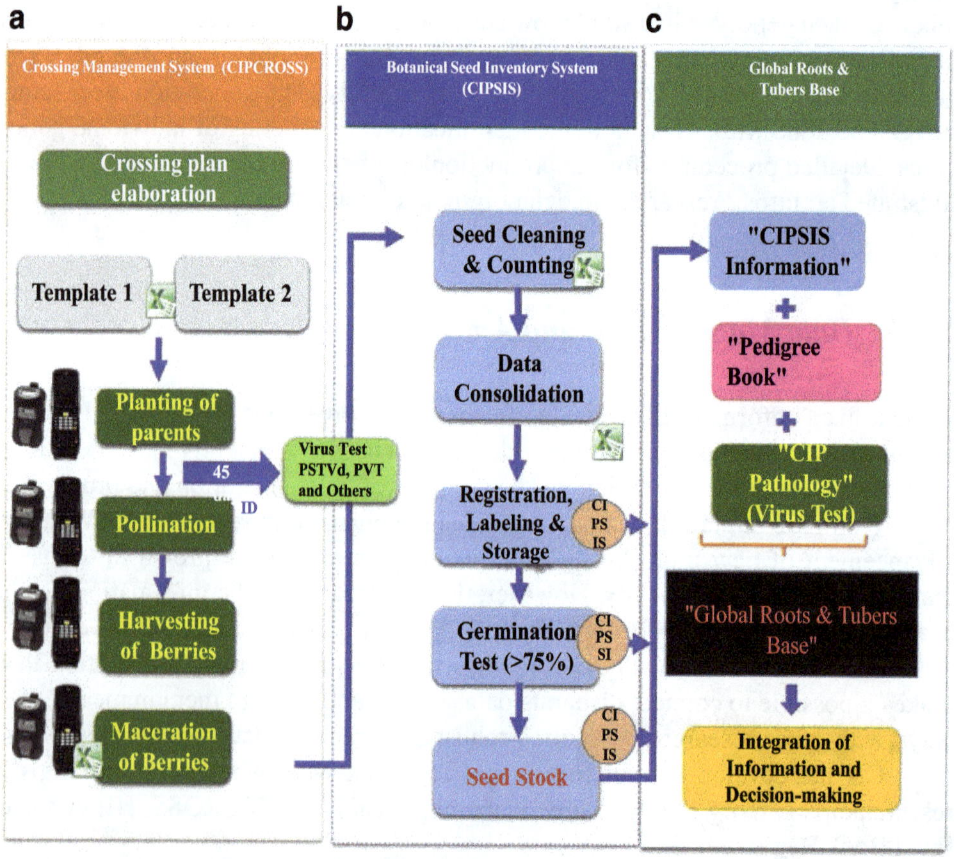

Fig. 6.7 CIPCROSS workflow: (**a**) The Crossing Management System (CIPCROSS Tool v1.1 for Pocket PC); (**b**) The Botanical Seed Inventory System (CIPSIS Tool v1.1) and c) Global Roots & Tubers Base as a centralize data repository; (**c**) Global Roots & Tubers Base as a centralize data repository

Data collection in potato breeding requires capacity to use and share standard protocols for the selection of clones from seedlings through to observational trials and on to preliminary and advanced yield trials, each accompanied by field books and structures for data collection, analysis, and reporting.

The highly integrated data analysis platform called HIDAP was developed by CIP's breeding program and Research Informatics Unit to facilitate and unify data collection, quality control, and data analysis for clonally propagated crops. HIDAP provides a single platform for use by potato and sweetpotato breeders. It supports compliance with Open Access, open standards such as the potato and sweetpotato crop ontologies and linkages with relevant corporate and community databases such as CIP's BioMart (https://research.cip.cgiar.org/gtdms/biomart) and SweetPotato Base (www.sweetpotatobase.org). HIDAP builds on the statistical platform R. This includes the R shiny tools, the knitr package, the Agricolae package also developed by CIP (https://cran.r-project.org/web/packages/agricolae/agricolae.pdf), and more than 100 other R packages. The R shiny package enables implementation of interactive web pages that are usable online and offline. The knitr package enables the creation of reproducible reports. Numerous statistical analyses can be performed using R and R functions developed at CIP. The software is available for download at https://research.cip.cgiar.org/gtdms/hidap/.

HIDAP is connected to the institutional pedigree and corporate database at CIP facilitating the tracking of clones and families generated through breeding. This connectivity enables verification and maintenance of the identity of clones across the different selection stages and tracing of pedigrees in selection and breeding. The HIDAP network enables researchers to share field books with colleagues, regional breeding programs, and/or partners. To use this network, users must register and create a login account. Once logged in, the user can share, download, and receive field books for different selection stages in a user-friendly interface. A download count helps to keep track of users and uses of this tool.

The Field Book Registry (https://research.cip.cgiar.org/cipfieldbookregistry/) facilitates updating field books in real time and viewing their status in the database. The data generated in CIP's potato breeding program is stored in the "*Global Roots & Tubers Base*" utilizing the free BioMart software https://research.cip.cgiar.org/gtdms/biomart/. This database has been structured for storage of phenotypic, genotypic, pedigree, geographical, and environmental data. Through the metadata and the search function using filters, the user can retrieve data from the experiments conducted by CIP or partner programs using CIP materials. The availability of the data is managed in conjunction with Dataverse following CGIAR open access guidelines (Fig. 6.8).

The use of the database and software tools enables analysis of phenotypic and genotypic data. A key goal is the identification of effective models that predict phenotypic traits and outcomes, elucidating important biomarkers and generating important insights into the genetic underpinnings of heritability.

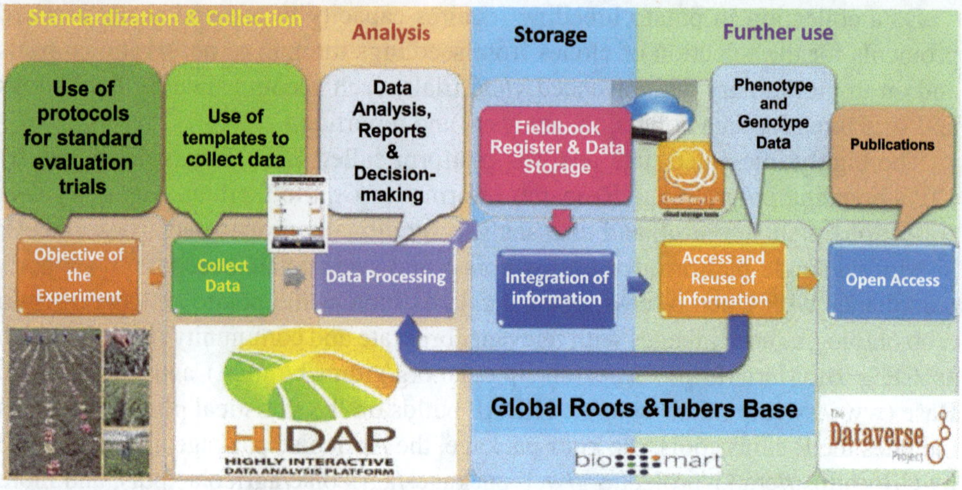

Fig. 6.8 Flowchart of CIP's breeding information system: Standardized evaluation protocols and ontologies implement templates to collect and analyze data with HIDAP software; Subsequent data storage through the field book register allows incorporation of breeding information into the Roots & Tubers Base in which phenotypic and genotypic data are integrated and made available for further use

6.4 Setting Objectives for Potato Breeding

Successful breeding relies fundamentally on having a clear set of prioritized breeding objectives. Key enabling information and tools for setting breeding objectives include: knowledge of the opinions and needs of stakeholder groups, knowledge of the production constraints and cropping systems of target populations and environments, and standardized means to measure and describe traits. Knowledge of the strengths and weaknesses of predominant varieties provides an important baseline for setting breeding objectives.

Having a good understanding of desired end-user traits requires ongoing, interdisciplinary communication and collaboration. This helps increase the chances that a new variety meets actual needs and is thus adopted, as well as minimizing the chance that new varieties (and the traits that define them, their management and marketing or use) might marginalize or disadvantage social or gender groups such as poor farmers or women, who often have access to different resources than men do. There is also a need to consider social and economic dimensions and client demography such as market access, youth, and urban/rural factors. A CGIAR system-wide initiative is currently underway to assess and encourage breeding that responds to social diversity including often-neglected or marginalized groups with emphasis on gender (http://www.rtb.cgiar.org/gender-breeding-initiative/?lang=en).

Potato breeding is characterized by having a large number of objectives, which include traits specific to the intended market and use of the crop (end-user traits), as well as traits related to productivity and protection against local diseases and insect pests (grower traits).

6.4.1 Targeting of New Potato Varieties

Building on CIP's experiences of a global breeding program that developed two agro-ecologically adapted populations (one for highland tropics and one for subtropical lowlands), increased attention is now given to prioritizing packages of traits for which relative values are determined with stakeholders in a more decentralized manner. This involves the cross-disciplinary estimation of trait values at global, regional, and local scales, the development of corresponding variety or product profiles with stakeholder involvement, and the setting of breeding priorities to meet those via selection decisions.

6.4.2 Case Study 5: RTB Priority Setting

An example of assessing trait values can be seen in the priority setting exercise conducted by the Root, Tubers and Bananas (RTB) Program of the CGIAR (http://www.rtb.cgiar.org/resources/impact-assessments/) Kleinwechter et al. (2014). Through a global survey carried out from 2012 to 2013, the RTB program sought to rank production constraints of each RTB crop (i.e., cassava, yam, banana, sweetpotato, and potato) and research options to alleviate them. This exercise was led by agricultural economists specialized in impact assessment. It provided an overview of problems affecting the potato sector and potential solutions including, but not limited to, breeding. The potato survey (Hareau et al. 2014) provided respondents with a list of 91 research options, organized around the areas of crop improvement, crop and resource management, seed management, genetic resources, value chains, postharvest utilization, and marketing, as well as socioeconomic research and extension.

Five of the top ten ranked research options for potato were directly related to breeding. Breeding for late blight resistance, drought tolerance, earliness, and high yield ranked second, third, fourth, and eighth of the 91 options provided, while the generic "germplasm enhancement" ranked seventh. An ex-ante assessment of the selected potato technologies revealed significant differences in terms of net present value (NPV) and internal rate of return (IRR) on investments across the different research options. Late blight- and virus-resistant varieties had the largest expected net benefits ($US 4.7 billion and $US 3.9 billion), and high rates of return of 87% and 104%, respectively, in the high-adoption scenario.

6.4.3 Case Study 6: Adjusting and Ranking Priority Traits in New York (NY) State

Staff of the breeding program in New York State U. S. meet many times each year with stakeholders, in both formal and informal settings, to continually discuss what the program's priorities should be. The frequent interaction helps stakeholders feel comfortable expressing their views, and allows the program to detect changes in industry priorities more quickly, as breeding goals always change with time.

Two examples can illustrate how stakeholder feedback has changed the NY program over the past 5 years. (1) NY growers recently made repeated requests to develop earlier maturing potatoes, to solve a problem the NY breeding program had unwittingly helped create. NY breeders have always put a high priority on yield, and selection for yield tends to select for later maturity. Unfortunately, a suite of late maturing varieties also shortened the timeframe in which NY growers could harvest their crop. The NY program now pays much more attention to maturity, and is prepared to select potatoes with lower yield if they mature early. (2) Potato chip factories recently began to ask for smaller chipping potatoes, as more and more of their product is now sold in small bags. In response, the program now selects smaller potatoes than it (or regional chipping factories) would have been willing to process in the not-so-distant past.

One approach that the NY breeding program has found useful when prioritizing traits is to compare what stakeholders ask for ("what stakeholders say") with the attributes of widely grown varieties ("what stakeholders do"). When there is an apparent disparity between words and action, there is an opportunity for deeper understanding. Fifteen years ago growers in NY kept asking for new varieties with resistance to common scab. What made the request unusual is that growers already had several resistant varieties to choose from, and that the variety they grew most each year was highly susceptible. What the NY program eventually realized is that the popular scab susceptible variety had two quality attributes—outstanding fry color out of cold storage and high specific gravity—that were far more important to the chip factories than the scab resistance growers kept asking for. As a result, NY re-ranked selection criteria, placing fry color and specific gravity above resistance to common scab.

6.4.4 Product Profiles

A product profile establishes a set of targeted attributes that a new plant variety or animal breed is expected to meet for successful release in a given market segment. Attributes must be understood as traits reaching a specified level; this level being defined either in absolute or relative terms (Ragot et al. 2018). Thus, a product profile may list yield (25 tons/ha or more; or 15% over variety V1 across a range of soil fertility conditions), tolerance to potato leaf roll virus (same as or better than variety V2), or dry matter content (no less than 18%, or no less than variety V3). The development of product profiles may best be done in collaboration with, for example, pathologists, agronomists, or nutritionists, as well as user communities who can contribute specific knowledge, tools, and approaches for setting trait levels, which in turn become the breeding targets within a profile.

6.4.5 Tools and Metrics

CIP's potato breeding program targets low input conditions by relying heavily on endemic disease pressure, poor soils and a series of intentional exposures to stress for screening and selection. The product profiles incorporate quantitative breeding objectives for productivity, protection, and utilization traits with emphasis on resilience, the setting of quantitative breeding objectives relies on knowledge of baseline variety characteristics and available genetic resources, the expected effects of changes in trait levels, and means to measure gains toward them.

Metrics for disease resistance: One special tool used by CIP is an interval scale for expressing potato resistance to late blight that indicates resistance levels required for satisfactory control of the disease in agro-ecologies with varying degrees of pathogen pressure. The use of this interval scale facilitates setting quantitative resistance breeding objectives in a robust manner within a breeding program's trait improvement framework. The late blight susceptibility scale of Yuen and Forbes (2009) enables assessment of resistance levels with a reduced coefficient of variation among trials as compared to other semi-quantitative metrics like AUDPC or rAUDPC. The scale uses reference cultivars in regression analysis and helps breeders to measure and describe resistance of genotypes independent of environment or inoculum level, which can vary from site to site and year to year. This approach gives breeders a simple numerical metric for quantitative traits that is useful for setting baselines, breeding goals, and calculating genetic gains.

Application of the susceptibility scale to setting breeding targets is based on the understanding that a variety with level 5 (more susceptible) would be sufficiently tolerant to provide the same level of protection in low-pressure agro-ecologies as a variety with level 2 (less susceptible) would provide in a agro-ecology with high disease pressure. The use of this scale is illustrated in Field assessment of resistance in potato to *Phytophthora infestans* at https://research.cip.cgiar.org/potatoknowledge/lateblight.ph.

Metrics for nutritional traits: To help reduce the health burden of iron and zinc deficiencies, quantitative targets for levels of these two elements in potato (a biofortification breeding goal) were set through collaboration between potato breeders at CIP and nutritionists of Harvest Plus. Consideration was given to (1) nutritional status of the target population, (2) dietary increments of iron and zinc known to be effective from other approaches (i.e., food fortification or nutrient supplementation), and (3) features affecting feasibility of nutrition impact such as consumption (potato intake), heritability, bioavailability, and retention of these minerals in potato prepared for use as food.

Nutritionists consider that dietary increments of 0.4 and 0.2 mg/day of iron and zinc, respectively, can have a positive biological effect on the health status of populations at risk of micronutrient malnutrition associated with deficiencies in these minerals. The provision of these increments through biofortified varieties is feasible when mineral concentration of a variety, consumption levels, bioavailability, and retention in the diet are sufficient. This food-based approach to reducing micronutrient malnutrition relies on variety change but not on modification of consumption patterns, i.e., high versus low potato intake. Logically, a population that consumes twice as much potato as another one would ingest twice as much iron from that source. And the percent of the Estimated Average Requirement (EAR) of the nutrients it provides would be double.

Quantification of these features for a given varietal and dietary context assumes conservative estimates of 10% and 25% bioavailability of iron and zinc, respectively, from potato, and that minerals are not lost in cooking. Figure 6.9 illustrates how different levels of iron and zinc concentrations in potato (referred to as biofortification levels: Base line, 1, 2, and 3, where base line represents current potato varieties) contribute to the EAR of iron and zinc for women of fertile age who consume 100, 200, or 400 g of potato per day. For reference, women in parts of Rwanda consume over 500 g of potato per day (Personal communication, Harvest Plus), while women in parts of Peru, may consume 800. The iron and zinc levels included in potato product profiles for populations of the Andes or the central African highlands are 45 mg/kg Fe and 35 mg/kg Zn (just over biofortification level 2 in Fig. 6.9) which can be expected to provide 50% of the EAR of both minerals for women consuming 400 g potato a day. This is a considerable increment over the baseline, but has been assessed as feasible by CIP's breeding program after evaluating genetic diversity, estimating heritability, and realizing significant gains (subsection 6.1.5) in cultivated potato germplasm. Achievement of such genetic gains would only provide 12–15% of the EAR for women who consume 100 g of potato per day or less, as in much of the subtropical lowlands of Asia. The inclusion of a biofortification breeding target to increase Fe or Zn intake from potato by 0.4 or 0.3 mg/day, or reach 50% EAR for this market segment, would require greater genetic gains than presently estimated to be feasible by the interploid breeding strategy undertaken by CIP. Nevertheless, high iron and zinc potatoes have been requested by the national programs of both India and Bangladesh. In such cases, a

A

Biofortification Level	Iron (mg/Kg DW)	Zinc (mg/Kg DW)
Baseline (non-biofortified)	20	16
1	27	24
2	42	32
3	50	40

B

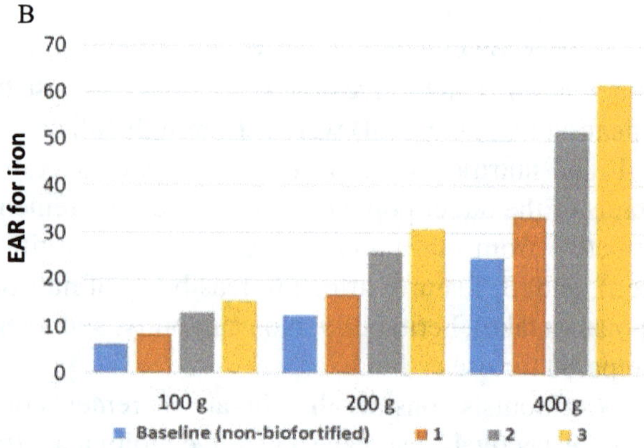

Fig. 6.9 Influence of Fe and Zn concentration and intake of potato on the estimated average requirement for women of fertile age. (**a**) Definition of baseline (non-biofortified) and incremental concentrations (biofortification levels 1, 2, and 3) of iron and zinc in potato tubers. (**b**) Percent of EAR for iron for women of fertile age met by consuming 100, 200, or 400 g/day of potatoes with concentration levels 0, 1, 2, and 3

food systems intervention seeking overall increments of micronutrients from co-staple crops should be considered.

Biofortification targets between level 2 and 3 in Fig. 6.9a have been set as part of the product profiles for table potatoes oriented to tropical highland agro-ecologies and populations with significant levels of anemia, who also consume potato as a main food or staple crop. The value of nutritional traits can be assessed with the disability adjusted life years (DALY metric). DALY is ordinarily used to assess the impact of public health burdens like human disease or illness. It extends the concept of potential years of life lost due to premature death to include equivalent years of "healthy" life lost by virtue of being in a state of poor health or disability. In so doing, mortality and morbidity are combined into a single, common metric (Meenakshi et al. 2007).

6.4.6 *Setting of Breeding Priorities*

Once product profiles are defined, breeding programs can proceed to refine trait lists and allocate resources to address those that are most feasible. While goals and priorities are important for breeding programs and the institutions that host and sponsor them, the list of traits a breeder can work with is limited, and in this light, he/she will select a subset of those in the profile for the day-to-day efforts of the program. Breeders may also be aware of a few "must have" traits that do not come up in stakeholder consultation, and since the inclusion of any trait in a breeding effort (whether to change or even maintain trait levels) requires investment, and thus has a cost, these should be made explicit and, ideally, added to the product profile.

Experience and knowledge of existing germplasm are applied to identify the set of attributes to be addressed in breeding. For each attribute, a quantified description of the desired result (maximize, reach specified level, maintain a certain minimum level, etc.) and a unique rank or priority is assigned. Ranking represents the relative effort required to develop the set of traits that will enable the new variety to respond to the opportunity or constraint identified in the targeting exercise and taken on by the breeding program. Relative effort required to address each objective are determined by the proximity of the program's genetic materials to the needed level of each trait, and the genetic control and relationships among various traits.

6.4.7 *Selection Decisions in Potato Breeding*

Because potato is highly heterozygous, a large number of traits segregate every time a cross is made. Every offspring receives a unique combination of desirable alleles at some loci and undesirable alleles at others. No offspring is ever "perfect." An issue of considerable practical importance, therefore, is deciding which phenotypes (clones; genotypes) are worth keeping, and which should be discarded at each stage of the selection scheme.

Selection decisions are the decisions a breeder makes to use, advance, or discard a selection unit (plant, progeny, progenitor, clone, experimental variety), guided by the breeding priorities, using specific decision-making tools (e.g., assessments of the targeted traits, estimates of breeding value, presence/absence of specific markers, selection indices, combining ability).

In early stages of selection, when the number of plants of each genotype is small, it is important to select stringently only for highly heritable traits. Meaningful assessments of yield, on the other hand, can only be performed in later stages. As selection progresses, and the individuals not meeting breeders' criteria have been eliminated, it becomes especially important to keep product profiles in mind, i.e., to select not just on the basis of individual traits, but on the suite of traits that collectively define a product, with attention to specific traits at specific stages of the selection scheme.

6.4.8 *Case Study 7: Profile and Selection Decisions for Chipping Potato for Northeast USA*

As nations develop there tends to be a shift in potato consumption, away from fresh and more towards processed potatoes, primarily as chips or French fries. Varieties intended for fries or chips ideally have higher levels of starch, so that they absorb less oil when fried. In addition, if potatoes need to be stored in the cold for a long time—typically to ensure a year-round supply of raw product—then it is useful if they are resistant to cold-induced sweetening, as the presence of high levels of glucose and fructose will cause potato slices to turn dark brown when fried.

To improve resistance to cold-sweetening, the New York program has adopted a low-tech approach that could be adopted anywhere. The essence of the approach is this: store potatoes at a temperature where only a small percent will fry to a light color. Intercross those that do, and after 5 years or so, once the population has reached a point where many progeny fry well, lower the cold storage temperature by 0.5–1.0 °C the following year. Intercross the few that fry well under the new, colder regime, and repeat the process. Recurrent selection has proven very effective at improving fry color in the New York program; many clones now fry well out of 3–9 months of storage at 6 °C. Ideally the breeding program will fry from slightly colder storages than those currently in use by regional industry, to help drive down the temperatures that industry can use over time. In general, the colder the storage, the longer that potatoes can be stored.

The two most important attributes in chipping potatoes are starch content (strongly correlated with specific gravity) and fry color, followed by tuber shape and size. About 70% of the New York breeding program effort is dedicated to developing new chipping varieties.

The New York program practices very little selection on seedling tubers, as the performance of plants grown in pots correlates poorly with performance of plants grown in the field. When the first field generation is harvested, the principal selection criteria are for tuber yield, tuber shape, and size (all visually assessed). The ideal chipping potato is round and the size of a baseball. Clones with low yield, or shapes and sizes too far from baseballs, are not selected.

The New York breeding program chips (out of cold storage) every clone that survives 1 or more years of selection in the field, discarding all that do not chip as well as, or better than, the industry standard chipping variety "Snowden." Absolute fry color can vary from season to season, although the ranking of fry color between clones does not vary greatly.

In the USA a specific gravity of 1.080 is deemed the minimum for processing. Nevertheless, because starch content can vary considerably from season to season (warm nights result in lower starch content than cool nights, all other things being equal), in New York the level of starch is always evaluated relative to that in the widely grown chipping variety, "Atlantic." The specific gravity of Atlantic is typically above 1.090 in New York, but in some seasons it is less. The New York breeding program will not continue to evaluate a chipping clone whose gravity averages 0.010 or more less than Atlantic, and prefers clones to average 0.005 or fewer points less. New York begins to evaluate specific gravity after a clone's second year in the field.

Replicated, multisite yield trials begin in the fourth field year, where clones that yield less than 90% of Atlantic are discarded. Clones that exhibit considerably more internal or external defects than "Atlantic" or "Snowden" in yield trials are also discarded, regardless of their yield, fry color, or specific gravity. Clones that survive several years of field trials are then evaluated on commercial farms, and processed in commercial chipping plants, where growers and chipping plants make the final decisions about which clones merit release as new varieties.

6.4.9 Selection Decisions: Marker-Assisted Selection

6.4.9.1 How the New York Program Uses Molecular Markers

Even though current NY breeder WDJ was trained as a molecular geneticist, the New York breeding program does not make extensive use of marker assisted selection, primarily because there are not yet many publicly available markers linked to traits of high priority.

One of the two markers currently used in New York is 57R (Finkers-Tomczak et al. 2011), tightly linked to the H1 gene, which confers resistance to race Ro1 of *Globodera rostochiensis*. Developing potatoes resistant to race Ro1 has been the highest priority of the New York breeding program for the past four decades; almost every cross made in NY has at least one Ro1-resistant parent. All offspring that survive 2 years of visual selection for appearance and yield in the field (about 200 clones, out of 20,000 initially planted) are screened with 57R over the winter. The vast majority of clones lacking the marker are discarded, the only exceptions being those that look especially promising for other high value traits (e.g., excellent chip color combined with high specific gravity).

RYSC3 (Kasai et al. 2000) is the other marker currently used in New York. This marker is tightly linked to Ry_{adg}, which confers immunity to potato virus Y (PVY), and like 57R, is used to screen clones that have survived 2 years of selection in the field. At present the presence/absence of the RYSC3 marker is just one of many data points NY uses to make a decision about each clone, with other data points including chip color, specific gravity, resistance to common scab, and visual assessments of yield and tuber appearance.

Although markers tightly linked to potato virus X (Gebhardt et al. 2006), wart (Gebhardt et al. 2006), verticillium wilt (Bae et al. 2008), Columbia root-knot nematode (Zhang et al. 2007), the pale cyst nematode (Sattarzadeh et al. 2006), and late blight resistance genes (e.g., Colton et al. 2006) have been reported, none of these traits are important enough in New York to influence selection decisions, and thus none of these markers are used routinely.

6.4.10 Breeding Objectives at CIP

Priority setting exercises (Fuglie 2007; Hareau et al. 2014), complemented by evolving requests from national programs for elite germplasm and candidate varieties, both contributed to CIP's profiles for new potato varieties tailored to the contrasting agroecologies, cropping, and food systems its program addresses. Two key products conceived in a recent exercise were (1) "Agile Potato" for intensification and diversification of cereal-based systems of Asia and, (2) mid-elevation tropical highlands late blight resistant potato.

CIP's major breeding populations are oriented to develop table and processing varieties for sustainable potato production in tropical highland and lowland areas faced with either of two basic sets of biotic and abiotic constraints: those encountered in the tropical and subtropical highlands, where late blight, cyst nematodes and frost are limiting, and farmers are increasingly faced with drought; and warmer, drier production areas, that are challenged by viruses, bacterial wilt, drought, root knot nematode and other pests, Increasing requests to support production in mid-elevation tropics of Sub Saharan Africa and Asia as well as temperate regions of Asia have led to the inclusion of heat tolerance and long day adaptation into our breeding objectives, requiring the introduction of new germplasm, screening methods, and trial sites. Present emphasis is on: earliness to enable potato's full potential to contribute to incomes and productive cropping systems; heat and drought tolerance to enable expansion to mid-elevation regions nearer to population centers, at the same time helping farmers adapt to climate change, and; nutritional value with specific attention to iron and zinc biofortification. CIP aims to reach new thresholds for these traits, while maintaining and broadening levels of disease resistance achieved in the past, and also responding to regional preferences for culinary characteristics and market opportunities.

6.4.10.1 Case Study 8: Agile Potato for Asia

Product Profile

The "Agile Potato" project seeks to develop potato varieties for fresh and processing use in short crop production windows of Asia. The product profile includes 70–90 day maturity; resistance to multiple viruses (PVY and PLRV, with PVX resistance desirable); stress tolerance (heat, drought, and salinity); resistance to late blight; and market-oriented quality (high dry matter and chip processing). The expectation is that varieties with short crop duration that resist disease, require less-frequent seed replacement, thrive in warm seasons, and store well under rustic conditions, can readily be inserted into diverse cropping patterns dominated by rice or wheat. Agile potato varieties would fit into cropping windows currently left fallow in the cereal-based systems of subtropical lowland and temperate regions of Asia. They should help increase system productivity through intensification as well as help meet growing demand for processed potatoes, thus contributing to income generation and employment. Breeding to meet this profile is backed up by trait research to enhance and accelerate breeding methods, and inter-disciplinary research to promote fast-track systems for variety identification and release. With this profile, CIP seeks to develop strategies for sustainable intensification of farming systems, raise awareness and expand consumer demand for potato while diversifying cereal-based diets with potato as a nutritious co-staple or vegetable food. A gamut of research on ecological and socioeconomic consequences of intensification is also planned in order to identify appropriate indicators of sustainability and monitor effects of benefits to intended sectors, including the poor whose livelihoods would be affected by the proposed intervention.

The target is to select 70–80 day heat-tolerant potatoes with competitive yields and more than 18% dry matter and resistance to potato viruses for subtropical lowlands of Southwest Asia. Commercial varieties and elite selections from previous cycles of the lowland tropics virus resistant population (LTVR) are comparators for performance in representative or target selection sites. In the following figures we show a group of selected early clones that performed significantly better than past generations and local varieties for yield and quality at 70 days after planting in warm environments, yielding in a range of 20–40 tons/ha, compared to the best local varieties such as Unica and Reiche (Fig. 6.10) and adequate dry matter content over 18% under the abiotic stress conditions of lowlands (Fig. 6.11).

Dry matter (DM) content is an important component to be taken into account in the breeding process especially when selection is focused on lowland tropical areas. In addition to the loss of fresh weight yield of tubers associated with high temperatures, the DM content of tubers are also reduced by 1% as the air temperature rises by 1 °C over a temperature variation average of 15–25 °C. This is due to the rapid

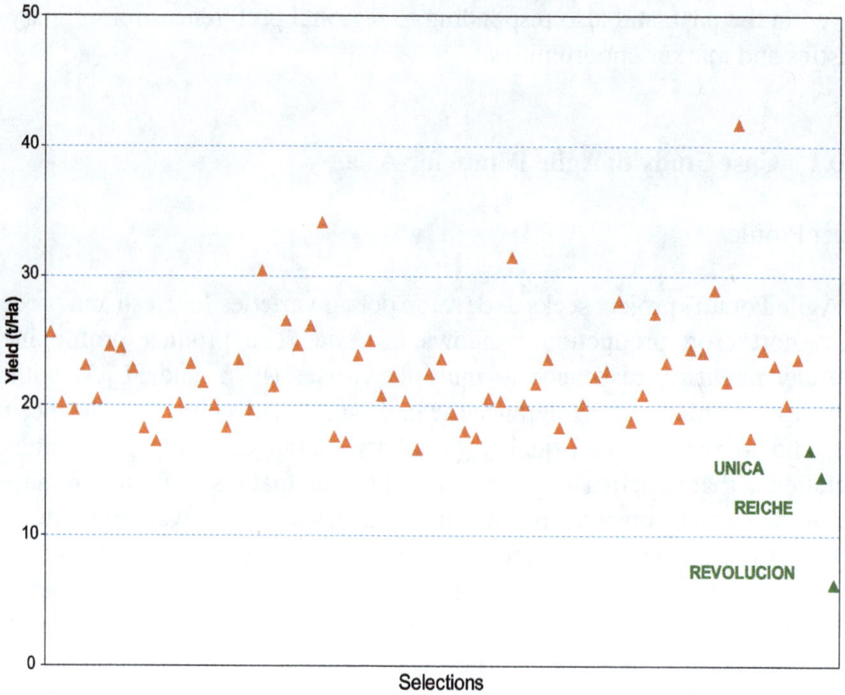

Fig. 6.10 Scatter plot showing yield of selected clones in a new generation of LTVR population that surpass levels of varieties and clones from previous generation grown under hot conditions of San Ramon, Junin, Peru, and harvested at 70 days after planting

decline of photosynthesis (Midmore and Rhoades 1988). This tendency of decrease in DM of clones in the LTVR population can be appreciated in Fig. 6.11, showing decreasing DM contents with decreasing altitude and thus increasing temperature of the evaluation sites. However, breeding by recurrent selection for yield as well as DM content in this population has maintained this important trait at between 11 and 19% even in the warmest season, and 15–21% in the high humid jungle site of San Ramon, Peru. This means that genetic diversity is sufficient to permit selection for heat tolerance and genotypes with good yield and adequate DM content can be identified in LTVR in spite of the critical negative correlation between temperature and DM in potato.

Crossing with long day adapted parents and evaluation under simulated long day conditions in Peru permitted the identification of LTVR families and advanced clones with adaptation to higher latitudes, for example of Vietnam where day lengths are intermediate, and Uzbekistan and Tajikistan where they are long combined with high summer temperatures. The same screening approach also helped provide pressure for improving the earliness (70–80 days) of LTVR at low latitudes. The compensatory effect of temperature on day length provided by warm mid-latitude conditions of southern Peru has also been key in breeding for early maturity, which can also be referred to as a relaxed requirement for short days. The identification of a moderate frequency of genotypes as well as superior parents with good performance under higher latitudes suggests that the LTVR breeding strategy

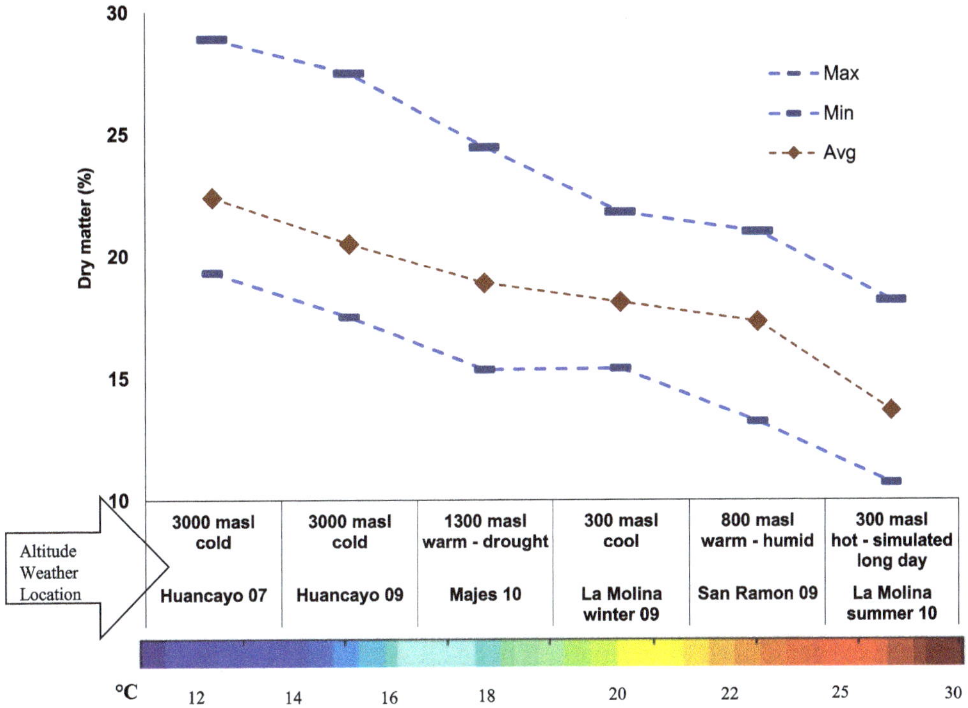

Fig. 6.11 Variation for dry matter content of potato selections in LTVR population across environments. Three hundred LTVR clones were grown in ten hill plots in six locations in Peru and evaluated for dry matter content within 10 days of harvest

is likely to be successful in developing the agile potato for both of its target environments—subtropical lowland and temperate low and mid-elevation regions of Asia.

Source Population LTVR

From the 1980s through the late 1990s, CIP endeavored to strengthen its lowland tropics (LT) population with multiple virus resistance developing the lowland tropics virus resistant population (LTVR). Development of the broad-based LTVR population used an open-recurrent selection strategy to incorporate foreign elite and advanced clones into its lowland tropics adapted breeding population. Lowland tropical environments were characterized as having day lengths of 10–14 h, minimum night-time temperatures of 18–20 °C, mean maximum temperatures greater than 25 °C and mean annual soil temperatures at 50 cm depth of 22 °C or more. LTVR population combines the heat tolerance and early bulking ability of *S. tuberosum* germplasm bred under the summer conditions of the northern Hemisphere (Cubillos and Plaisted 1976) on one hand, with virus resistance from native *Andigenum Group* germplasm and long day adapted *Andigenum Group* germplasm, namely Neotuberosum, on the other (Plaisted 1987; Mendoza 1990). The Neotuberosum and *Andigenum Group* germplasm contributed major genes for extreme resistance to potato virus Y (PVY) and potato virus X (PVX) (Muñoz et al. 1975; Gálvez et al. 1992). Early generation screening was routinely performed to discard PVX and PVY susceptible individuals in any trial in which LTVR families were evaluated.

Testing sites included CIP's experimental stations in Peru (see Section 6.4.6) which are relatively dry, humid, hot, and hot and humid lowland environments. Experiments comparing the yield stability of diverse clones across these locations revealed that hybrids from *S. tuberosum* × Neotuberosum crosses performed significantly better than those from *S. tuberosum* × *S. tuberosum or S. tuberosum* × *Andigenum Group*. It is possible that the higher level of heterozygosity together with prior selection of the divergent parental materials for adaptation accounted for the better performance of crosses between *S. tuberosum* and Neotuberosum clones (Mendoza and Estrada 1979). These authors also reported that *S. tuberosum* × neotuberosum hybrids are adapted to hot and humid conditions and able to produce tubers at 60–70 days after planting, albeit under short days. These hybrids made a strong contribution to the LTVR population, and thus may account in part for its adaptation to warm conditions. The subsequent cycles of genotypic recurrent selection led to the selection of a number of multiplex PVY and/or PVX extreme resistant progenitors (Mendoza et al. 1996) some of which have since been shown to also possess good GCA for yield and dry matter content. Multiplex refers to the genetic constitution of a polyploid individual having more than one copy of an allele, in this case RRrr or RRRr, assuring the transmission of resistance conferred by a major gene to nearly 100% of its cross progeny. In parallel to population improvement and parental line development, high yielding clones with excellent tuber quality were selected during each recurrent selection cycle.

A second stage of the breeding process aimed to incorporate foreign elite and advanced clones by means of an open-recurrent selection strategy. Assessment of genetic parameters in a North Carolina Design II estimated heritability for resistance to PLRV infection to be 0.54–0.69. Higher estimates were obtained when greenhouse-reared infective aphids were used to conduct controlled inoculations with sprouted tubers before tuber families were planted in the field, suggesting that controlled inoculations minimize errors or escapes because of a more even distribution of inoculum. Additive genetic effects were shown to contribute the most to resistance variability (74%), followed by dominance effects (10%) and additive × environment (location) effects (5%), suggesting that breeding advances could be expected in further cycles of recurrent selection (Salas 2002).

Multilocation Testing and Selection Decisions

A stage plan can be used to communicate selection decisions made in early and late stages of breeding. The stage-gate process portrays stepwise decisions with thresholds for each characteristic that govern decisions to keep or discard individuals in the course of a given breeding cycle. In this process, the "gates" describe the collection of thresholds or criteria that must be met if a clone is to pass from one stage of selection to another along the clonal selection scheme outlined in Fig. 6.12, presents a stage plan for selection in the LTVR population to meet the Agile Potato profile.

Yield and processing quality traits of LTVR clones generated during open-recurrent selection were evaluated in replicated experiments in a range of lowland environments of Peru. Dry matter content was consistently higher in the cool arid lowlands of the coast under irrigation (average = 21%) than in the warm humid area of the central jungle (average = 19%), highlighting the detrimental effect of high temperatures on this trait. As expected, significant G × E interaction was encountered for dry matter content, but also for glucose content and chip color. High yielding clones with good table or processing quality combining virus resistances and heat tolerance have been released by national programs and public institutions in target countries. Varieties of LTVR origin include: Tacna, Costanera, Unica, Reiche, Maria Bonita, and Maria Tambeña released in Peru; Kinga, Meva, Kinigi, Muziranzara, Muruta, Yayla Kizi, Baseko, Enfula, Chamak, Dheera, and IRA-92 in Africa; and Raniag, which was named in the Philippines. Evidence of the wide adaptation of bred materials from this population is provided by the success of the variety "Tacna" in the lowland subtropics of China where it is known as "Jizhangshu 8" and planted on an estimated 133,215 ha in 2015 from which 70,533 ha in Hebei, 26,000 ha in Shanxi, 22,667 in Inner Mongolia, and further 3000 has in the increasingly drought-prone northwest province of Gansu with more than 3000 ha (Gatto et al. 2018). Additional evidence of broad adaptation comes from the variety Unica that is now grown in at least six countries.

Breeding Priorities

Following CIP's multidisciplinary conceptualization of the agile potato, the potato breeding program refreshed its strategy by setting priorities for combining the needed traits in subgroups of the LTVR population. The inclusion of resistance to PVY in the LTRV population and selection for highland tropical as well as subtropical lowland conditions to enable seed production has had a cost in terms of earliness even after three to four cycles of recurrent selection. The infusion of Tuberosum germplasm and selection under heat and mid-long days, for plant type and stability from early generations is strategy to improve earliness in the LD (long day) and HT (heat tolerant) groups. In this interest, tuber quality becomes an issue and attention is needed to keeping dry matter high and glycoalkaloids, which provide a bitter flavor to potato and can be toxic, low.

Often unrecognized and rarely controlled, the accumulation of viruses in seed and ware tubers is perhaps the most serious constraint to economical potato production in the tropics. Breeding for resistance to the world's most important potato virus PVY is possible due to the availability of major genes (Ry_{adg}, Ry_{sto}) that confer extreme resistance, and for which selectable markers are available. On the other hand, developing effective levels of resistance to PLRV is one of the more challenging components of the Agile Potato profile. Most sources of resistance to PLRV are quantitatively inherited and provide partial resistance to infection and/or accumula-

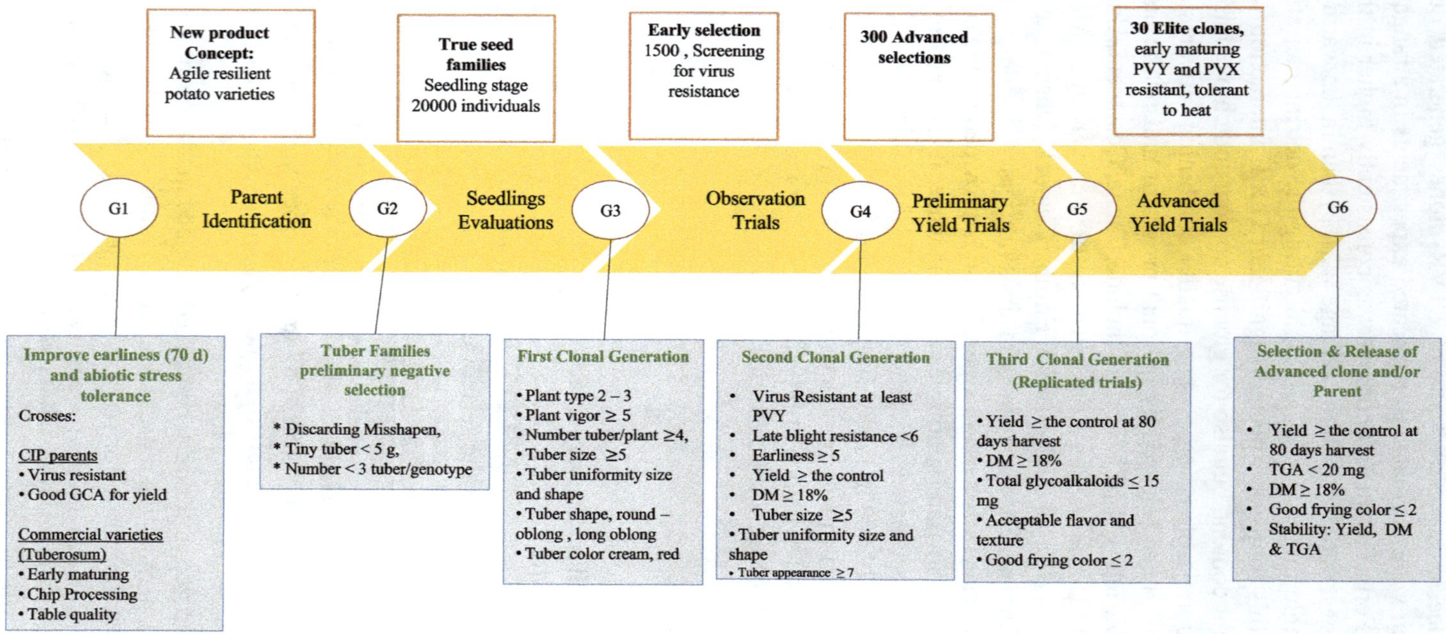

Fig. 6.12 Stage plan for selection in the LTVR population to meet the Agile Potato profile

tion of the virus. Despite moderate heritability, the frequency of resistant clones in advanced populations has been low. Thus, a relatively high priority is given to incorporating a new, highly heritable source of high levels of resistance to PLRV, designated Rl_{adg} recently identified in Andigenum Group germplasm into advanced populations that already carry extreme resistance to PVY at high frequency.

New sources of PLRV resistance, namely, accessions LOP-868, HUA-332 and OCH-7643, confirmed following mass selection in *S. tuberosum* Andigenum Group germplasm held at CIP, were shown to have exceptionally high GCA for this trait (Mihovilovich et al. 2007). Further characterization of these sources has been the basis of supporting trait research, the benefits of which are expected to boost levels and frequencies of PLRV resistance in the LTVR and other potato populations. Using a population of dihaploid individuals derived from LOP-868, Rl_{adg} was mapped to potato chromosome V, with high heritability demonstrated to be due to its multiple resistance mechanisms and a duplex allelic conformation of a major dominant gene in the donor accession (Velásquez et al. 2007). A PCR based marker (RGASC850) located 2 cM from the gene has been developed to facilitate the introgression and tracking of this major gene in advanced breeding populations (Mihovilovich et al. 2014). This marker has been incorporated into a multiplex PCR system for assisted selection of PLRV and PVY resistance using previously available information on Ry_{adg} from Kasai et al. (2000).

Table 6.2 presents the attributes, profile and breeding priorities contributing to the Agile Potato program and outlines some of the key research that supports the objective. Breeding priorities represent the relative effort dedicated to each attribute of the profile, given present status of the LTVR population.

Multi Trait Selection for Increased Earliness and Tolerance to Abiotic Stress

Most recently, emphasis in improving CIP's LTVR population toward the agile potato has targeted increased earliness and tolerance to abiotic stress. A new earlier maturing and abiotic stress tolerant generation that combined superior LTVR progenitors with tuberosum varieties was generated. Accelerated selection based on multilocation trials and simultaneous screening for abiotic stress (heat, drought and simulated long days) in divergent and stressful environments as well as screening for PVY resistance was implemented in Peru. The scheme requires only 4 years of selection and evaluation from the true seed progeny generation until the identification of advanced clones (see Fig. 6.6).

Evaluation and Selection Sites

CIPs proximity to significant agroecological diversity in Peru is an asset for the identification of patterns and trends of potato genotypes' response to environments. Potato is grown in coastal, high jungle and highland ecologies ranging from sea level to 3500 masl, providing wide variation in weather patterns including tempera-

Table 6.2 Agile potato profile, breeding objectives, and supporting research

Attribute	Product profile	Breeding priority
1. Yield 2. Earliness 3. Table quality 4. PVY resistance 5. PLRV resistance 6. Heat tolerance 7. Good storability 8. Late blight resistance 9. PVX resistance 10. Water productivity 11. Cold chipping ability 12. Bacterial wilt resistance	1. >25 tons/ha 2. 70 day maturity 3. TBD (flavor appealing to consumers, more than 17% dry matter, <20 ppm glycoalkaloid content) 4. Extreme resistance to PVY 5. Tuberization at >22 °C night temperature; bulking under warm day temperatures 6. Drought tolerance-TBD (ratio of fresh tuber yield to applied water expressed as kg/ha/mm) 7. Bacterial wilt resistance-TBD % plants wilted to degree *x* and % latent infection 8. Resistance to PLRV as high or higher than Granola 9. TBD (*x* days to sprouting, low rottage, low water loss at given temperature) 10. Late blight resistance <6 on susceptibility scale 11. 20% selections with chip color score 1–2 after cold storage; dry matter >18% 12. Extreme resistance to PVX	1. Earliness 2. High and stable yield across day lengths 3. PLRV resistance 4. Heat tolerance 5. Flavor and dry matter 6. PVY resistance 7. Water productivity 8. Storage quality 9. Chip quality 10. PVX resistance 11. Bacterial wilt resistance

Supporting research:

Earliness: implement protocol and validate markers to develop MAS for bulking-based maturity. Complete GWAAS for day neutrality in subset of Trait Observation network panel across environments

Introgress and increase frequency of Rl_{adg} using MAS for PLRV resistance

Implement proximal sensing phenotyping for drought tolerance (water productivity). Develop inexpensive assay for glycoalkaloid contents

Validate pre-bred *S. commersonii* source of bacterial wilt resistance in Kenya and Ethiopia

Refine genome estimated breeding values (GEBV) for early bulking and heat tolerance extend pilot study of genomic selection for early bulking to LTVR in Asia

Optimize field trials and statistical analysis to make full use of trial data; genetic gains and heterotic groups at tetraploid and diploid levels

ture, humidity and rainfall. Most environments have short daylengths of around 11–12 h but the southern extreme at 17°S, provide daylength of about 14 h, which considering high summer temperatures approach threshold conditions for long day adaptation. Three experiment stations (San Ramon, Huancayo and La Molina) and additional sites accessed by local agreements (Table 6.3 provide evaluation conditions representative of the tropical lowland and highland potato environments targeted by agile potato. Indeed, potato seed for the subtropical lowlands is often produced at higher, vector-free, locations, requiring adaptation of the same variety to both highland and lowland conditions. Moving potato populations through the biotic and abiotic stresses present among these sites, and particularly the combination of temperature and day length, permit the identification of potato genotypes with wide adaptability and stability for yield.

Table 6.3 Location and conditions of major selection sites used for potato breeding by CIP in Peru

Location	Season	Climate	Latitude (°S)	Altitude (masl)	Geography	Main traits assessed	Agroecology
San Ramon (SR), Junín	Winter	Warm, humid	11	800	Central jungle	Heat tolerance; glycoalkaloid accumulation	Subtropical lowland humid
La Molina (LM), Lima	Spring-summer	Warm, semi-arid	12	133	Central coast	Yield, drought and heat tolerance, processing	Subtropical lowland arid
Tacna (Tac),	Summer	Hot, saline soils	18	500	Southern coast	Drought, soil salinity and adaptation to long days	Subtropical lowland arid
Huancayo, Junín	Winter–autumn	Cold	11	3200	Central highland	Yield potential, processing quality	Tropical highland cold
Oxapampa	Spring	Rainy, warm	10	1000	Central high jungle	Late blight	Tropical highland humid
Majes	Spring–summer	Warm, arid	16	1300	Southern coast	Drought tolerance, glycoalkaloid accumulation	Subtropical lowland arid

Trait Research in Support of the Agile Potato

Proximal sensing for efficient development of the agile potato: Remote sensing technologies are increasingly being used to assess crops by breeders and physiologists. These technologies have the potential to simplify evaluation of large breeding populations and improve selection efficiency over traditional field measurements. Different vegetative indices combined together may, for example, improve yield prediction (Rodriguez Junior et al. 2014). CIP's work in this area has aimed to explore genetic variation for the canopy temperature differential (CTD) and Normalized Difference Vegetation Index (NDVI) measurements and their relations with tuber yield components using 27 segregating families of the LTVR population under water stress, toward identifying families with tolerance to drought at early stages of selection. A high negative correlation (−0.66) of NDVI with Harvest index-fresh weight was observed, as were a high positive correlation (0.74) with plant wilting, a moderately high (0.50) correlation with plant vigor under drought conditions, and a significant negative correlation (−0.5) of CTD with tuber yield (tons/ha). Remote sensing techniques may provide a way to save resources and accelerate genetic gains through the early identification of families with tolerance to abiotic stress in a breeding program.

Trait research for understanding of complex, underground development: In breeding for a short vegetative cycle, CIP emphasizes bulking-based maturity, as opposed to foliar maturity or senescence. Tuber bulking results from two basic processes, tuber initiation and tuber growth (Radley et al. 1961). Like Andigenum Group potatoes, most varieties and clones from CIP's program yield significantly before physiological maturity, and development and harvest date cannot be assessed on foliar senescence alone. Indeed, if CIP-bred varieties are harvested when foliage has senesced, they will generally be well beyond the optimum time for harvest. Timing and duration of bulking vary with location, environmental factors, and genetics. Early tuber initiation and growth are necessary for acceptable production in areas or varieties in which potatoes are often harvested prior to physiological maturity.

To help provide early-maturing, resilient potatoes to farmers in medium and high latitudes of Asia, a panel of 162 CIP breeding lines was characterized for bulking behavior under four environmental conditions in field plots in La Molina, Peru. The environmental conditions for this study were: simulated long (16 h), and short (12 h) day lengths, in each of two seasons—cool winter and hot summer. The traits evaluated in the four field plots include yield (number and weight of tubers on two harvest dates) and dry matter content, in addition to which tuber induction was assessed by the cutting method of Ewing (1992). Tuber bulking ratio was calculated as the number of medium-sized and large tubers over the total number of tubers (tiny + small + medium + large) multiplied by the ratio of tuberized plants. Phenotypic assessment according to CIP's Protocol for tuber bulking maturity assessment of elite and advanced potato clones (https://research.cip.cgiar.org/potatoknowledge/bulking.php) enabled the classification of genotypes by maturity through cluster analysis according to Evanno and Regnaut (2005), and thereby, the identification of clones for which tuberization and bulking were more or less influenced by heat and photoperiod, including a small set of selections classified as day neutral.

Genotyping of the panel with version 1 of the SolCAP infinium array (8303 single nucleotide polymorphism (SNP) markers) enabled the identification of genomic regions and SNP markers influencing potato tuberization and bulking related traits affected by photoperiod and high temperatures. Prerequisites for this genome-wide association study (GWAS) included the calculation of the population membership matrix and SNP-based kinship relationship matrix and the estimation of linkage disequilibrium (LD decay to $r^2 = 0.2$ over 0.3 Mb) for the panel. GWAS-associated SNP jointly explained up to 60% of the phenotypic variation for the various traits, Therefore, MAS could be efficiently used for pre-screening of seedlings before further field evaluation, and contribute to shift the mean of the selection population for adaptation to warm, long day environments.

The same information was also used in a pilot study of genomic selection (GS). The underlying concept in GS is to model the entire complement of quantitative trait loci (QTL) effects across the genome to produce a genomic estimated breeding value (GEBV) from which progenitors or clones can be selected after genotyping. This pilot study tested several models proposed by Endelman (2011) to assess the feasibility of applying GS to improve selection of parental clones for early tuberization and bulking under long day-length. Prediction accuracies of ≥0.3 can be regarded as sufficiently high to merit implementation of GS (Heffner et al. 2010). Sufficiently high values were attained for tuberization induction, stolon length, bulking ratio at 75 and 90 days, and marketable tuber number at 75 days. As all markers are simultaneously fitted to capturing most of the trait heritability, GS can be expected to help reduce the length of selection cycles aimed at improving traits for which accurate phenotyping can be performed.

6.5 Prospects for True Hybrid Potato Breeding

Several countries and institutions have conducted research proposing the production of potato by sexual or botanical seed. Since the vegetative (tuber) planting material of clonally propagated potatoes is typically called seed potato, the term true potato seed (TPS) was introduced for tetraploid potato varieties produced from botanical seed. TPS aimed to overcome difficulties associated with production of vegetative seed including slow multiplication rate, accumulation of pathogens, and the bulky and perishable nature of tuber seed (Kidane-Mariam et al. 1985; Golmirzaie et al. 1994; Almekinders et al. 2009). As potato is a cross-fertilizing crop developed from heterozygous parents that do not "breed true," TPS is heterogeneous and thus variable for a large number of characteristics. While this variability is the source of variation for a clonal variety program, it is generally undesirable in varieties per se.

True hybrid potato breeding, on the other hand, aims at developing uniform true seed (TS) hybrid varieties from complementary superior inbred stocks developed at the diploid level. It would combine the advantages of TPS, (large numbers of disease-free propagules), diploid breeding (simplified genetics), and homogeneity (by way of combining inbred instead of heterozygous parents) in a new potato breeding system akin to that of maize.

Preliminary reports of productive diploid hybrids (de Vries et al. 2016), and the advantages of "re-inventing potato" as an inbred line-based crop (Jansky et al. 2016) make this effort one of the most intriguing developments in potato breeding over the past decade. Diploid inbreeding-tolerant potatoes would allow for considerably more systematic incorporation of new genes and traits, such as by backcrossing to fixed genotypes, as well as the possibility of substantial yield gains by crossing between well-defined heterotic groups. Working with diploid potatoes would also greatly facilitate the use of desirable recessive alleles. In principle, the low multiplication rate of clonally propagated potato can be circumvented by moving to diploid hybrids, where true seed from inbred × inbred crosses can produce vast numbers of genetically identical offspring in a short period of time.

Several steps need to be taken before diploid potato inbred line breeding might become routine. Most important is eliminating the self-incompatibility system, so that inbred lines can be readily developed. Recently, the dominant self-incompatibilty locus inhibitor allele (*Sli*) identified in certain self-compatible (SC) variants of the diploid wild potato *Solanum chacoense* (Hosaka and Hanneman 1998) has been used to overcome self-incompatibility (SI) and develop SC inbreds directly from the original source or by introgression to SI 2x cultivated germplasm (Phumichai and Hosaka 2006; Jansky et al. 2014). Additional research may be needed in order to manage SC as a trait in potato breeding, to improve the vigor of potato seedlings and to assure male fertility in diploid male parents.

Investment in inbreeding seems to be greater so far in the private than in the public sector. Nevertheless, some public breeding programs including CIP's have begun to seek some of the advantages that this novel strategy can offer to breeding and genetics for potato improvement. CIP has undertaken research to develop (1) a self-compatible) 2x breeding platform for broadening the genetic base of potatoes by hybridization across genepools, and (2) inbred genetic stocks for gene discovery and trait capture. Initial stocks and experience gained in the process of this research will contribute to longer term goals of developing complementary superior diploid inbred parental lines that will produce uniform, heterotic true seed 2x varieties when combined by crossing.

6.6 Variety Testing and Seed Links

6.6.1 How the Breeding Program Is Integrated with Seed Production in the State of New York in the United States

In addition to a public sector breeding program, the state of New York also maintains a public sector nuclear seed farm (the Uihlein Farm, in Lake Placid NY), which produces clean seed for commercial seed growers in the region. Each year the NY breeding program identifies up to five promising selections—typically clones that have been evaluated for at least 6 years in the field—and sends them to the Uihlein Farm. The Uihlein Farm first tests each clone for viruses, removing any that are present, and then, after a few years, depending on the multiplication scheme chosen, offers minitubers or limited-generation field grown tubers of these selections for sale.

Over time the breeding program has learned that it is important not to discuss promising clones with local industry until seed will soon be, or already is, available. If a clone elicits excitement, and growers ask for seed yet find none, they quickly lose interest, and it is very difficult to rekindle interest once seed does become available. When a clone is discussed before seed is available from the Uihlein Farm, the breeding program strives to provide up to 1000 kg of seed for commercial ware and chipstock growers to evaluate on their farms in the interim.

The NY seed farm bears some risk in this arrangement, as the breeding program typically discovers, over the course of additional years of evaluation, serious weaknesses in some of the clones being multiplied, rendering the clone unmarketable. On the other hand, the NY seed farm also benefits when it can be the first supplier of a new variety.

6.6.2 Regional Trials for Variety Assessment

Procedures, interfaces and responsibilities for trialing elite potato clones for variety release vary in degree of organization or formality across and sometimes within Countries.

As the aim of *plant breeding* is to produce new, improved *varieties, breeders should be aware of and anticipate meeting the several steps required for variety recognition, release and registration.*

Depending on program structure and target territory) and the complexity of the selection criteria, potato variety breeding typically involves a minimum of 6–7 years (and often more) of crossing and selection work, followed by a year or 2 of early testing to identify superior clones. The next step is the entry of candidate varieties into official variety registration recommendation trials. Programs in which breeding is done in the territory of release are more straight-forward in this respect than international programs in which additional time is required for quarantine exchange and bulking of seed in a new location. Depending on access to infrastructure and communication about objectives, this can take 2–5 years. The public and private sector differ in this respect since while public programs generally provide elite materials as true seed families or a small number of in vitro plants for research and breeding that may lead to local selection, breeding companies provide a reasonable quantity of seed of known varieties along with support and incentives for their testing and release.

6.6.3 Case Study 9: Regional Trials for Potato Variety Development in the United States

Northeast Regional Potato Trials—trialing system is a network for testing elite clones as candidate varieties. At present public-sector potato breeding in the US is divided into four regions—east, central, west and south. Each region has a regional trialing network, in which the agronomic performance of advanced clones is evaluated across a wide geographical range. The trial serves both to identify those relatively uncommon clones with broad adaptation, which are most likely to become commercially acceptable, and to identify varieties that may serve the needs of narrow geographic areas that have no breeding program of their own. Taking the eastern US as an example: the system consists of four public sector breeding programs—2 based in Maine, one in North Carolina, and 1 in New York—and 11 evaluation sites, 2 in NY and 1 each in Maine, North Carolina, Pennsylvania, Florida, Virginia, Ohio, Maryland and occasionally the Canadian provinces of New Brunswick and Prince Edward Island. Evaluators at each site rate clones using standardized scoring scales for yield, specific gravity, shape, internal and external defects, maturity, texture of tuber skin, and overall appearance, and upload the data to a common web portal (potatoes.ncsu.edu/NE.html). The originating breeding programs use the data to help make selection decisions, while regional evaluators use it to help inform local growers about which new varieties will perform best in their region. Breeders and evaluators also meet once a year to discuss, in person, the performance of clones in the trial the preceding year, and to make incremental adjustments in how the trial data is collected, analyzed and reported. Over the past decade separate national programs in the US, focused on chip and french fry variety development, have built upon the standardized framework of the eastern region trial network, with that data housed at potatoesusa.mediusag.com.

6.6.4 How Does the New York Program Decide When to Release a Variety?

In many countries a new potato variety cannot be legally grown until it has been formally evaluated in registration trials, found to be distinct, uniform and stable (DUS), and shown to have "value for cultivation and use." Decisions on all of these characters are made by an independent regulatory body, not the originating breeder.

In contrast, in the United States there is no legal requirement for comparable tests before a potato variety can be grown. (Although there is a requirement for DUS data, if a US breeder wants to procure plant breeders rights.) While some US universities have panels of breeders that evaluate candidate varieties, and can deny release of a variety by a colleague, others, like Cornell University in New York, have no formal potato variety release process at all.

Is either of these approaches preferable? Coauthor Walter de Jong is comfortable to operate in an unregulated environment. In his view the uniformity and stability tests are essentially meaningless for a clonally propagated crop. In addition, while he recognizes that evaluating distinctness is important to prevent anyone from re-releasing varieties developed by others, as long as two highly heterozygous autotetraploids have been crossed, all offspring will be distinct, as so many loci segregate. In other words, it is trivial to create a distinct potato. Variety release decisions in New York are made, instead, by gauging stakeholder interest. Prior to formal release, seed growers are permitted to grow candidate quantities under an evaluation license. When a candidate variety reaches a threshold, typically 10 ha or more of seed across all seed growers, and the seed growers ask for the variety to receive a formal (nonexperimental) name, it is released. One specific situation where the New York program will not release a variety, even if there is industry interest, is if tuber glycoalkaloids are too high (>20 mg/100 g tuber fresh weight).

6.6.5 *Case Study 10: All India Coordinated Crop Improvement Project and Release of Kufri-Lima from LTVR Population*

In India, state and national systems for verity testing and release coexist, and to some degree depend on each other, and these policies serve as a model for other countries in south Asia. Formal testing of a candidate variety by the All India Coordinated Crop Improvement Projects (AICCIP), and the subsequent recommendation of the variety for release by the AICCIP are required before variety can be released by the Central Sub-Committee that permits notification and thereby enables seed production and trade. In the case of potato, a minimum of 3 years of multilocational trials and assessment for Value for Cultivation and Use (VCU) take advantage of 16 locations across the country. These locations use standardized design and data templates and report on performance as well as meteorological conditions. Experiments are designed for Rabi (winter) season for irrigated areas such as Central, Northern, and Eastern regions (Indo-Gangetic Plains). Plateau regions (700–1200 masl) and hilly regions (1500–3000 masl) are represented in the multilocational trial series.

A new variety should be suitable for specified agro-climatic and soil conditions, have an ability to withstand typical stress conditions, and have tolerance/resistance to pests and diseases. It should also show distinct advantages over the existing equivalent released varieties, a process facilitated by standard formats for data collection. The breeders or Principal Investigators and Zonal Co-ordinators attend the meeting to provide wider information on the variety. The Director of Crop Development Programme is invited to provide information on the response of farmers to promotion and demonstration trials if they were conducted. The release proposal proforma requires the breeder to ensure availability of enough seed stock for seed multiplication on at least 10 ha.

An interesting feature of the All India trials project is that it reports on potato varieties as components of cropping systems, and not only on potato as a commodity. The pertinent proforma includes information on the productivity and profitability of potato and each of different crops in intercropping systems. Long-term experiments address issues of nutrient cycling and crop management that offer valuable methodology and possibly valuable data sets and recommendations for CIP's concern for the resilience of its Agile Potato proposition, proposing intensification of cereal-based systems with attention to sustainability.

Recent emphasis of collaboration between CIP and the Central Potato Research Institute (CPRI) of India have been on identifying early-maturing, heat tolerant, virus-resistant varieties suitable for intensifying and diversifying rice-based production systems. CIP introduces elite clones as in vitro plants as well as true seed from crosses between parents with good GCA for features of performance in subtropical lowland environments for used in breeding or direct identification of varieties for the vast Indo-Gangetic Plains and hilly regions of southwest Asia. The release of Kufri-Lima in October 2017, bred as CIP397065.28, is an exciting result of this collaboration. Potato virus are a serious problem for farmers in India, particularly in Gujarat, Rajasthan, Madhya Pradesh, Chhattisgarh, Odisha, West Bengal, and East Uttar Pradesh, where temperatures are higher in the winter season. Crop losses due to viruses range between 20 and 50%, and farmers from these areas traditionally need to import seed from Northern India (Punjab). "Farmers usually have to wait until temperatures drop to plant potatoes, but because of its tolerance to heat, Kufri Lima can be planted a full 20–30 days earlier than other local varieties." Earlier planting means earlier harvests, giving Kufri Lima farmers the ability to sell their potatoes at a premium price before other varieties hit the market. Farmers can expect to be paid 40–50% higher prices than those who harvest potatoes during the normal season. Heat tolerance and early maturity mean that Kufri Lima farmers can invest their earnings into a second winter harvest helping them to improve the overall productivity of the cropping system, while virus resistance may enable seed production closer to home. https://www.potatopro.com/companies/central-potato-research-institute-cpri.

Kufri-Lima was introduced to CPRI as CIP397065.28 in 2006 after initial selection in Peru. It was evaluated for adaptation, quality, and resistance to degeneration along with several other selections from the LTVR population with the support of ICAR (International Council of Agricultural Research), Central Potato Research Institute, GIZ/BMZ and USAID.

6.6.6 Variety Release and Registration

Keen awareness of timelines, procedures and actors involved in decisions on the release of varieties from elite clones and of policies and practices that permit or enable seed production is essential for the uptake of new varieties and subsequent benefit of their component and composite traits. Country specific systems for variety testing, release and dissemination of seed to farmers are governed by sets of national laws, scientific guidelines, norms, and standard practices which together can be termed "Regulatory Frameworks."

Registration is often required for seed to be multiplied and marketed in a country or territory or for exportation outside the territory. Release is intended to make a newly developed variety available to the public for general cultivation in the regions for which it is adapted. Even though they are not directly responsible, breeders should be aware of the designated service responsible for recommending the registration of new varieties in their target locations, as well as the Basic information required for application. Additional efficiencies can be realized when breeding programs are aware of and contribute to the availability of Basic material required for application and variety release. Knowledge of the agency with which a reference sample has to be deposited, up to date reference to the format for information required and the quantity of seed to be available at time of release vary be country or state. Depending on the territory in question, the designated service responsible for approving the registration of new varieties is usually a ministry or department of agriculture, which may have crop specific rules and regulations for variety development. In India and Bangladesh, for example, potato is among the crops for which a special category called "notified" is required for varieties to enter seed production schemes.

Additional policies, processes and agencies are those responsible for demonstration or promotion trials for pre-released and released varieties where required; maintenance of certification standards; and the production and storage of foundation seed and provision of seed of new varieties. Policy regulating seed trade is of less direct concern, though also important to breeding programs. An excellent reference for procedures and policy for variety testing and release can be found at http://www.coraf.org/wasp2013/wp-content/uploads/2013/07/ECOWAS_VAR_REGIST_MANUAL_SEP_081.pdf.

6.6.7 *Regional Frameworks for Variety Release*

Beyond national seed regulatory systems, several regional frameworks are being developed or implemented across countries. Advantages of regional frameworks include improved access by farmers to seed of improved varieties at affordable prices, avoidance of repetitive national testing, making seed trade easier, faster, and less expensive for introduction of new varieties, timely availability of quantity and high quality of seed and choice of variety needed. Regional frameworks encourage individual countries to adopt national seed legislation and establish national variety catalogues that contribute to growers' awareness of new varieties. Regional registration can result in a larger market share for varieties and reduced time and cost for national variety release when second and third country releases are simplified by admissibility of available performance trial data.

Fast track options for variety registration at regional level rely on policies and communication practices that recognize prior registration in one member country and one season testing as sufficient for variety release in a second one. Similarly relevant are federal and state or provincial procedures that complement each other, enabling the benefits of new varieties to be realized across states or countries. Mechanisms are also needed whereby release proposal on the basis of data from state and farmers' field trials could be used for the zonal release of a variety, across an agro-ecological zone that covers more than one state or one country (Tripp et al. 1997). These same authors conclude that adjustment to seed regulatory frameworks is necessary because of significant changes in national seed systems, including: reductions in budget for public agricultural research; the failure of many seed parastatals; increasing concern about plant genetic diversity; pressure for the establishment of plant variety protection; the increasing contributions of commercial seed enterprises; and the emergence of innovative local level variety development and seed production initiatives.

A report of test cases in Africa supported by Syngenta documented the inclusion of three new potato varieties in the COMESA (Common Market for Eastern & Southern Africa) regional catalogue in 2016 (https://www.syngentafoundation.org/

sites/g/files/zhg576/f/seeds_policy_regional_variety_release_test_cases_317.pdf). Notably, these are the first non-maize varieties to be included in the region's list.

6.7 Concluding Remarks

Potato breeding is at a crossroad. Gene editing and large scale genotyping are at hand, and true hybrid breeding appears feasible, even for this complex polyploid and traditionally clonally propagated crop. Together with pre- and post-Mendelian methods, these advances promise to help accelerate genetic gain to nourish a growing population, help preserve the environment, and confront economic and climate constraints. Sustained breeding and supporting trait research will assure potato's role in a more comprehensive and integrated agriculture in which quality and resilience, in addition to the historical emphasis on yield, are major goals. Community efforts including the development of cross-species tools and research aimed at understanding the principles underlying crop performance will be needed to meet growing demands on agriculture. The programs illustrated here use crop ontologies to support understanding of traits and attempt to exploit heterosis, inter-ploidy breeding, and accelerated selection schemes to reach their goals. Beyond this, research on root systems, G × E interaction and photosynthetic efficiency merit attention. The lengthy 12–20 years required to develop and release a new potato variety, coupled with climate variability, call for intensified research on selection strategies and choice of breeding method. The use of marker-assisted and genomic selection, the possibilities offered by inbreeding coupled with gametophyte selection and hybrid development, "big data" and efficient real-time phenotypic data collection and analysis are most compelling at present. Finally, collaboration across sites and disciplines and the utilization of statistical models for assessing genotypic adaptation and breeding value are critical to the efficient development and deployment of better potato varieties.

Acknowledgements We gratefully acknowledge Elisa Mihovilovich and Thiago Mendes for valuable discussions toward this chapter.

References

Almekinders CJM, Chujoy E, Thiele G (2009) The use of true potato seed as pro-poor technology: the efforts of an international agricultural research institute to innovating potato production. Potato Res 52:275–293

Bae J, Halterman D, Jansky S (2008) Development of a molecular marker associated with Verticillium wilt resistance in diploid interspecific potato hybrids. Mol Breed 22:61–69

Bernardo R (2002) Breeding for quantitative traits in plants. Stemma Press, Woodbury

Bonierbale M, Plaisted RL, Tanksley S (1993) A test of the maximum heterozygosity hypothesis using molecular markers in tetraploid potatoes. Theor Appl Genet 86:481–491

Bonierbale M, Amoros W, Burgos G, Salas E, Suarez H (2007) Prospects for enhancing the nutritional value of potato by plant breeding. African Potato Association conference proceedings, Alexandria, Egypt, vol 7, pp 26–46

Butler DG, Cullis BR, Gilmour AR, Gogel BJ (2009) ASReml-R reference manual. v. 3. Queensland Department of Primary Industries & Fisheries, Brisbane

Chase SS (1963) Analytic breeding in *Solanum tuberosum* L.—a scheme utilizing parthenotes and other diploid stocks. Can J Genet Cytol 5:359–363

Colton L, Groza H, Wielgus S, Jiang J (2006) Marker-assisted selection for the broad-spectrum potato late blight resistance conferred by gene RB derived from a wild potato species. Crop Sci 46:589–594

Comai L (2005) The advantages and disadvantages of being polyploid. Nat Rev Genet 6:836–846

Comstock RE, Robinson HF (1948) The components of genetic variance in populations of biparental progenies and their use in estimating the average degree of dominance. Biometrics 4:254–266

Cubillos A, Plaisted R (1976) Heterosis for yield in hibrids between *S. tuberosum* spp. *Tuberosum* and *S. tuberosum* spp. *Andigena*. Am Potato J 53:143–150

De Jong H, Rowe PR (1971) Inbreeding in cultivated diploid potatoes. Potato Res 14:74–83

De Nettancourt D (1977) Incompatibility in angiosperms. Springer, Berlin

Endelman JB (2011) Ridge regression and other kernels for genomic selection with R package rrBLUP. Plant Genome 4:250–255

Evanno G, Regnaut S (2005) Detecting the number of clusters of individuals using the software STRUCTURE: a simulation study. Mol Ecol 14:2611–2620. https://doi.org/10.1111/j.1365-294X.2005.02553.x

Ewing EE (1992) Critical photoperiods for tuberization: a screening technique with potato cuttings. Am Potato J 55:43–53

Finkers-Tomczak A, Bakker E, de Boer J, van der Vossen E, Achenbach U, Golas T, Suryaningrat S, Smant G, Bakker J, Goverse A (2011) Comparative sequence analysis of the potato cyst nematode resistance to locus H1 reveals a major lack of co-linearity between three haplotypes in potato (*Solanum tuberosum* ssp.). Theor Appl Genet 122:595–608

Frankel R, Galun E (1977) Pollination mechanisms, reproduction, and plant breeding. Springer, New York

Fuglie K (2007) Priorities for potato research in developing countries: results of a survey. Am J Potato Res 84:353–365

Gaiero P, Mazzella C, Vilaró F, Speranza P, de Jong H (2017) Pairing analysis and in situ hibridisation reveal autopolyploid-like behaviour in *Solanum commersonii* × *S. tuberosum* (potato) interspecific hybrids. Euphytica 213:137. https://doi.org/10.1007/s10681-017-1922-4

Gallais A (2004) Quantitative genetics and selection theory in autopolyploid plants. INRA, Paris

Gálvez R, Mendoza HA, Fernández-Northcote EN (1992) Herencia de la inmunidad al virus Y de la papa (PVY) en clones derivados de *Solanum tuberosum* spp. Andigena. Fitopatología 27:8–15

Gardner CO, Heberhart SA (1966) Analysis and interpretation of the variety cross diallel and related populations. Biometrics 22:439–452

Gatto M, Hareau G, Pradel W, Suarez V, Qin J (2018) Release and adoption of improved potato varieties in Southeast and South Asia. International Potato Center (CIP) Lima, Peru, 42p. Social Sciences Working Paper No. 2018-2. ISBN 978-92-9060-501-0

Gebhardt C, Bellin D, Henselewski H, Lehmann W, Schwarzfischer J, Valkonen J (2006) Marker-assisted combination of major genes for pathogen resistance in potato. Theor Appl Genet 112:1458–1464

Golmirzaie A, Malagamba P, Pallais N (1994) Breeding potatoes based on true seed propagation. In: Bradshaw J, MacKay G (eds) Potato genetics. CAB International, Wallingford, pp 499–512

Griffing B (1956) Concepts of general and specific combining ability to relation to diallel crossing system. Anst J Biol Sci 9:463–493

Grüneberg WJ, Mwanga R, Andrade M, Espinoza J (2009) Chapter 5: Selection methods Part 5: Breeding clonally propagated crops. In: Ceccarelli S, Guimarães EP, Weltzien E (eds) Plant breeding and farmer participation. FAO, Rome, pp 275–322

Hareau G, Kleinwechter U, Pradel W, Suarez V, Okello J, Vikraman S (2014) Strategic assessment of research priorities for potato. CGIAR Research Program on Roots, Tubers and Bananas (RTB), Lima (Peru). RTB Working Paper 2014-8. Available online at: www.rtb.cgiar.org

Harlan IR, de Wet IMI (1971) A rational classification of cultivated plants. Taxon 20(4):509–517

Hayman BI (1954) The theory and analysis of diallel crosses. Genetics 39:789–809

Heffner EL, Lorenz AJ, Jannink J l, Sorrells ME (2010) Plant breeding with genomic selection: Gain per unit time and cost. Crop Sci 50(5):1681–1690

Hirsch C, Hisrch CD, Felcher K, Coombs J, Zarka D, VanDeynze A, DeJong W, Veilleux RE, Jansky S, Bethke P, Douches DS, Buell CR (2013) Retrospective view of North American potato (*Solanum tuberosum* L.) breeding in the 20th and 21st centuries. G3 Genes Genomes. Genetics 3:1003–1013

Hirsch CD, Hamilton J, Childs K, Cepela J, Crisovan E, Vaillancourt B, Hirsch CN, Habermann M, Neal B, Buell CR (2014) Spud DB: a resource for mining sequences, genotypes, and phenotypes to accelerate potato breeding. Plant Genome 7(1):1–12

Hosaka K, Hanneman RE (1998) Genetics of self-compatibility in a self-incompatible wild diploid potato species *Solanum chacoense*. 1. Detection of an S locus inhibitor (Sli) gene. Euphytica 99:191–197. https://doi.org/10.1023/A:1018353613431

Jansky SH, Chung YS, Kittipadukal P (2014) M6: a diploid potato inbred line for use in breeding and genetics research. J Plant Regist 8:195–199. https://doi.org/10.3198/jpr2013.05.0024crg

Jansky SH, Charkowski AO, Douches DS, Gusmini G, Richael C, Bethke PC, Spooner DM, Novy RG, De Jong H, De Jong WS, Bamberg JB, Thompson AL, Bizimungu B, Holm DG, Brown CR, Haynes KG, Sathuvalli VR, Veilleux RE, Miller JC, Bradeen JM, Jiang JM (2016) Reinventing potato as a diploid inbred line–based crop. Crop Sci 56:1–11

Kasai K, Morikawa Y, Sorri VA, Valkonen JP, Gebhardt C, Watanabe KN (2000) Development of SCAR markers to the PVY resistance gene Ryadg based on a common feature of plant disease resistance genes. Genome 43:1–8

Kempthorne O, Curnow RN (1961) The partial diallel cross. Biometrics 17:229–250

Kidane-Mariam HM, Mendoza H, Wissar R (1985) Performances of true potato seed families derived from intermating tetraploid parental lines. Am Potato J 62:643–652. https://doi.org/10.1007/BF02853473

Kleinwechter U, Hareau G, Suarez V (2014) Prioritization of options for potato research for development—results from a global expert survey. CGIAR Research Program on Roots, Tubers and Bananas (RTB), Lima (Peru). RTB Working Paper 2014-7. Available online at: www.rtb.cgiar.org\

Levin DA (1983) Polyploidy and novelty in flowering plants. Am Nat 122:1–25

Lynch M, Walsh B (1998) Genetics and analysis of quantitative traits. Sinauer Associates, Sunderland

Meenakshi JV, Johnson N, Manyong VM, De Groote H, Javelosa J, Yanggen D, Naher F, Gonzalez C, Garcia J, Meng E (2007). How cost-effective is biofortification in combating micronutrient malnutrition? An ex-ante assessment HarvestPlus Working Paper No.2/. HarvestPlus, Washington, DC. www.harvestplus.org

Mendiburu AO, Peloquin SJ (1977) Bilateral sexual polyploidization in potatoes. Euphytica 26:573–583

Mendoza HA (1990) Mejoramiento poblacional. Una estrategia para la utilización del germoplasma de papa. En: Mejoramiento de papa en el Cono Sur, pp 47–62

Mendoza HA, Estrada NR (1979) Breeding potatoes for tolerance to stress: heat and frost. In: Mussel H, Staples RR (eds) Stress physiology in crop plants. Wiley, New York, pp 227–262

Mendoza HA, Sawyer RL (1985) The breeding program at the International Potato Center (CIP). In: Russell GE (ed). Progress in plant breeding, 1. London, Butterworths, 325 p

Mendoza HA, Mihovilovich EJ, Saguma F (1996) Identification of triplex (YYY) potato virus Y (PVY) immune progenitor derived from *Solanum tuberosum* spp. Andigena. Am Potato J 73(1):13–19. ISSN 0003-0589

Midmore DJ, Rhoades RE (1988) Applications of agrometeorology to the production of potato in the warm tropics. In: Rijks D, Stigter CJ (eds) Agrometeorology of the potato crop. International Society for Horticultural Science (ISHS); International Potato Center (CIP); World Meteorological Organization (WMO); European Association for Potato Research; Royal Netherlands Meteorological Institute, Wageningen, pp 103–136. Acta Horticulturae. no. 214

Mihovilovich E, Alarcón L, Pérez AL, Alvarado J, Arellano C, Bonierbale M (2007) High levels of heritable resistance to potato leafroll virus (PLRV) in *Solanum tuberosum* subsp. Andigena. Crop Sci 47(3):1091–1103

Mihovilovich E, Aponte M, Linqvist-Kreuze H, Bonierbale M (2014) An RGA-derived SCAR marker linked to PLRV resistance from *Solanum tuberosum* ssp. Andigena. Plant Mol Biol Report 32:117–128. https://doi.org/10.1007/s11105-013-0629-5

Muñoz FJ, Plaisted RL, Thurston HD (1975) Resistance to potato virus Y in *S. tuberosum* spp. Andigena. Am Portao J 52:107–115

Oakey H, Verbyla AP, Cullis BR, Wei XM, Pitchford WS (2007) Joint modeling of additive and non-additive (genetic line) effects in multi-environment trials. Theor Appl Genet 114:1319–1332. https://doi.org/10.1007/s00122-007-0515-3

Ortiz R (1998) Potato breeding via ploidy manipulations. Plant Breed Rev 16:15–85

Paget M, Amoros W, Salas E, Eyzaguirre R, Alspach P, Apiolaza L, Noble A, Bonierbale M (2014) Genetic evaluation of micronutrient traits in diploid potato from a base population of Andean Landrace Cultivars. Crop Sci 54:1949–1959

Phumichai C, Hosaka K (2006) Cryptic improvement for fertility by continuous selfing of diploid potatoes using Sli gene. Euphytica 149:251–258. https://doi.org/10.1007/s10681-005-9072-5

Piepho H-P, Möhring J, Melchinger AE, Büchse A (2008) Blup for phenotypic selection in plant breeding. Euphytica 161:209–222

Plaisted RL (1987) Advances and limitations in the utilization of Neotuberosum in potato breeding. In: Jellis GJ, Richardson DE (eds) The production of new potato varieties. Cambridge University Press, Cambridge, pp 186–196

Plaisted RL, Sanford L, Federer WT, Kehr AE, Peterson LC (1962) Specific and general combining ability for yield in potatoes. Am Potato J 39:185–197

Pushkarnath P (1942) Studies on sterility in potatoes. 1. The genetics of self- and cross-incompatibilities. Indian J Genet and Pl Br 2:11–36

Radley RW, Taha MA, Bremner PM (1961) Tuber bulking in the potato crop. Nature 191:782–783

Ragot M, Bonierbale M, Weltzien E (2018) From market demand to breeding decisions—a framework. RTB Working Paper.

Rodriguez Junior FA, Ortiz-Monasterio I, Zarco-Tejada PJ, Ammar K, Gérard B (2014) Using precision agriculture and remote sensing techniques to improve genotype selection in a breeding program. Conference Paper.International Conference on Precision Agriculture. http://quantalab.ias.csic.es/pdf/Rodrigues%20Jr.%20et%20al.%202014.pdf

Salas E (2002) Estudio de la variabilidad genética de la resistencia al Virus del Enrollamiento de la papa (PLRV) en una población tetraploide. Tesis para optar el Grado de Magister Scientiae. Universidad Nacional Agraria—La Molina. Lima, Peru

Sanford JC, Hanneman RE (1982) Intermating of potato haploids and spontaneous sexual polyploidization—effects on heterozygosity. Am J Potato Res 59:407–414

Sattarzadeh A, Achenback U, Lubeck J, Strahwald J, Tacke E, Hofferbvert H, Rothsteyn T, Gebhardt C (2006) Single nucleotide polymorphism (SNP) genotyping as basis for developing a PCR-based marker highly diagnostic for potato varieties with high resistance to Globodera pallida pathotype Pa2/3. Mol Breed 18:301–312

Slater AT, Wilson GM, Cogan NO, Forster JW, Hayes BJ (2014) Improving the analysis of low heritability complex traits for enhanced genetic gain in potato. Theor Appl Genet 127(4):809–820

Spooner DM, Ghislain M, Simon R, Jansky SH, Gavrilenko T (2014) Systematics, diversity, genetics, and evolution of wild and cultivated potatoes. Bot Rev 80:283–383. https://doi.org/10.1007/s12229-014-9146. —Toward a Rational Classification of Cultivated Plants—Jstor https://www.jstor.org/stable/1218252

Sprague GF, Tatum A (1941) General vs. specific combining ability in single crosses of corn. J Am Soc Agron 34:923–932

Tripp R, Louwaars NP, Van Der Burg WJ, Virk DS, Witcombe JR (1997) Alternatives for Seed regulatory reform: an analysis of variety testing, variety regulation and seed quality control, Agricultural Research and Extension Network Paper, no. 69. http://hdl.handle.net/10535/4581

Ugent DL, Pozorski S, Pozorski T (1982) Archaeological potato tuber remains from the Casma Valley of Peru. Econ Bot 36:182–192

Ugent DL, Dillehay T, Ramirez C (1987) Potato remains from a Late Pleistocene settlement in southcentral Chile. Econ Bot 41:17–27

Uitdewilligen JGAML, Wolters A-MA, D'hoop BB, Borm TJA, Visser RGF, van Eck HJ (2013) A next-generation sequencing method for genotyping-by-sequencing of highly heterozygous autotetraploid potato. PLoS One 8(5):e62355. https://doi.org/10.1371/journal.pone.0062355

Velásquez AC, Mihovilovich E, Bonierbale M (2007) Genetic characterization and mapping of major gene resistance to potato leafroll virus in *Solanum tuberosum* ssp. *Andigena*. Theor Appl Genet 114(6):1051–1058

De Vries MI, ter Maat M, Pim Lindhout P (2016) The potential of hybrid potato for East-Africa. Open Agric. 1: 151–156

Wricke G, Weber WE (1986) Quantitative genetics and selection in plant breeding. de Gruyter, Berlin

Xu X et al (2011) Genome sequence and analysis of the tuber crop potato. Nature 475(7355):189–195

Yuen J, Forbes G (2009) Estimating the level of susceptibility to *Phytophthora infestans* in potato genotypes. The American Phytopathological Society. Phytopathology 99(6):2

Zhang LH, Mojtahedi H, Kuang H, Baker B, Brown CR (2007) Marker-assisted selection of Columbia root-knot nematode resistance introgressed from *Solanum bulbocastanum*. Crop Sci 47:2021–2026

7

Potato Genetics and Cytogenetics: Germplasm Enhancement and Breeding

Rodomiro Ortiz and Elisa Mihovilovich

Abstract Tetraploid potato (*Solanum tuberosum* L.) is a genetically complex, polysomic tetraploid ($2n = 4x = 48$), highly heterozygous crop, which makes genetic research and utilization of potato wild relatives in breeding difficult. Notwithstanding, the potato reference genome, transcriptome, resequencing, and single nucleotide polymorphism (SNP) genotyping analysis provide new means for increasing the understanding of potato genetics and cytogenetics. An alternative approach based on the use of haploids ($2n = 2x = 24$) produced from tetraploid *S. tuberosum* along with available genomic tools have also provided means to get insights into natural mechanisms that take place within the genetic load and chromosomal architecture of tetraploid potatoes. This chapter gives an overview of potato genetic and cytogenetic research relevant to germplasm enhancement and breeding. The reader will encounter findings that open new doors to explore inbred line breeding in potato and strategic roads to access the diversity across the polyploid series of this crop's genetic resources. The text includes classical concepts and explains the foundations of potato genetics and mechanisms underlying natural cytogenetics phenomena as well as their breeding applications. Hopefully, this chapter will encourage further research that will lead to successfully develop broad-based potato breeding populations and derive highly heterozygous cultivars that meet the demands of having a resilient crop addressing the threats brought by climate change.

7.1 Introduction

The most grown potatoes are tetraploid ($2n = 4x = 48$), but farmers in the Andes grow diploid ($2n = 2x = 24$), triploid ($2n = 3x = 36$), and pentaploid ($2n = 5x = 60$) cultivars (Watanabe 2015). The basic chromosome number of these tuber-bearing

R. Ortiz (✉)
Swedish University of Agricultural Sciences (SLU), Alnarp, Sweden
e-mail: rodomiro.ortiz@slu.se

E. Mihovilovich
Independent Consultant, Lima, Peru

Solanum species is 12. Diploid cultivars along with tetraploid cultivars are used in potato breeding through ploidy manipulations with haploids and $2n$ gametes (or gametes with the sporophytic chromosome number), and chromosome engineering using aneuploidy (Ortiz 1998). Improvement of cultivated potato is challenged by its high heterozygosity (Bradshaw et al. 2006; Hirsch et al. 2013) and complex polysomic tetraploid inheritance (Howard 1970; Ortiz and Peloquin 1994; Ortiz and Watanabe 2004). The tetraploid *Solanum tuberosum* has four homologues, which include 12 unique chromosomes each, thus showing tetrasomic inheritance (Bradshaw 2007). Epistasis and heterozygosity are key for succeeding in $4x$ potato breeding because multi-allelic quantitative trait loci showing high-order genic interactions, while additivity also contributes to quantitative traits with high heritability.

The genetics of tetrasomic potato depends on four sets of homologous chromosomes (instead of two as in diploid potato). Three genotypes (*AA*, *Aa*, *aa*) are expected after selfing a heterozygous diploid (*Aa*), while the selfing of a comparable tetraploid (*AAaa*) gives five different genotypic classes in its offspring: *AAAA* (quadruplex), *AAAa* (triplex), *AAaa* (duplex), *Aaaa* (simplex), and *aaaa* (nulliplex). Double reduction—related to tetrasomic inheritance—occurs when two chromosomes in a gamete derive from two sister chromatids; i.e., the sister chromatids end in same gamete. Quadrivalent formation, a single crossing over between the centromere and the locus to allow sister chromatids to attach to two different centromeres, that these centromeres with sister chromatids move to the same pole in anaphase I, and sister chromatids go the same pole in anaphase II are necessary for double reduction. The probability of double reduction to occur is noted as α, which is equalto $\frac{qea}{2}$, where q is the quadrivalent frequency, e is the frequency of equational separation that depends on the gene–centromere map distance, and a is the frequency of non-disjunction (often $\frac{1}{3}$). Chromosome segregation arises when $\alpha \neq 0$,thus indicating that a locus of interest lies close to the centromere; while if $\alpha = \frac{1}{7}$ or $\frac{1}{6}$ then chromatid segregation or maximal equational division (MED), respectively, occurs. MED is rarely found because of the requirements for its occurrence (Burham 1984). DNA-aided marker analysis confirmed the occurrence of double reduction and that it increases with distance from the centromeres (Bourke et al. 2015). As noted by Gálvez et al. (2017), the potato reference genome and transcriptome, research on both gene expression and regulatory motif, plus resequencing and SNP genotyping analyses, provide new means for increasing the understanding of potato genetics. For example, DNA resequencing allows assembling a genome reference for each cultivar or landrace, which may provide useful knowledge regarding structural differences between the various potato groups. In this regard, the resequencing of diversity panel including wild species, landraces, and cultivars demonstrated that a limited gene set accounts for early improvement of the potato cultigen, while distinct loci seems to be involved on the adaptation *S. tuberosum* group Andigenum (upland potato) and *S. tuberosum* groups Chilotanum and Tuberosum (lowland potato) populations (Hardigan et al. 2017). Signatures of selection in genes regulating pollen development/gametogenesis reduced fertility. Introgression of truncated alleles of wild species, particularly *S. microdontum*, was noted in long day cultivars, thus showing how wild tuber-bearing *Solanum* species are key sources of variation for breeding.

7.2 Haploids and Disomic Inheritance

Tetraploid potato shows significant inbreeding depression (De Jong and Rowe 1971) because polyploidy and heterozygosity mask deleterious recessive mutations and buffer genomic imbalance (Comai 2005; Henry et al. 2010; Tsai et al. 2013). These characteristics led to an alternative breeding approach based on the use of haploids ($2n = 2x = 24$) produced from tetraploid *S. tuberosum*. The homozygosity/heterozygosity of methylated DNA may be, however, involved in inbreeding depression/heterosis in self-compatible diploid potatoes because DNA methylation may suppress gene expression (Nakamura and Hosaka 2010).

The induction of haploid plants is generally referred to as "haploidization." There are two main pathways by which haploid formation can be induced in potato: androgenesis and gynogenesis. Androgenesis is through in vitro culturing of whole anthers or free microspores on a nutrient rich medium to induce plantlet regeneration from single gametic cells or haploid calli (Veilleux 1996), while gynogenesis is haploidization via the "maternal" or seed parent's genome. Potato haploids are routinely obtained by gynogenesis, a process in which specific *S. tuberosum* Group Phureja ($2n = 2x = 24$) selections, known as "haploid inducers," contribute the paternal gametes for pollination of the desired haploid progenitor. The formation of a haploid embryo begins when the egg is either induced into parthenogenesis (Hermsen and Verdenius 1973) or when the zygote experiences spontaneous abortion of the pollen donor's set of chromosomes (Clulow et al. 1991). Evidence for the latter has been the identification of Phureja-specific molecular markers in aneuploids ($2n = 2x + 1 = 25$) among the offspring of some haploid induction crosses (Clulow et al. 1991; Clulow and Rousselle-Bourgeois 1997; Samitsu and Hosaka 2002; Straadt and Rasmussen 2003; Ercolano et al. 2004). Ortiz et al. (1993a) suggested that the genetics of the ability to induce haploids is relatively simple.

Putative haploids are identified firstly by the lack of a dominant morphological marker for anthocyanin pigmentation on developing embryos (embryo spots) or on seedling shoots (nodal bands). This marker that allows early haploid selection is present in homozygosis in certain pure *S. tuberosum* Group Phureja clones or has been bred to homozygosity in its derived hybrids. The ploidy of the resultant seedlings is confirmed by counting chromosomes in mitotically dividing root cells (Sopory 1977), counting chloroplasts in stomatal guard cell pairs (Singsit and Veilleux 1991) or through flow cytometric analysis (Owen et al. 1988). The haploid-inducing clones often used owing to their relatively superior haploid-induction frequency and homozygosity for the seed marker "embryo-spot" are the following

Table 7.1 Haploid induction ability of "IVP 35," "IVP 101," and "PL-4"

Character[a]	"IVP 35"	"IVP 101"	"PL-4"
Number of haploids per 100 berries	69	62	96
Number of haploids per 1000 seeds	52	85	103

[a]Pooled data per haploid inducer from 13 seed parents

Group Phureja clones: "IVP 35," "IVP 48", and "IVP 101" (Ross 1986). "IVP 101" has been derived from the cross [(G609 × "IVP 48") × ("IVP 10" × "IVP 1")]. G609 is a haploid from Group Tuberosum cultivar "Gineke" that combines its own haploid induction ability with a high degree of male fertility, profuse flowering, and vigor.

The efficiency of haploid production is determined by both, production ability of the tetraploid seed parent and induction ability of the diploid pollinator (Hougas et al. 1964; Frandsen 1967; Hermsen and Verdenius 1973). However, interaction between seed parents and pollinators were also noted (Frandsen 1967). Despite this interaction, haploid induction ability of "IVP 101" has proved to be higher than "IVP 35" and "IVP 48" (Hutten et al. 1993). By the end of the 1990s a promising haploid inducer named "PL-4" (CIP596131.4) was selected at the International Potato Center (CIP, Lima, Perú) as a transgressive genotype from the cross between "IVP 35" × "IVP 101" due to its highest haploid inducer ability, degree of flowering, shedding and pollen viability relative to its parents (M. Upadhya and R. Cabello, CIP, unpublished data). Historical data accumulated from 2001 to 2009 from haploid induction crosses between 37, 4*x* breeding lines with both "IVP 101" and "PL-4" showed that the latter produced twice the amount of seeds without embryo spot (putative haploids) of "IVP 101." A more comprehensive study to determine the haploid inducer ability of "PL-4" relative to their parents was performed during 2015 and 2016 at CIP. This involved haploid induction of 13, 4*x* breeding clones with the three haploid inducers. Haploid confirmation was made by counting chloroplasts in stomatal guard cell pairs and flow cytometric analysis in seedlings grown from seeds without embryo spot. "PL-4" produced a significantly higher number of haploids than its parents. Meanwhile, "IVP 101" outperformed "IVP 35" in number of haploid per 1000 seeds (Table 7.1). There were also differences in haploid production ability between seed parents (Table 7.2). Two breeding clones, CIP 300056.33 and CIP 392820.1, showed the highest number of haploids.

7.2.1 Further Research and New Evidence on Haploid Origin

Previous research in potato haploids originated by gynogenesis detected aneuploids ($2n = 2x + 1 = 25$ and $2n = 2x + 2 = 26$) instead of the expected 24-chromosome karyotypes with concurrent appearance of Group Phureja-specific molecular markers (Clulow et al. 1991, 1993; Waugh et al. 1992; Wilkinson et al. 1995; Clulow and Rousselle-Bourgeois 1997; Ercolano et al. 2004). Moreover, there was one case in which translocation of a Group Phureja chromosomal segment to the Group Tuberosum genome was detected by genomic in situ hybridization (GISH; Wilkinson

Table 7.2 Haploid production ability of 13, 4*x* breeding clones from CIP potato breeding program

CIP-number[a]	Breeding code	Number of haploids per 100 berries	Number of haploids per 1000 seeds
CIP 300056.33	LR00.014	141	169
CIP 300072.1	LR00.022	68	101
CIP 300093.15	LR00.027	82	58
CIP 301023.15	C01.020	40	63
CIP 388615.22	C91.640	75	141
CIP 388676.1	Y84.027	40	32
CIP 390478.9	C90.170	2	2
CIP 390637.1	93	76	107
CIP 391931.1	458	11	17
CIP 392780.1	C92.172	9	21
CIP 392820.1	C93.154	258	175
CIP 397073.16	WA.104	39	30
CIP 397077.16	WA.077	100	68
Average		72	76

[a]Pooled data per breeding clone from three haploid inducers

et al. 1995). The cause, frequency, and nature of these introgressions remain unknown, though they may affect performance of the haploids (Allainguillaume et al. 1997). A similar phenomenon has been observed in CenH3-based haploid induction in *Arabidopsis thaliana* (Ravi and Chan 2010; Ravi et al. 2014; Tan et al. 2015). In this system, mis-segregation of the haploid inducer chromosomes leads to genome elimination. In a fraction of the haploid progeny, one or few of the haploid inducer chromosomes were retained, resulting in aneuploid progeny. This DNA introgression was identified readily by low-pass sequencing and single nucleotide polymorphism (SNP) analysis (Tan et al. 2015). These findings added evidence that DNA introgression from a haploid inducer is expected to involve large contiguous segments, and often whole chromosomes. Lately, K.R. Amundson et al. (unpublished) surveyed a haploid segregating population for aneuploids by low-pass sequencing. The population was previously developed at CIP for tetraploid genetic mapping of a major gene controlling *Potato leaf roll virus* (PLRV) resistance in the Group Andigena cultivar "Alca Tarma" (Velásquez et al. 2007). They identified 19 haploids (11.4%) that displayed elevated relative sequence read coverage of a single chromosome consistent with 25-chromosome karyotypes in root tip metaphase spreads in these putatively aneuploid clones. By sequencing parental genotypes to higher depth (40–66×) and identifying homozygous SNP between "Alca Tarma" and either of the two haploid inducers, "IvP-101" or "PL-4," plus assuming to have sired each haploid (Velásquez et al. 2007), they found nearly 0% haploid inducer

SNP for all chromosomes of these aneuploids. Thus, the additional chromosomes observed likely did not originate from the haploid inducer genome, but were maternally inherited. This lack of Group Phureja SNP in haploids was previously reported in another study that employed DNA markers (Samitsu and Hosaka 2002). The production of aneuploid gametes is a common property of polysomic polyploids (Comai 2005), and "Alca Tarma" was not an exception. Admusson et al. (unpublished) concluded that for haploid inducers "IvP-101" and "PL-4," either the mechanism of haploid induction does not involve egg fertilization or genome elimination in "Alca Tarma" was very efficient.

7.3 Relevance of Haploids in Plant Breeding and Genetics

Haploids showed disomic inheritance, which means that each chromosome paired with its homolog, thus providing means for simplifying genetic research in potato. They can also be efficiently used for research on chromosome pairing and natural mutation accumulated at the tetraploid level. Initially, potato haploids were envisioned as a tool to simplify the breeding of *S. tuberosum* cultivar production by reducing tetraploid germplasm to a diploid breeding level (Chase 1963). A second early reason for the production of haploids was to acquire a "genetic bridge" between the various genomes of *Solanum* species. Ploidy barriers between the cultivated and wild *Solanum* species could be circumvented by crossing haploids to the wild diploid *Solanum spp.* and novel hybrid germplasm incorporated back into tetraploid breeding programs through $4x \times 2x$ crosses using $2n$ gamete formation, or by colchicine-doubling of the novel diploid hybrid (Ross 1986). In addition to their use in breeding, diploid potato hybrids represent a powerful tool for genetic analysis due to its much simpler segregation ratios compared to tetraploid cultivated potatoes (Ortiz and Peloquin 1994). Thus, diploid potato has been used to determine the inheritance of economically important traits such as tuber shape (De Jong and Burns 1993; Van Eck et al. 1994b), tuber flesh and skin pigmentation (De Jong 1987; Van Eck et al. 1994a), and tuber skin texture (De Jong 1981). The genetic basis of some physiological mutants has also been analyzed with the use of diploids (De Jong et al. 1998, 2001). Haploids have been convenient for trait mapping (Kotch et al. 1992; Pineda et al. 1993; Freyre et al. 1994; Simko et al. 1999; Naess et al. 2000; Velásquez et al. 2007) and in the development of an online catalogue of amplified fragment length polymorphisms (AFLP) covering the potato genome (Rouppe van der Voort et al. 1998). On the other hand, many breeders have extracted haploids from superior parents and also maintain a diploid gene pool composed of hybrids between haploids and diploid wild species carrying specific quality and host plant resistance genes not found in cultivars (Carputo and Barone 2005; Ortiz et al. 2009).

Currently, there are some ongoing efforts towards genetically restructuring potato as a diploid inbred line-based crop (Jansky et al., 2016; Lindhout et al. 2011). Here, the vision is a diploid potato crop composed of a series of inbred lines that capture the favorable genetic diversity available in the potato cultigen. This diploid genepool with a broad suite of traits represent a valuable stock for fixing desirable gene combinations and realized breeding gains. Last but not least, haploids have been regarded as a tractable ploidy state for copy number variation (CNV) analysis in tetraploid potatoes as these variants are meiotically transmissible and diploid gametes can likely shield deficiencies such as recessive lethal and dosage sensitive loci (Comai 2005; Lovene et al. 2013; Henry et al. 2015). CNV is defined as stretches of DNA from 1 kilobase (kb) to several megabases (Mb) that display different copy numbers in populations (Feuk et al. 2006). Analyses of multiple genotypes in *Arabidopsis* and maize suggest that CNV may play a significant role in phenotypic diversity and hybrid heterosis in plant species (Swanson-Wagner et al. 2010; Cao et al. 2011). Moreover, they can affect phenotype impacting important agronomic and host plant resistance traits (Maron et al. 2013; Díaz et al. 2012; Zhu et al. 2014; Cook et al. 2012). Direct CNV detection in tetraploid potato is challenging. The dosage increase associated with a duplication is subtler in tetraploids (25% increase) than in diploids (50% increase), and high heterozygosity impedes haplotype assembly (Potato Genome Sequencing Consortium 2011). As a consequence, CNV analysis is often limited to few loci of interest and is costly to be practical in breeding programs. Hence, sampling the gametophyte genome of tetraploids by ploidy reduction through haploidy is an alternative approach.

7.4 2*n* Gametes

Gametes with the sporophytic chromosome number should be named as 2*n* gametes and not as "unreduced" gametes as wrongly dubbed. They result from pre-meiotic, meiotic, or post-meiotic abnormalities during gametogenesis. The modes of formation are pre-meiotic doubling, first division restitution (FDR), chromosome replication during meiotic interphase, second division restitution (SDR), post-meiotic doubling, and apospory (diploid sac formed from nucellus or integument cell). FDR and SDR mechanisms are the most common for 2*n* pollen and 2*n* egg formation in potato (Ortiz 1998). Heterozygous 2*x* parents transmit 80% and 40% of their heterozygosity to their 4*x* hybrid offspring after sexual polyploidization with FDR or SDR 2*n* gametes, respectively.

The parallel orientation of the spindles in the second meiotic division accounts frequently for FDR 2*n* pollen, while omission of the second division after a normal first division seems to be often involved in SDR 2*n* eggs. The abnormal meiosis leading to these 2*n* gametes are under the genetic control of recessive mutants: *ps* for 2*n* pollen and *os* for 2*n* egg, both of which appear to be ubiquitous in *Solanum* species. The finding of genes whose mutations led to a high frequency of 2*n* gametes in the model plant species *Arabidopsis thaliana* provided further means for understanding their formation in plants (Brownfield and Kölher 2011). It appears to be very likely that a mechanism related to a loss of protein function leads to the formation of 2*n* gametes.

The frequency of $2n$ gametes may be affected by incomplete penetrance and variable expressivity, which is under minor modifier genes and influenced by plant age and the environment. Phenotypic recurrent selection could be effective for increasing the frequency of FDR $2n$ pollen, while recurrent selection with progeny testing may raise SDR $2n$ egg expressivity.

There are synaptic mutants affecting gametogenesis in haploids, *Solanum* species and haploid-species hybrids. They may cause poor pairing, reduced chiasma, or both, thus reducing recombination. For example, the synaptic mutant sy_3—found in Group Phureja–haploid hybrids—along with *ps* produces FDR $2n$ pollen without crossing over (FDR-NCO), while the desynaptic mutant *ds-1* generates sterile *n* eggs and fertile FDR $2n$ eggs owing to a direct equational division of univalent chromosomes at anaphase I, i.e., pseudohomotypic division. Desynaptic gametes may transfer about 95% of the $2x$ genotype to their $4x$ hybrid offspring.

7.5 Cytoplasm Diversity and Male Sterility

There are six distinct cytoplasmic genome types in potato, namely, M, P, A, W, T, and D. Many clones bred worldwide show a genetic bottleneck in cytoplasmic diversity due to the continuous use of cytoplasmic-based male sterility "lineages" derived from *S. demissum* or *S. stoloniferum* that are often used as sources of host plant resistance. For example, T (45%), D (38%), and W (11%) are the most frequent types in CIP bred germplasm (Mihovilovich et al. 2015); while the most popular among EU cultivars and breeding clones are T (59%), D (27%), and W (12%) (Sanetomo and Gebhardt 2015), and cultivars and breeding lines from Japan plus a sample of landraces and foreign cultivars show 73.9% T, 17.4% D, and 2.4% W (Hosaka and Sanetomo 2012).

Cytoplasmic factors and nuclear alleles are involved in indehiscence, shriveled microspores, sporad formation, anther-style fusion, ventral-styled anthers, and thin anthers (Grun et al. 1977). Group Andigena and its ancestors, Group Stenotomum and Phureja, share most plasmon factors, of which many differ from those found in Group Tuberosum. Hence, cytoplasmic-genetic male sterility is often noted in hybrid offspring among some tuber-bearing *Solanum* species because interactions between sensitive factors in the cytoplasm of one species and nuclear genes from the other species. Hybrids derived from crossing Group Tuberosum haploids as female and Group Phureja or Stenotomun as males are very often male sterile, but the reciprocal cross show male fertile offspring. Male sterility ensues from the interaction of a dominant gene (*Ms*) from the Group Phureja or Stenotomun with Group Tuberosum sensitive cytoplasm. Diploid (2EBN) wild species such as *S. chacoense, S. berthaultii,* and *S. tarijense* do not carry genes that interact with the Group Tuberosum cytoplasm as shown by the high male fertile hybrid offspring between them.

The frequency of male fertile offspring in hybrids between Group Tuberosum and groups Stenotomum or Phureja may vary because some tetraploid cultivars bear a dominant male fertility restorer (*Rt*) gene (Iwanaga et al. 1991b). The *Ms* and *Rt* genes, which are independently inherited, are very distal from the centromere, thus showing both loci chromatid segregation (Ortiz et al. 1993b). The *Rt* gene allows to partially bypass male sterility using crossbreeding, e.g. by crossing Group Tuberosum haploids bearing *Rt* with 2× species carrying *Ms* because ½ (*rt*/*rt* × *Ms*/*ms*) ¾ (*Rt*/*rt* × *Ms*/*ms*) or 100% (*Rt*/*Rt* × *Ms*/*Ms* or *Rt*/*Rt* × *Ms*/*ms*) of the resulting hybrid offspring will be male fertile.

7.6 Self-Incompatibility and *s* Locus Inhibitor Mechanism

Diploid potatoes and their related wild tuber-bearing *Solanum* species are self-incompatible due to a gametophytic self-incompatibility (GSI) system controlled by the interaction of a pollen *S* gene with pistil *S* gene(s). GSI inhibits fertilization by self-pollen or pollen from closely related (sibling) plants (Hanneman Jr 1999). In this system, compatibility is controlled by the *S* locus, which consists of two closely linked genes: *S-RNase* and *S-Locus F-box* (*SLF/SFB*) that control the female and male specificity, respectively. S-RNase-based self-incompatibility systems are widespread mechanisms for controlling selfing (Hancock et al. 2003). *S* locus variants, currently known as S-haplotypes, determine self-incompatible pollen rejection when there is a match between the single S-haplotype in the haploid pollen and either of the two haplotypes in the diploid system. S-RNases are the determinants of *S*-specificity in the pistil and act in pollen recognition as well as in direct pollen growth inhibition by degrading pollen RNA in incompatible pollinations (McClure et al. 1990). A distinct S-RNase protein is expressed from each functional S-haplotype and upon recognition the protein enters intact pollen tubes retaining its potentially cytotoxic enzyme activity (Gray et al. 1991). Conversely, pollen RNA is stable in compatible pollinations as interaction between S-RNase and *SLF* confers resistance to the cytotoxic effects of S-RNase (Fig. 7.1; Golz et al. 2001). The *SLF/SFB* is a family of F-box protein genes whose most well-known role is ubiquitin-mediated protein degradation. Pollen modifier genes encoding proteins that form complexes with SLF provides for ubiquitylation and degradation of non-self S-RNase as a necessary step to overcome the cytotoxicity of S-RNase. On the other hand, self S-RNases fail to bind productively and thus escape degradation (Hua et al. 2008; Zhang et al. 2009).

Modifier genes encoding putative pistil self-incompatibility factors were found in potato wild relatives. *HT-A* and *HT-B* are two similar genes expressed in the genus *Solanum*, but only *HT-B* has shown to be strongly suppressed in self-incompatibility breakdown in *S. chacoense*. HT-B proteins appear to be degraded in pollen tubes after compatible pollination, while this is not the case in incompatible pollen tubes where substantial amounts of HT-B reactive protein were found (Fig. 7.1; Goldraij et al. 2006).

Self-compatibility can be obtained by converting a self-incompatibility diploid (e.g., S_1S_2) to a tetraploid (de Nettancourt 1977). Here the defect occurs only in the pollen due to the so-called heteroallelic pollen (HAP) effect. Thus, $S_1S_1S_2S_2$ pistils

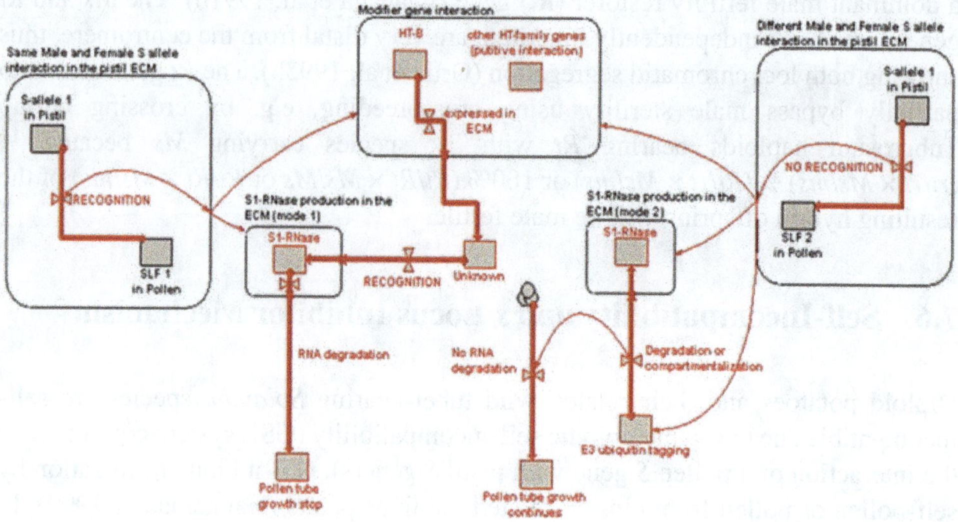

Fig. 7.1 A model for S-RNase-based self-incompatibility. When the pollen tube enters the extracellular matrix (ECM) the proteins HT-B, S-RNase, and 120K (not represented) are expressed. On a positive recognition of self-pollen (or a self *SLF* gene) the pollen tube RNA is broken down by the production of S-RNase and fertilization is unlikely. In the opposite reaction, when an unrecognizable match is made, S-RNase is degraded after being tagged by an E3 ubiquitin ligase complex. (Courtesy: Dr. Philippe Kear, International Potato Center)

reject S_1- and S_2-pollen normally, but diploid pollen is not rejected. Self-incompatibility breakdown only occurs in the HAP case, S_1S_2. Pollen *S* functions to provide resistance to S-RNase. Self-compatible variants have often been described among genotypes of self-incompatible potato species, such as, *inter alia*, *S. chacoense*, *S. kurtzianum*, *S. neohawkesii*, *S. pinnatisectum*, *S. raphanifolium*, *S. sanctaerosae*, *S. tuberosum* Groups Phureja and Stenotomum (Cipar et al. 1964), and *S. verrucosum* (Eijlander 1998). SC variants were also noted in haploids ($2n = 2x = 24$ chromosomes) from Group Tuberosum (De Jong and Rowe 1971; Olsder and Hermsen 1976) and in hybrids between Group Phureja and haploids ($2n = 2x = 24$ chromosomes) of Group Andigenum (Cipar 1964).

Self-compatible variants in *S. chacoense* show a single dominant gene "*Sli*" that is expressed in a sporophytic fashion (Hanneman 1985; Hosaka and Hanneman 1998a). The existence of self-incompatible progeny segregating from *S. chacoense* selfed plants having *Sli* gene in a heterozygous condition shows that fully functional S-haplotypes are transmitted through pollen even when the *Sli* factor is not. *Sli* was mapped to the end of chromosome 12 (Hosaka and Hanneman 1998b). Since the *S* locus has been localized on chromosome 1 (Gebhardt et al. 1991; Jacobs et al. 1995; Rivard et al. 1996), it is evident that the *Sli* gene is independent of the *S* locus. *Sli* can be regarded as a dominant gain-of-function (GOF) pollen-part mutant (PPM) that interacts in some way with the GIS system in pollen and results in self-compatible plants. This pollen side gene may inhibit S-RNase uptake, break down the pollen-stigma recognition system, overcome the cytotoxic activity of S-RNase independent of its pollen *S*-genotype or through the action on a non-S-specific factor (Hosaka and Hanneman 1998b; McClure et al. 2011). Likewise, self-compatible variants in *S. verrucosum* (*ver*) has been assumed to be a pistil side nonfunctional S-RNase haplotype (*Sv* allele) that allows *ver* plants be pollinated by its own pollen as well by the pollen of other self-incompatible potato species (Eijlander 1998). Absence of pistillate S-RNases seems to be a characteristic feature of this species (Makhan'ko 2011).

7.6.1 *Self-Compatibility in Breeding*

Most cultivated tetrasomic polyploid or self-incompatible diploid potatoes have not realized breeding gains due to low recombination, long generation cycles, polyploidy, inbreeding depression, and poor adaptation of wild potato germplasm (Visser et al. 2009; Lindhout et al. 2011). New breeding methods that involve the development of diploid inbred lines in potato were proposed as a strategy to address many perceived limitations faced by potato breeders (Birhman and Hosaka 2000; Phumichai et al. 2005). In the last few years a trend emerged in a group of potato breeders to reconsider the crop as a diploid species composed of a series of inbred lines that capture the favorable genetic diversity available in cultivated and wild potatoes. Inbreeding due to selfing may be efficient for organizing the whole gene pool into various favorably interacting and stable epistatic systems (Allard 1999).

The self-incompatibility inhibitor (*Sli*) gene opens new doors to explore inbred line breeding in potato (Lindhout et al. 2011). Highly inbred *S. chacoense* lines such as M6, which has been self-pollinated for seven generations, are vigorous and fertile (Jansky et al. 2014). CIP has incorporated *Sli* into diploid cultivars of *S. tuberosum* groups Stenotomum and Phureja and selected a panel of 20 *Sli* bearing self-compatible hybrids to provide a more desirable self-compatibility source than wild *S. chacoense* for the development of inbred lines. These self-compatible hybrids denoted BSLi, are being used to incorporate novel diversity from wild species and take full advantage of modern genetics and genomics tools to generate inbred genetic resources such as recombinant inbred lines (RILs), for fundamental gene discovery and gene mapping.

Self-compatibility has been identified in five diploid cultivated potatoes of *S. tuberosum* Phureja (phu) and Stenotomum (stn) Groups held in CIP's genebank. Selfing three of these self-compatible diploid cultivars; i.e., CIP705468 (goniocalyx), CIP703320 (stn), and CIP701165 (stn) yielded progenies that segregated for self-compatible and self-incompatible individuals. The segregation ratio 2 self-compatible: 1 self-incompatible was significantly skewed from the expected ratio of 3 self-compatible: 1 self-incompatible for a mutant factor in heterozygosis because of the small population size ($N < 60$) analyzed in each selfed cultivar. Self-compatible plants due to pistil side mutations that compromised S-RNase or HT factors produce only self-compatible plants. The same is true for GOF mutations of the *SLF* gene

(McClure et al. 2011). On the other hand, self-compatible plants with a *Sli* mutation in heterozygocity produce 3 self-compatible: 1 self-incompatible plants regardless of its haplotype condition in the *SLF* locus; i.e., *SxSx*, *SxSy*, *SySy*. Hence, the presence of self-compatible plants in the offspring of self-compatible 2*x* cultivars suggests a dominant pollen-side mutation similar to the *Sli* gene since this is the only scenario that yields self-incompatible offspring. Further research will be required to elucidate whether this *Sli*-like phenotype is a novel pollen-side GOF factor or *Sli* gene variant. Whatever its nature, self-compatible 2*x* cultivars would provide a more desirable self-compatible source than *S. chacoense* as they will avoid the undesirable linkage drag associated with the use of a wild species in the development of 2*x* inbred lines.

7.6.2 Interspecific Crosses and Incompatibility

Interspecific reproductive barriers (IRBs) complicate using wild germplasm species for crop improvement (Zamir 2001; Jansky et al. 2013). Therefore, the rich trait diversity available in CIP's extensive collection of tuber-bearing *Solanum* accessions requires overcoming IRBs to be introgressed into cultivated *Solanum* taxa. These barriers include incompatibility between pollen and pistil, male sterility resulting from interactions between nuclear and cytoplasmic genes, and endosperm failure (Camadro et al. 2004). Post-zygotic IRBs are due to EBN incompatibilities that lead to endosperm failure, whereas pre-zygotic IRBs are typically associated with pollen tube growth inhibition (Camadro and Peloquin 1981; Fritz and Hanneman 1989; Novy and Hanneman 1991; Camadro et al. 1998; Erazzú et al. 1999; Hayes et al. 2005).

Genetic and molecular research shows that some self-incompatibility factors also function in prezygotic IRBs. However, self-incompatibility and IRBs differ in terms of specificity and the precise factor requirements. IRBs show broad specificity, and a single S-RNase can cause rejection of pollen from species or groups of species (Murfett et al. 1996; Tovar-Méndez et al. 2014). *S-RNase* and *HT* genes dual roles in self-incompatibility and interspecific pollen rejection points out pleiotropic effects and hence linkage between these two mechanisms.

IRB mechanisms' complexity is such that multiple redundant mechanisms can contribute to interspecific incompatibility, even between a single pair of species (Murfett et al. 1996; McClure et al. 2000). This may complicate experiments because defects in one rejection mechanism do not necessarily result in compatibility. For example, HT-proteins previously found implicated only in S-RNase-dependent self-incompatibility and IRBs have also been involved in S-RNase-independent pollen rejection in tomato (McClure et al. 2011).

Selective pressures like reproductive assurance make self-incompatible to self-compatible mating system transitions (MSTs) common in nature (Barrett 2002; Goldberg et al. 2010; Goldberg and Igic 2012). Self-compatibility has not been extensively investigated in the potato clade and few self-compatible species have been recognized. Loss of S-RNase function is a common route to self-compatibility. This is the case of self-compatible variants of *S. verrucosum* in which pollen tubes of other potato species, including those having 1EBN can grow without inhibition in their pistils reaching ovules in great quantity (Hermsen and Ramanna 1976; Makhan'ko 2011). Furthermore, dominant gain-of-function (GOF) pollen-part mutants such as *Sli*, which results in self-compatible variants in *S. chacoense*, increase seed set in interspecific crosses to *S. pinnatisectum* in addition to suppressing self-incompatibility (Sanetomo et al. 2014).

Phumichai et al. (2006) were able to introduce after crossing the *S*-locus inhibitor gene (*Sli*), which can inhibit gametophytic self-incompatibility, in diploid potatoes and alter self-incompatible to self-compatible plants, into 32 diploid genotypes. *Sli* has been also successfully introduced to diploid cultivars from *S. tuberosum* Phureja and Stenotomum Groups using SC *S. chacoense* variants as male parents at CIP. Assuming that this pollen side mutant acts either breaking down pollen-stigma recognition system or overcoming S-RNAse cytotoxic activity (Hosaka and Hanneman 1998b), then a change in interspecific compatibility may occur after crossing *Sli*-bearing self-compatible hybrids as pollen parents and self-incompatible sources of late blight resistance from wild *S. piurae* and *S. chiquidenum* species. Previous attempts at CIP to cross these wild species with 2*x* *S. tuberosum* cultivars produced very few seeds with *S. chiquidenum* and no seed set at all with *S. piurae*. Embryo rescue was often required to save the few hybrids produced from *S. chiquidenum*. Linkage between self-incompatibility and IRBs results in MSTs with significant implications in germplasm enhancement programs particularly when decisions between different crop wild relatives have to be made in interspecific crosses.

7.7 Unilateral Compatibility

This is a very common IRB pattern that refers to crosses that are compatible in only one direction (Lewis and Crowe 1958). Most IRBs conform to the self-incompatible × self-compatible rule, in which pollen from the self-compatible species is rejected on pistils of related self-incompatible species but the reciprocal pollination is compatible (Nathan Hancock et al. 2003; Bedinger et al. 2011). In potato only some IRBs conform to this rule (Hermsen and Ramanna 1976; Eijlander et al. 2000). For example, crosses between the self-compatible 2*x* (1EBN) species *S. pinnatisectum* and the 2*x* (1EBN) self-incompatible species *S. cardiophyllum* were successful only when self-compatible *S. pinnatisectum* was used as the female parent (Chen et al. 2004). A similar pattern was observed in crosses between the self-compatible 2*x* (1EBN) species *S. commersonii* and the self-incompatible 2*x* (2EBN) species *S. chacoense* (Summers and Grun 1981). Although it is not absolute, the consistency of the "self-incompatible × self-compatible" rule suggests a link between inter- and intraspecific pollen rejection. The suggested linkage is that the *S*-locus controls unilateral incompatibility as well as self-incompatibility. Conversely, there are many exceptions to the self-incompatible × self-compatible rule in potato such as the occurrence of both unilateral and bilateral self-incompatible × self-incompatible conflicts (Camadro et al. 1998, 2004; Kuhl et al. 2002; Raimondi et al. 2003). Genetic systems entirely independent of the *S*-locus have been proposed to explain cross incompatibility, which is a term used to emphasize potato clade cross distinctive complexity. Cross incompatibility includes both post- and pre-zygotic mechanisms.

An important phenomenon worth mentioning is compatibility encountered in crosses between self-compatible *S. verrucosum* plants (2*x*, 2EBN) and a wide range of accessions of various 1EBN potato species. Success on producing these novel sexual hybrids was achieved with several 1EBN diploid species, such as *inter alia*, *S. bulbocastanum*, *S. pinnatisectum*, *S. polyadenium*, *S. commersonii*, and *S. circaeifolium* (Yermishin et al. 2014). Absence of S-RNases in pistils in self-compatible *S. verrucosum* allowed growth of pollen tubes from these 1EBN wild species and fertilization of egg cells. In addition, "rescue pollination" was used to improve hybridization effectiveness. "Rescue pollination" also known as "double pollination" is a technique that reduces premature fruit drop and involves the application of pollen from the incompatible species, followed a day or 2 later by that of a compatible species, denoted as "mentor pollen" (Singsit and Hanneman Jr 1990). The "mentor pollen" fertilizes several ovules, stimulating the development of fruit and pollen tubes from the incompatible parent to reach the ovules and effect fertilization (Singsit and Hanneman Jr 1990; Yermishin et al. 2014). Pollen of Group Phureja pollinators are used as "mentor pollen" because of their typical dominant seed spot marker, so its offspring can be visually identified and eliminated (Brown and Adiwilaga 1991; Iwanaga et al. 1991a).

A feature of inter-EBN interspecific hybridization using self-compatible *S. verrucosum* as "bridge species" is male sterility in most resulting hybrids (Abdalla and Hermsen 1972–Abdalla and Hermsen 1973; Yermishin et al. 2014). Cytoplasmic male sterility factors (CMS) from *S. verrucosum* have been assumed to interact with dominant nuclear genes from the 1EBN male parents, resulting in male sterility of hybrids. CMS has also been suggested to account for male sterility of hybrids produced when 2*x* (2EBN) Group Tuberosum haploids were used as male parents in crosses with cultivated 2*x* (2EBN) Phureja and Stenotomum Groups (Grun et al. 1962; Ross et al. 1964; Carroll 1975) as well as with a wide range of wild species (Hermundstad and Peloquin 1985; Tucci et al. 1996; Santini et al. 2000). Male fertile hybrids obtained when the haploids were the male parent corroborated this assumption (Tucci et al. 1996; Novy and Hanneman 1991). Consequently, this type of barrier can be overcome by carrying out reciprocal crosses.

Male sterility of hybrids from *S. verrucosum* did not represent a drawback since crosses with Tuberosum haploids were successful when the (*S. verrucosum* × 1EBN) hybrids were used as females (Yermishin et al. 2014). However, male fertile Tuberosum haploids are not widely available (Makhan'ko 2008). Breeders may overcome this drawback by extracting haploids from 4*x S. tuberosum* cultivars or breeding clone bearing the male fertility dominant restorer gene *Rt* that gives fertility to plants that contain the dominant male sterility gene *Ms* in the presence of sensitive cytoplasm (Iwanaga et al. 1991b). Selection of Group Tuberosum haploids carrying a restorer of fertility (*Rt*) gene can be used to pollinate (*S. verrucosum* × 1EBN) hybrids and produce male fertile hybrids for further backcrossing with cultivated potato.

An alternative appealing approach for creating opportunities to solve the problem of the prezygotic interspecific incompatibility with a number of 1EBN wild species was proposed by Polyukhovich et al. (2010). These investigators transferred the nonfunctional S-RNase *Sv* haplotype from self-compatible *S verrucosum* directly to Group Tuberosum haploids using some rare pollen receptive haploids. Further development of homozygote *SvSv* hybrids were achieved after selfing or sib-mating of F_1 self-compatible hybrids with high functional pollen fertility. These *SvSv* hybrids were identified based on their good penetration of pollen tubes in the styles of 1EBN *S. bulbocastanum* and *S. pinnatisectum* species. These *SvSv* Tuberosum hybrids may produce fertile hybrids in direct crosses with 1EBN wild diploid potato species.

According to available knowledge, 2*x* (2EBN) self-compatible *S. verrucosum*, which is used as a "bridge species", provides potato breeders an ideal route in germplasm enhancement aimed at introgressing valuable genes from 1EBN wild diploid *Solanum* species into breeding populations. Efficiency of hybrid production using self-compatible *S. verrucosum* is greater than other methods based on ploidy manipulations such as somatic hybridization or embryo rescue, which require considerable experience and expenditure of time and resources (Jansky 2006).

Group Tuberosum haploids are the most promising recipients for introgression of novel genes for traits of interest from diploid wild species germplasm (Peloquin et al. 1989; Jansky et al. 1990). Smaller population size in comparison with tetraploids is needed for selecting recombinants that meet breeding requirements due to their disomic inheritance. Accumulation of desirable genes and elimination of undesirable ones flows faster at this ploidy level. In addition, the presence of naturally occurring meiotic mutations in the potato gene pool, which leads to the production of 2*n* gametes, allows chromosome doubling for the transfer of valuable gene combinations to the tetraploid level (Carputo et al. 2000; Yermishin et al. 2014).

7.8 Endosperm Balance Number (EBN) and Interspecific Reproductive Barriers

Various genetic mechanisms account for the success or failure of seed development in flowering plants, which undergo double fertilization during sexual reproduction (Kinoshita 2007). Failure for producing triploids after tetraploid × diploid crosses—also known as "triploid block"—led to studying as a reproductive barrier the endosperm, which provides nourishment to the seed embryo. A first concept was to consider a 2:3:2 ploidy balance between maternal tissue, endosperm and embryo but further research demonstrated that normal endosperm development depends on having a 2:1 maternal to paternal genome dosage in the endosperm (Ehlenfeldt and

Ortiz 1995 and references therein). This endosperm dosage system for both intra- and inter-specific crossing seems to be multigenic in *Solanum* species, in which is known as the endosperm balance number (EBN) that also explains some aspects of species evolution therein. For example, Ortiz and Ehlenfeldt (1992) indicated the role of EBN in the origin of both diploid and polyploid potato species or how it becomes a hybridization barrier for speciation among sympatric *Solanum* species with same ploidy.

The EBN is a unifying concept that may predict endosperm function in intraspecific, interploidy, and interspecific crosses in potato and wild crop relatives (Johnston et al. 1980). Each species carries an EBN value that is constant across interspecific crossing, thus determining the effective ploidy in the endosperm, which must be in a 2 maternal:1 paternal ratio that is a necessary for successful endosperm development. The EBN, which is in itself an arbitrary value, is given to a species based on its crossing behavior with known EBN standards. The EBN does not reflect directly the ploidy of a species (Hanneman Jr 1999). For example, there are 2*x* (1EBN), 2*x* (2EBN), 3*x* (2EBN), 4*x* (2EBN), 4*x* (4EBN), 5*x* (4EBN), and 6*x* (4EBN) *Solanum* species.

The EBN concept was useful to elucidate the nature of the pollinator effect in haploid extraction (Peloquin et al. 1996). Haploid embryos appear to be associated with hexaploid endosperms as a result of having the union of 2-chromosome sets from the pollinator with the polar nuclei and lack of fertilization of the egg, thus having a 2 maternal: 1 paternal EBN in the endosperm that normal seed development requires.

The "triploid block" is a reproductive barrier resulting from endosperm malfunction due to the epigenetic phenomenon of genomic imprinting (Ehlenfeldt and Ortiz 1995; Köhler et al. 2009). Evidence shows that the endosperm dosage systems are imprinted within the gametes, thus the same gene being functionally different in maternal and paternal chromosomes. The maternally and paternally imprinted genes often carry a DNA methylation or histone modification in their vicinity (Kinoshita 2007). The position of these epigenetic modifications determines how the gene will be expressed; i.e., either from the maternal or paternal inherited allele (Köhler et al. 2012). Maternally imprinted genes appear to repress endosperm proliferation, which seems to be promoted by paternally imprinted genes. Abnormal endosperm development results from the imbalance of these imprinted genes. This "parental conflict" is consistent with the differential maternal and paternal genome effects in interploidy mating (Köhler et al. 2009). Products of imprinted PcG genes such as *Medea* (*MEA*) and *Fertilization Independent Seed2* (*FIS2*) could be limiting factors affecting endosperm and seed development. For example, FIS PcG proteins repress fertilization-independent seed formation and restrict endosperm proliferation (Köhler and Makaverich 2006).

The EBN provides useful prior knowledge for germplasm transfer from 2*x* (1EBN), 2*x* (2EBN), 4*x* (2EBN) and 6*x* (4EBN) species into 4*x* (4EBN) potato (Ortiz and Ehlenfeldt 1992). For example, chromosome engineering using 4*x* (2EBN) and 2*x* (2EBN) addition lines will allow to introduce specific chromosomes bearing the target allele into the 4*x* cultigen pool for further use in breeding.

Likewise, 2*x* chromosome addition lines from may result using 2*x* (1EBN) and 2*x* (2EBN) *Solanum* species. The first step will be to get a 3*x* (2EBN) interspecific hybrids after crossing both 2*x* species and thereafter using these hybrids by crossing them with 2*x* (2EBN) species to obtain 2*x* (2EBN) chromosome addition lines, which can be further used for breeding at 2*x* level. Haploids from 5*x* (4EBN) clones may be another method for producing 2*x* (2EBN) offspring with traits from the 2*x* (1EBN) parent because haploid extraction will allow screening for the right balance in the endosperm.

7.9 Trait Genetic Research: A Summary Prior to DNA Markers

Swaminathan and Howard (1953) provided the first summary of the inheritance of most important traits in potato, which was further updated by Howard (1960, Howard 1970), Bradshaw and Mackay (1994 and chapters therein), and Tiemens-Hulscher et al. (2013). Table 7.3 give some details about trait genetics and gene symbols as noted in some of these publications.

Further research using the candidate gene approach led to identifying diagnostic DNA-based markers for genes involved, *inter alia*, in quantitative resistance to late blight (*R1* gene family in chromosome V) or cyst nematode (major QTL in same resistance "hot spot" on potato chromosome V), and both chip color (e.g., co-localized with a cold-sweetening QTL in chromosome IX and other sugar QTL in chromosome X) and tuber starch content in tetraploid potato cultivars (Gebhardt et al. 2007). Other annotated loci in the potato genetic map are *Y* (yellow flesh color) in chromosome III, *H3* of *Gpa4* in chromosome IV, Rx_2 (extreme resistance to *Potato Virus X*) and Nx_{tbr} (hypersensitivity to *Potato Virus X*) in chromosome V, Nx_{phu} in chromosome IX, *Ro* (tuber shape) and anthocyanin (including flower and skin color) in chromosome X, Ry_{sto}, and *Ry-adg* in chromosome XI, and both *Gpa 2* (resistance to potato cyst nematode) plus Rx_1 in chromosome XII.

7.10 Cytogenetics for Crossing, Scaling Up and Down Ploidy, and Chromosome Engineering

Germplasm enhancement (mistakenly replaced by some as pre-breeding, Ortiz 2002) is the early component of sustainable plant breeding that includes identifying a useful character, "capturing" its genetic diversity, and putting those genes into a "usable" form (Peloquin et al. 1989). Potato provides an interesting example of using crop wild relatives in germplasm enhancement. The wild species *S. demissum* (6*x*, 4EBN), *S. stoloniferum* (4*x*, 2EBN), and *S. vernei* contributed host plant resistance genes to late blight, *Potato Virus Y*, and cyst nematode, respectively; while

Table 7.3 Gene symbols and their genetics of key traits for potato breeding

Trait	Gene symbol	Genetics (disomic inheritance unless indicated otherwise)
Plant growth type		Upright dominant to prostrate, and intermediate procumbent recessive to upright but unknown to prostrate
Dwarfism		Recessive phenotype producing compact, dark green, rosette plant
Flower color	*D* red	*F* involved in anthocyanin expression in flowers
	P blue	Purple = *D_P_F_*, Red to rose *D_ppF_*, Blue = *ddP_F_*, White = *D_P_ff* or *ddppF_*
Skin color	*D* red	*I* engaged in anthocyanin expression in tuber skin
	P blue	Purple = *D_P_I_*, Red *D_ppI_*, Blue = *ddP_I_*, White to yellow to brown = *D_P_ff* or *ddppF_*
Flesh color	*Y* yellow, *y* white	Yellow caused by carotenoids dominant to white
	Or allele at *Y* locus	Orange depends on two genes: one determining production and other (recessive) accounting for accumulation of zeazhantin. Orange dominant to yellow
	D red	Purple or red due to anthocyanins regulated by *B*.
	P blue	Purple = *D_ P_ B_*, Red = *D_ pp B_*, Blue = *dd P_ B_* (2*x*)
Tuber color pattern		Splashed and spotted only in eyes while speckled (dominant) reverses; i.e., eyes are either white or yellow
Tuber shape	*Ro* round	Round (*Ro__*) dominant to long (*roro*)
	ro long	
Russet skin		Three independent loci having an additive effect to each other; i.e., *AABBCC* show more russet skin that *AaBbCc*
Stem pubescence		Single gene, being pubescent dominant to glabrous stem
Early maturity		Earliness due to dominant allele with additive effects; i.e., duplex (*AAaa*) earlier than simplex (*Aaaa*) in 4*x*
Tuber dormancy		Dominant early sprouting with short dormancy
Tuberization under long days		High heritability but influenced by day length, light intensity, and temperature

(continued)

Table 7.3 (continued)

Trait	Gene symbol	Genetics (disomic inheritance unless indicated otherwise)
Chip color		3-Gene hypothesis for both reversion resistance (ability to produced light-colored chips after harvest or short storage) and reconditioning (controlled tuber warming after storing fall crop for 4–6 months) to eliminate reducing sugars: A dominant allele in each of three loci for good chipping, being one or two loci common to both traits
Late blight resistance	12 known *R* genes derived from *S. demissum*, *Rpi-blb3* from *S. bulbocastanum*, and *Rpi-abpt* derived from quadruple hybrid involving *S. acaule*, *S. bulbocastanum*, plus groups Phureja and Tuberosum	Dominant *R* for race-specific host resistance corresponding to virulence genes in oomycete *Phytophthora infestans* (*R8* and *R9* seem to be "durable") Many genes are involved in partial host plant resistance that slow down the development of all *Phytophthora infestans* races, thus being likely more durable than race-specific resistance
Early blight resistance		High heritability for partial resistance, thus additivity being the most important gene action
Potato leaf roll virus resistance	N_L for hypersensitivity	Dominance of resistance but major gene does not provide enough durable resistance to a cultivar
Potato virus X resistance	Rx_i for extreme resistance, while Nx_i controls hypersensitivity	Dominant inheritance
Potato virus Y resistance	Ny_{chc}, Ny_{dms}, $Ry_{sto}{}^{n1}$ and $Ry_{sto}{}^{n2}$ gives hipersensitivity, whereas *Ry*-adg, *Ry-chc*, *Ry-hou* and *Ry-sto* are for extreme resistance	Dominant extreme resistance from *S. stoloniferum*, *S. hougasii*, and Group Andigena, while multigenic for Group Phureja
Potato virus A	*Na* protects for infection through hypersensitivity	
Wart resistance		Dominant alleles in two loci necessary for resistance
Verticillium wilt		2 genes
Bacterial wilt resistance		3–4 dominant genes required
Black leg and bacterial soft rot resistance		Minor genes with additive effects increase resistance
Potato tuber moth resistance		Simple inheritance due to additivity

(continued)

Table 7.3 (continued)

Trait	Gene symbol	Genetics (disomic inheritance unless indicated otherwise)
Root-knot nematode resistance		High narrow-sense heritability for resistance in Group Phureja (susceptible)—*S. sparsipilum* (likely three major complementary genes in heterozygous state) segregating population, but reciprocal effects suggest maternal effects of *S. sparsipilum* or cytoplasmic-nuclear gene interaction
Potato cyst nematode resistance	*H1* (from Group Andigena), *H2* (from CPC2602), *Fa* and *Fb* (from *S. spegazzinii*), plus *B* and *C* from *S. verneii* provide host plant resistance against *Globodera rostochiensis*, while *Gpa2*, *Gpa5*, *GPa1*, *H2*, and *H3* of *Gpa4* give resistance against *G. pallida*	*H1* gives host plant resistance against Ro-1 and Ro-4 pathotypes

processing quality was derived from S. *chacoense* (2*x*, 2EBN) in various cultivars (Jansky et al. 2013).

Introgression and incorporation are the two approaches for using wild species in plant breeding (Simmonds 1993). Introgression refers to transferring one or a few alleles from wild germplasm to breeding populations that lack them, while a large-scale program for developing breeding populations using wild germplasm to broaden its genetic base is known as incorporation. Figure 7.2 illustrates both in potato breeding. "Bridge" species, double pollination, embryo rescue, 2*n* gametes, and the EBN knowledge allows using chromosome engineering, while haploids, wild 2*x* (2EBN) species, and 2*n* gametes are used in ploidy manipulations.

CIP released in the mid-1990s diploid bred-germplasm generated from using haploids from tetraploid cultivars and breeding clones, plus diploid landraces and wild species (Watanabe et al. 1994). These genetic resources are of value for potato breeding due to their genetic diversity, crossability with the 4*x* cultigen pool facilitated by 2*n* gametes (mostly FDR 2*n* pollen), and high host plant resistance to pathogens and pests derived mostly from wild relatives. Ploidy manipulations at CIP led to the transmission of host plant resistance to cyst and root-knot nematodes, bacterial wilt, early blight, and potato tuber moth, as well as producing high-yielding 4*x* (4EBN) breeding clones also having yield stability over environments (Ortiz et al. 1994). Burundi released in the 1990s the tetraploid potato cultivar "Nemared," which derived from this diploid breeding population (Fig. 7.3), because of both its host plant resistance to root knot nematode and desired agronomic traits.

Potato, as shown above, is the model crop species for germplasm enhancement of polysomic polyploids. Crop wild relatives and landraces are the diversity sources, while haploid derived from the tetraploid cultigen pool "capture" this diversity after crossing them with the former. The haploid-species hybrids producing 2*n* gametes

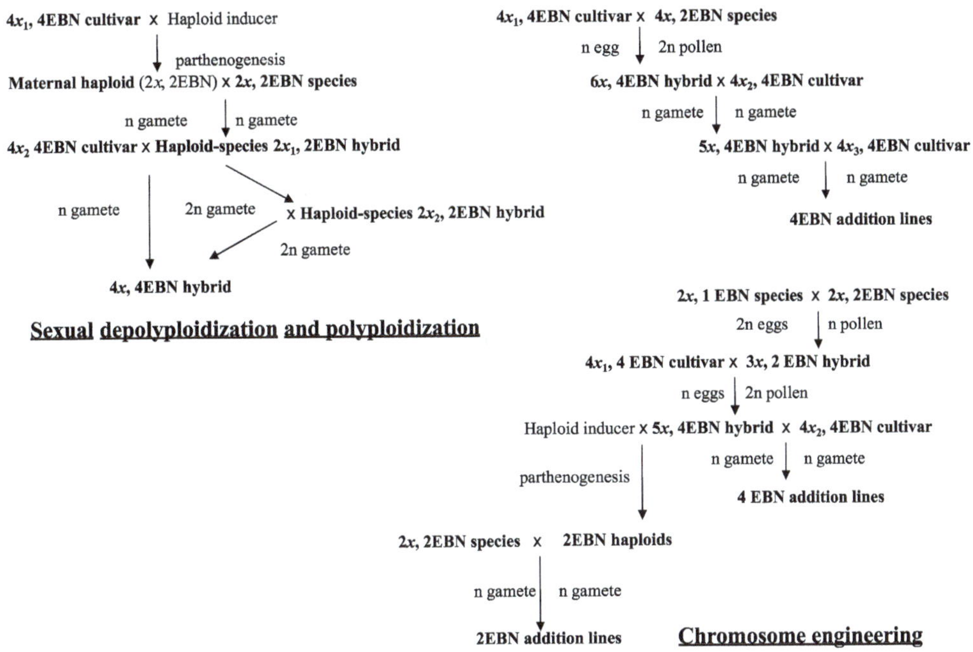

Fig. 7.2 Germplasm enhancement approaches in potato breeding: introgression through chromosome engineering and incorporation using sexual depolyploidization and polyploidization

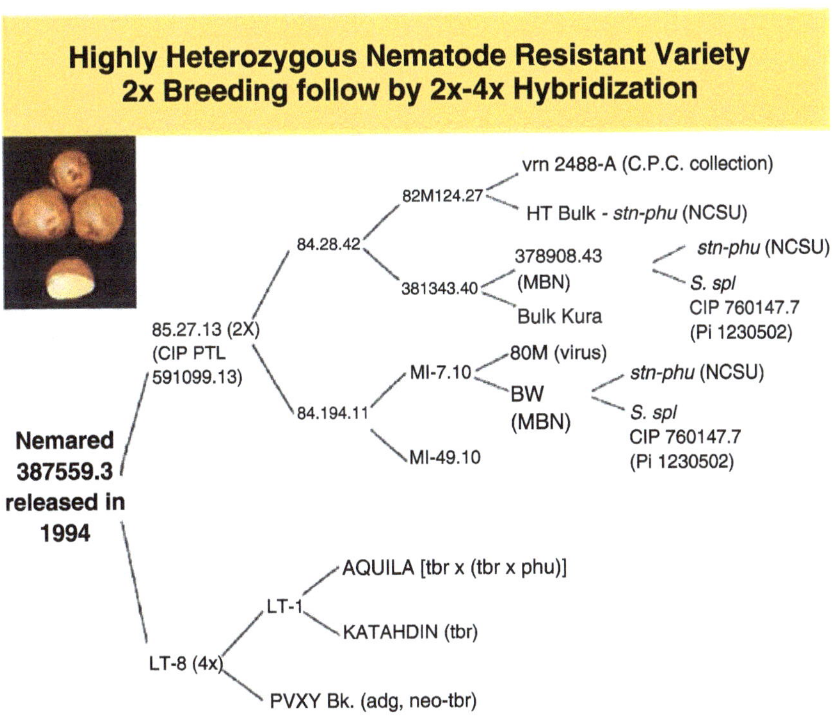

Fig. 7.3 Pedigree of tetraploid cultivar potato "Nemared" resulting from ploidy manipulations

transfer this diversity to the tetraploid breeding pool through sexual polyploidization in which the EBN ensures the resulting ploidy of the hybrid offspring.

7.11 Concluding Remarks

Despite its the biological characteristics which made genetic improvement more complex in potato than in other crops, potato breeders have at their disposal such a powerful and effective approach as the unique ability to conduct across ploidy and species-wide crosses in order to introgress relevant genetic variation from its genetic resources into potato breeding programs. Progress in the field of cytogenetics of potato enables a more effective transfer of relevant genetic variation from wild *Solanum* accessions kept in genebanks or through in situ approaches. The increasing demands to develop resilient potato varieties able to withstand the threats brought by climate change, as well as the substantial progress recently achieved with the development of potato hybrid cultivars at the 2*x* level, underpin the increasing contribution of cytogenetics to the genetic improvement of the potato crop. The increasing availability of vast amounts of genomic information, as well as the continuous reduction of expenses associated with the sequencing of whole genomes are expected to further increase the contribution of potato cytogenetics, in terms of accelerating the development of varieties which provide not only superior adaption to changing production environments but also increased food and nutrient security to the millions of people to whom potato represents a staple crop.

References

Abdalla MMF, Hermsen JGT (1972) Plasmons and male sterility types in *Solanum verrucosum* and its interspecific hybrid derivatives. Euphytica 21:209–220

Abdalla MMF, Hermsen JGT (1973) An evaluation of *Solanum verrucosum* Schlechtd. For its possible use in potato breeding. Euphytica 22:19–27

Allainguillaume J, Wilkinson MJ, Clulow SA et al (1997) Evidence that genes from the male parent may influence the morphology of potato dihaploids. Theor Appl Genet 94:241–248

Allard RW (1999) History of plant population genetics. Ann Rev Genet 33:1–27

Barrett SCH (2002) The evolution of plant sexual diversity. Nat Rev Genet 3:274–284

Bedinger PA, Chetelat RT, McClure B et al (2011) Interspecific reproductive barriers in the tomato clade: opportunities to decipher mechanisms of reproductive isolation. Sex Plant Reprod 24:171–187

Birhman RK, Hosaka K (2000) Production of inbred progenies of diploid potatoes using an S-locus inhibitor (Sli) gene and their characterization. Genome 43:495–502

Bourke PM, Voorrips RE, Visser RGF, Maliepaard C (2015) The double-reduction landscape in tetraploid potato as revealed by a high-density linkage map. Genetics 201:853–863

Bradshaw JE (2007) The canon of potato science: 4 tetrasomic inheritance. Potato Res 50:219–222

Bradshaw JE, Mackay GR (eds) (1994) Potato genetics. CAB International, Wallinford

Bradshaw JE, Bryan GJ, Ramsay G (2006) Genetic resources (including wild and cultivated *Solanum* species) and progress in their utilization in potato breeding. Potato Res 49:49–65

Brown CR, Adiwilaga KD (1991) Use of rescue pollination to make a complex interspecific cross in potato. Am J Potato Res 68:813–820

Brownfield L, Kölher C (2011) Unreduced gamete formation in plants: mechanisms and prospects. J Exp Bot 62:1659–1668

Burham C (1984) Discussion in cytogenetics. Burgess Publishing, Minneapolis

Camadro EL, Peloquin SJ (1981) Cross-incompatibility between two sympatric polyploid *Solanum* species. Theor Appl Genet 60:65–70

Camadro EL, Verde LA, Marcellán ON (1998) Pollen-pistil incompatibility in a diploid hybrid potato population with cultivated and wild germplasm. Am J Potato Res 75:81–85

Camadro EL, Carputo D, Peloquin SJ (2004) Substitutes for genome differentiation in tuber-bearing Solanum: interspecific pollen-pistil incompatibility, nuclear-cytoplasmic male sterility, and endosperm. Theor Appl Genet 109:1369–1376

Cao J, Schneeberger K, Ossowski S et al (2011) Whole-genome sequencing of multiple *Arabidopsis thaliana* populations. Nat Genet 43:956–963

Carputo D, Barone A (2005) Ploidy level manipulations in potato through sexual hybridization. Ann Appl Biol 146:71–79

Carputo D, Barone A, Frusciante L (2000) 2n gametes in the potato: essential ingredients for breeding and germplasm transfer. Theor Appl Genet 101:805–813

Carroll CP (1975) The inheritance and expression of sterility in hybrids of dihaploid and cultivated diploid potatoes. Genetica 45:149–162

Chase SS (1963) Analytic breeding in *Solanum tuberosum* L.—a scheme utilizing parthenotes and other diploid stocks. Can J Genet Cytol 5:359–363

Chen Q, Lynch D, Platt HW et al (2004) Interspecific crossability and cytogenetic analysis of sexual progenies of Mexican wild diploid 1EBN species *Solanum pinnatisectum* and *S. cardiophyllum*. Am J Potato Res 81:159–169

Cipar MS (1964) Self compatibility in hybrids between Phureja and haploid Andigena clones of *Solanum tuberosum*. Eur Potato J 7:152–160

Cipar MS, Peloquin SJ, Hougas R (1964) Variability in the expression of self-incompatibility in tuber-bearing diploid *Solanum* species. Amer Potato J 41:155–162

Clulow SA, Rousselle-Bourgeois F (1997) Widespread introgression of *Solanum phureja* DNA in potato (*S. tuberosum*) dihaploids. Plant Breed 116:347–351

Clulow SA, Wilkinson MJ, Waugh R et al (1991) Cytological and molecular observations on Solanum phureja-induced dihaploid potatoes. Theor Appl Genet 82:545–551

Clulow SA, Wilkinson MJ, Burch LR (1993) *Solanum phureja* genes are expressed in the leaves and tubers of aneusomatic potato dihaploids. Euphytica 69:1–6

Comai L (2005) The advantages and disadvantages of being polyploid. Nat Rev Genet 6:836–846

Cook DE, Lee TG, Guo X et al (2012) Copy number variation of multiple genes at *Rhg1* mediates nematode resistance in soybean. Science 338:1206–1209

De Jong H (1981) Inheritance of russeting in cultivated diploid potatoes. Potato Res 24:309–313

De Jong H (1987) Inheritance of pigmented tuber flesh in cultivated diploid potatoes. Amer Potato J 64:337–343

De Jong H, Burns VJ (1993) Inheritance of tuber shape in cultivated diploid potatoes. Am Potato J 70:267–283

De Jong H, Rowe PR (1971) Inbreeding in cultivated diploid potatoes. Potato Res 14:74–83

De Jong H, Kawchuk LM, Burns VJ (1998) Inheritance and mapping of a light green mutant in cultivated diploid potatoes. Euphytica 103:83–88

De Jong H, Kawchuk LM, Coleman WK et al (2001) Development and characterization of an adapted form of droopy, a diploid potato mutant deficient in abscisic acid. Am J Potato Res 78:279–290

de Nettancourt D (1977) The genetic basis of self-incompatibility. In: Incompatibility in angiosperms. Monographs on theoretical and applied genetics, vol 3. Springer, Berlin

Díaz A, Zikhali M, Turner AS et al (2012) Copy number variation affecting the Photoperiod-B1 and Vernalization-A1 genes is associated with altered flowering time in wheat (*Triticum aestivum*). PLoS One 7:e33234

Ehlenfeldt MK, Ortiz R (1995) On the origins of endosperm dosage requirements in *Solanum* and other angiosperma genera. Sex Plant Reprod 8:189–196

Eijlander R (1998) Mechanisms of self–incompatibility and unilateral incompatibility in diploid potato (*Solanum tuberosum* L.). Dissertation, Wageningen University

Eijlander R, Ter Laak W, Hermsen JGT et al (2000) Occurrence of self-compatibility, self-incompatibility and unilateral incompatibility after crossing diploid *S. tuberosum* (SI) with *S. verrucosum* (SC): I. Expression and inheritance of self-compatibility. Euphytica 115:127–139

Erazzú LE, Camadro EL, Clausen AM (1999) Pollen-style compatibility relations in natural populations of the wild diploid potato species *Solanum spegazzinii* Bitt. Euphytica 105:219–227

Ercolano MR, Carputo D, Li J, Monti L et al (2004) Assessment of genetic variability of haploids extracted from tetraploid ($2n = 4x = 48$) *Solanum tuberosum*. Genome 47:633–638

Feuk L, Carson AR, Scherer SW (2006) Structural variation in the human genome. Nat Rev Genet 7:85–97

Frandsen NO (1967) Haploidproduktion aus einem Kartoffelzuchtmaterial mit intensiver Wildarteinkreuzung. Der Zuchter 37:120–134

Freyre R, Warnke S, Sosinski B et al (1994) Quantitative trait locus analysis of tuber dormancy in diploid potato (*Solanum* spp.). Theor Appl Genet 89:474–480

Fritz NK, Hanneman RE (1989) Interspecific incompatibility due to stylar barriers in tuber-bearing and closely related non-tuber-bearing Solanums. Sex Plant Reprod 2:184–192

Gálvez JH, Tai HH, Barkley NA et al (2017) Understanding potato with the help of genomics. AIMS Agric Food 2:16–39

Gebhardt C, Ritter E, Barone A et al (1991) RFLP maps of potato and their alignment with the homeologous tomato genome. Theor Appl Genet 83:49–57

Gebhardt C, Li L, Pajerowska-Mukthar K et al (2007) Candidate gene approach to identify genes underlying quantitative traits and develop diagnostic markers in potato. Crop Sci 47:S106–S111

Goldberg EE, Igic B (2012) Tempo and mode in plant breeding system evolution. Evolution 66:3701–3709

Goldberg EE, Kohn JR, Lande R et al (2010) Species selection maintains self-incompatibility. Science 330:493–495

Goldraij A, Kondo K, Lee CB et al (2006) Compartmentalization of S-RNase and HT-B degradation in self-incompatible Nicotiana. Nature 439(7078):805–810

Golz JF, Oh HY, Su V et al (2001) Genetic analysis of Nicotiana pollen-part mutants is consistent with the presence of an S-ribonuclease inhibitor at the S-locus. Proc Natl Acad Sci U S A 98:15372–15376

Gray JE, McClure BA, Bonig I et al (1991) Action of the style product of the self-incompatibility gene of Nicotiana alata (S-RNase) on in vitro-grown pollen tubes. Plant Cell 3:271–283

Grun P, Aubertin M, Radlow A (1962) Multiple differentiation of plasmons of diploid species of *Solanum*. Genetics 47:1321–1333

Grun P, Ochoa C, Capage D (1977) Evolution of cytoplasmic factors in tetraploid cultivated potato (Solanaceae). Am J Bot 64:412–420

Hancock CN, Kondo K, Beecher B, McClure B (2003) The S-locus and unilateral incompatibility. Phil Trans R Soc Lond B 358:1133–1140

Hanneman RE Jr (1999) The reproductive biology of the potato and its implication for breeding. Potato Res 42:283–312

Hanneman RE (1985) Self-fertility in *Solanum chacoense*. Am Potato J 62:428–429. (abstract)

Hardigan MA, Parker F, Laimbeer E et al (2017) Genome diversity of tuber-bearing Solanum uncovers complex evolutionary history and targets of domestication in the cultivated potato. Proc Natl Acad Sci USA. https://doi.org/10.1073/pnas.1714380114

Hayes RJ, Dinu II, Thill CA (2005) Unilateral and bilateral hybridization barriers in interseries crosses of 4x 2EBN *Solanum stoloniferum*, *S. pinnatisectum*, *S. cardiophyllum*, and 2x 2EBN S. tuberosum haploids and haploid-species hybrids. Sex Plant Reprod 17:303–311

Henry IM, Dilkes BP, Miller ES, Burkart-Waco D et al (2010) Phenotypic consequences of aneuploidy in *Arabidopsis thaliana*. Genetics 186:1231–1245

Henry IM, Zinkgraf MS, Groover AT et al (2015) A system for dosage-based functional genomics in poplar. Plant Cell 27:2370–2383

Hermsen JGT, Ramanna MS (1976) Barriers to hybridization of *Solanum bulbocastanum* Dun. and *S. verrucosum* Schlechtd. and structural hybridity in their F1 plants. Euphytica 25:1–10

Hermsen JT, Verdenius J (1973) Selection from *Solanum tuberosum* group Phureja of genotypes combining high-frequency haploid induction with homozygosity for embryo-spot. Euphytica 22:244–259

Hermundstad SA, Peloquin SJ (1985) Male fertility and 2n pollen production in haploid-wild species hybrids. Am Potato J 62:479–487

Hirsch CN, Hirsch CD, Felcher K et al (2013) Retrospective view of North American potato (*Solanum tuberosum* L.) breeding in the 20th and 21st centuries. G3 3:1003–1013

Hosaka K, Hanneman RE (1998a) Genetics of selfcompatibility in a self-incompatible wild diploid potato species *Solanum chacoense*. I. Detection of an S locus inhibitor (*Sli*) gene. Euphytica 99:191–197

Hosaka K, Hanneman RE (1998b) Genetics of self-compatibility in a selfcompatible wild diploid potato species *Solanum chacoense*. 2. Localization of an S-locus inhibitor (Sli) gene on the potato genome using DNA markers. Euphytica 103:265–271

Hosaka K, Sanetomo R (2012) Development of a rapid identification method for potato cytoplasm and its use for evaluating Japanese collections. Theor Appl Genet 125:1237–1251

Hougas RW, Peloquin SJ, Gabert AC (1964) Effect of seed-parent and pollinator on frequency of haploids in *Solanum tuberosum*. Crop Sci 4:593–595

Howard HW (1960) Potato cytology and genetics, 1952-59. Biblphia Genet 19:87–216

Howard HW (1970) Genetics of the potato *Solanum tuberosum*. Logos Press, London

Hua ZH, Fields A, Kao TH (2008) Biochemical models for S-RNase-based self-incompatibility. Mol Plant 1:575–585

Hutten RCB, Scholberg EJMM, Huigen DJ et al (1993) Analysis of dihaploid induction and production ability and seed parent x pollinator interaction in potato. Euphytica 72:61–64

Iwanaga M, Freyre R, Watanabe K (1991a) Breaking the crossability barriers between disomic tetraploid *Solanum acaule* and tetrasomic tetraploid *S. tuberosum*. Euphytica 52:183–191

Iwanaga M, Ortiz R, Cipar M, Peloquin SJ (1991b) A restorer gene for genetic-cytoplasmic male sterility in cultivated potatoes. Am Potato J 68:19–28

Jacobs JME, Van Eck HJ, Arens P et al (1995) A genetic map of potato (*Solanum tuberosum*) integrating molecular markers, including transposons, and classical markers. Theor Appl Genet 91:289–300

Jansky S (2006) Overcoming hybridization barriers in potato. Plant Breeding 125:1–12

Jansky S, Peloquin SJ, Yerk GL (1990) Use of potato haploids to put 2x wild species germplasm in usable form. Plant Breed 104:290–294

Jansky S, Dempewolf H, Camadro EL, Simon R, Zimnoch-Guzowska E, Bisognin DA, Bonierbale M (2013) A case for crop wild relative preservation and use in potato. Crop Sci 53:746–754

Jansky S, Chung YS, Kittipadukal P (2014) M6: a diploid potato inbred line for use in breeding and genetics research. J Plant Reg 8:195–199

Jansky SH, Charkowski AO, Douches DS, Gusmini G, Richael C, Bethke PC, Spooner DM, Novy RG, De Jong H, De Jong WS, Bamberg JB, Thompson AL, Bizimungu B, Holm DG, Brown CR, Haynes KG, Sathuvalli VR, Veilleux RE, Miller JC, Bradeen JM, Jiang J (2016) Reinventing potato as a diploid inbred line–based crop. Crop Sci 56:1412–1422

Johnston SA, den Nijs TP, Peloquin SJ, Hanneman RE Jr (1980) The significance of genic balance to endosperm development in interspecific crosses. Theor Appl Genet. 57:5–9

Kinoshita T (2007) Reproductive barrier and genomic imprinting in the endosperm of flowering plants. Genes Genet Syst 82:177–186

Köhler C, Makaverich G (2006) Epigenetic mechanisms governing seed development in plants. EMBO Rep 7:1223–1227

Köhler C, Scheid OM, Erilova A (2009) The impact of the triploid block on the origin and evolution of polyplpid plants. Trends Genet 26:142–148

Köhler C, Wolff P, Spillane C (2012) Epigenetic mechanisms underlying genomic imprinting in plants. Annu Rev Plant Biol 63:331–352

Kotch GP, Ortiz R, Peloquin SJ (1992) Genetic analysis by use of potato haploid populations. Genome 35:103–108

Kuhl JC, Havey MJ, Hanneman RE (2002) A genetic study of unilateral incompatibility between diploid (1EBN) Mexican species *Solanum pinnatisectum* and *S. cardiophyllum* subsp. *cardiophyllum.* Sex Plant Reprod 14:305–313

Lewis D, Crowe LK (1958) Unilateral interspecific incompatibility in flowering plants. Heredity 12:233–256

Lindhout P, Meijer D, Schotte T et al (2011) Towards F_1 hybrid seed potato breeding. Pot Res 54:301–312

Lovene M, Zhang T, Lou Q, Buell CR et al (2013) Copy number variation in potato—an asexually propagated autotetraploid species. Plant J 75:80–89

Makhan'ko OV (2008) Interspecific incompatibility in diploid potato breeding. Zemlyarobstva i Akhova Raslin 1:11–14

Makhan'ko OV (2011) Incompatibility in interspecific and intraspecific hybridization of diploid potato and ways of its overcoming. Dissertation, Institute of Genetics and Cytology Belarusian National Academy of Sciences

Maron LG, Guimarães CT, Kirst M et al (2013) Aluminum tolerance in maize is associated with higher *MATE1* gene copy number. Proc Natl Acad Sci USA 110:5241–5246

McClure B, Cruz-Garcia F, Romero C (2011) Compatibility and incompatibility in S-RNase based systems. Ann Bot 108:647–658

McClure BA, Gray JE, Anderson MA et al (1990) Self-incompatibility in Nicotiana alata involves degradation of pollen rRNA. Nature 347:757–760

McClure BA, Cruz-Garcia F, Beecher BS, Sulaman W (2000) Factors affecting inter- and intraspecific pollen rejection in Nicotiana. Ann Bot 85:113–123

Mihovilovich E, Sanetomo R, Hosaka K et al (2015) Cytoplasmic diversity in potato breeding: case study from the International Potato Center. Mol Breed 35:137. https://doi.org/10.1007/s11032-015-0326-1

Murfett J, Strabala T, Zurek D et al (1996) S RNase and interspecific pollen rejection in the genus Nicotiana: multiple pollen-rejection pathways contribute to unilateral incompatibility between self-incompatible and self-compatible species. Plant Cell 8:943–958

Naess SK, Bradeen JM, Wielgus SM et al (2000) Resistance to late blight in *Solanum bulbocastanum* is mapped to chromosome 8. Theor Appl Genet 101:697–704

Nakamura S, Hosaka K (2010) DNA methylation in diploid inbred lines of potatoes and its possible role in the regulation of heterosis. Theor Appl Genet 120:205–214

Nathan Hancock C, Kondo K, Beecher B et al (2003) The S locus and unilateral incompatibility. Phil Trans R Soc Lond B 358:1133–1140

Novy RG, Hanneman RE (1991) Hybridization between Gp. Tuberosum haploids and 1EBN wild potato species. Am Potato J 68:151–169

Olsder J, Hermsen JT (1976) Genetics of selfcompatibility in dihaploids of Solanum tuberosum L. 1. Breeding behaviour of two selfcompatible dihaploids. Euphytica 25:597–607

Ortiz R (1998) Potato breeding via ploidy manipulations. Plant Breed Rev 16:15–86

Ortiz R (2002) Germplasm enhancement. In: Engels JMM, Ramanatha Rao V, Brown AHD, Jackson MT (eds) Managing plant genetic diversity. CAB International, Wallingford—International Plant Genetic Resources Institute, Rome, pp 275–290

Ortiz R, Ehlenfeldt M (1992) The importance of endosperm balance number in potato breeding and the evolution of tuber bearing Solanums. Euphytica 60:105–113

Ortiz R, Peloquin SJ (1994) Use of 24-chromosome potatoes (diploids and dihaploids) for genetic analysis. In: Bradshaw JE, Mackay GR (eds) Potato genetics. CAB International, Wallinford, pp 133–153

Ortiz R, Watanabe KN (2004) Genetic contributions to breeding polyploid crops. Recent Res Devel Genet Breed 1:269–286

Ortiz R, Iwanaga M, Camadro EL (1993a) Utilizaci6n potencial de progenie autofecundada de IvP-35 como inductor de haploides en papa por cruzamientos $4x$-$2x$. Revista Latinoamericana de Papa 5/6:46–53

Ortiz R, Iwanaga M, Peloquin SJ (1993b) Male sterility and $2n$ pollen in $4x$ progenies derived from $4x \times 2x$ and $4x \times 4x$ crosses in potatoes. Potato Res 36:227–236

Ortiz R, Iwanaga M, Peloquin SJ (1994) Breeding potatoes for the developing countries using wild tuber bearing *Solanum* species and ploidy manipulations. J Genet Breed 48:89–98

Ortiz R, Simon P, Jansky S et al (2009) Ploidy manipulation of the gametophyte, endosperm and sporophyte: a tribute to Professor Stanley J. Peloquin (1921–2008). Ann Bot 104:795–807

Owen HR, Veilleux RE, Haynes FL et al (1988) Photoperiod effects on $2n$ pollen production, response to anther culture, and net photosynthesis of a diplandrous clone of *Solanum phureja*. Am Potato J 65:131–139

Peloquin SJ, Yerk GL, Werner JE, Darmo E (1989) Potato breeding with haploids and 2n gametes. Genome 31:1000–1004

Peloquin SJ, Gabert AC, Ortiz R (1996) Nature of "pollinator" effect in potato haploid production. Ann Bot 77:539–542

Phumichai C, Mori M, Kobayashi A et al (2005) Toward the development of highly homozygous diploid potato lines using the self-compatibility controlling Sli gene. Genome 48:977–984

Phumichai C, YIkeguchi-Samitsu Y, Fujimatsu M et al (2006) Expression of S-locus inhibitor gene (Sli) in various diploid potatoes. Euphytica 148:227–234

Pineda O, Bonierbale MW, Plaisted RL et al (1993) Identification of RFLP markers linked to the H_1 gene conferring resistance to the potato cyst nematode *Globodera rostochiensis*. Genome 36:152–156

Polyukhovich YV, Makhan'ko OV, Savchuk AV et al (2010) Production of bridge lines for overcoming interspecific incompatibility in potato. Vestsi Natsiyanalnai Akademii Navuk Belarusi. Seryia Biyalagichnykh Navuk 2:51–58

Potato Genome Sequencing Consortium (2011) Genome sequence and analysis of the tuber crop potato. Nature 475:189–195

Raimondi JP, Sala RG, Camadro EL (2003) Crossability relationships among the wild diploid potato species *Solanum kurtzianum*, *S. chacoense* and *S. ruiz-lealii* from Argentina. Euphytica 132:287–295

Ravi M, Chan SW (2010) Haploid plants produced by centromere-mediated genome elimination. Nature 464:615–618

Ravi M, Marimuthu MPA, Tan EH et al (2014) A haploid genetics toolbox for *Arabidopsis thaliana*. Nat Commun 5:5334

Rivard SR, Cappadocia M, Landry BS (1996) A comparison of RFLP maps based on anther culture derived, selfed, and hybrid progenies of *Solanum chacoense*. Genome 39:611–621

Ross H (1986) Potato breeding-problems and perspectives. Verlag Paul Parey, Berlin

Ross RW, Peloquin SJ, Hougas RW (1964) Fertility of hybrids from *Solanum phureja* and haploid *S. tuberosum* matings. Eur Potato J 7:81–89

Rouppe van der Voort JR, Van Eck HJ, Draaistra J et al (1998) An online catalogue of AFLP markers covering the potato genome. Mol Breed 4:73–77

Samitsu Y, Hosaka K (2002) Molecular marker analysis of 24-and 25-chromosome plants obtained from *Solanum tuberosum* L. subsp. *andigena* ($2n = 4x = 48$) pollinated with a *Solanum phureja* haploid inducer. Genome 45:577–583

Sanetomo R, Gebhardt C (2015) Cytoplasmic genome types of European potatoes and their effects on complex agronomic traits. BMC Plant Biol 15:162. https://doi.org/10.1186/s12870-015-0545-y

Sanetomo R, Akino S, Suzuki N et al (2014) Breakdown of a hybridization barrier between *Solanum pinnatisectum* Dunal and potato using the S locus inhibitor gene (Sli). Euphytica 197:119–132

Santini M, Camadro EL, Marcellán ON et al (2000) Agronomic characterization of diploid hybrid families derived from crosses between haploids of the common potato and three wild Argentinean tuber-bearing species. Am J Potato Res 77:211–218

Simko I, Vreugdenhil D, Jung CS et al (1999) Similarity of QTLs detected for in vitro and greenhouse development of potato plants. Mol Breed 5:417–428

Simmonds NW (1993) Introgression and incorporation: strategies for the use of crop genetic resources. Biol Rev 68:539–562

Singsit C, Hanneman RE Jr (1990) Isolation of viable inter-EBN hybrids among potato species using double pollinations and embryo culture. Am Potato J 67:578

Singsit C, Veilleux RE (1991) Chloroplast density in guard cells of leaves of anther-derived potato plants grown in vitro and in vivo. HortScience 26:592–594

Sopory SK (1977) Differentiation in callus from cultured anthers of dihaploid clones of *Solanum tuberosum*. Z Pflanzenphysiol 82:88–91

Straadt IK, Rasmussen OS (2003) AFLP analysis of *Solanum phureja* DNA introgressed into potato dihaploids. Plant Breed 122:352–356

Summers D, Grun P (1981) Reproductive isolation barriers to gene exchange between *Solanum chacoense* and *S. commersonii* (Solanaceae). Am J Bot 68:1240–1248

Swaminathan MS, Howard HW (1953) The cytology and genetics of the potato (*Solanum tuberosum*) and related species. Biblphia Genet 16:1–192

Swanson-Wagner RA, Eichten SR, Kumari S et al (2010) Pervasive gene content variation and copy number variation in maize and its undomesticated progenitor. Genome Res 20:1689–1699

Tan EH, Henry IM, Ravi M et al (2015) Catastrophic chromosomal restructuring during genome elimination in plants. Elife 4:e06516

Tiemens-Hulscher M, Delleman J, Eising J, Lammerts van Beuren ET (2013) Potato breeding. Aardappelwereld, The Hague

Tovar-Méndez A, Kumar A, Kondo K et al (2014) Restoring pistil-side self-incompatibility factors recapitulates an interspecific reproductive barrier between tomato species. Plant J 77:727–736

Tsai H, Missirian V, Ngo KJ et al (2013) Production of a high-efficiency TILLING population through polyploidization. Plant Physiol 161:1604–1614

Tucci M, Carputo D, Bile G et al (1996) Male fertility and freezing tolerance of hybrids *involving Solanum tuberosum* haploids and diploid Solanum species. Potato Res 39:345–353

Van Eck HJ, Jacobs JM, Van den Berg PM et al (1994a) The inheritance of anthocyanin pigmentation in potato (*Solanum tuberosum* L.) and mapping of tuber skin colour loci using RFLPs. Heredity 73:410–421

Van Eck HJ, Jacobs JM, Stam P et al (1994b) Multiple alleles for tuber shape in diploid potato detected by qualitative and quantitative genetic analysis using RFLPs. Genetics 137:303–309

Veilleux RE (1996) Haploidy in important crop plants—potato. In: Mohan Jain S, Sopory SK, Veilleux RE (eds) In vitro haploid production in higher plants, vol 3. Kluwer Academic, Dordrecht, pp 37–39

Velásquez AC, Mihovilovich E, Bonierbale M (2007) Genetic characterization and mapping of major gene resistance to potato leafroll virus in *Solanum tuberosum* ssp. *andigena*. Theor Appl Genet 114:1051–1058

Visser RGF, Bachem CWB, Boer JM et al (2009) Sequencing the potato genome: outline and first results to come from the elucidation of the sequence of the world's third most important food crop. Am J Potato Res 86:417–429

Watanabe K (2015) Potato genetics, genomics, and applications. Breed Sci 65:53–68

Watanabe K, Orrillo M, Iwanaga M et al (1994) Diploid potato germplasm derived from wild and landrace germplasm resources. Am Potato J 71:599–604

Waugh R, Baird E, Powell W (1992) The use of RAPD markers for the detection of gene introgression in potato. Plant Cell Rep 11:466–469

Wilkinson MJ, Bennett ST, Clulow SA et al (1995) Evidence for somatic translocation during potato dihaploid induction. Heredity 74:146–151

Yermishin AP, Polyukhovich YV, Voronkova EV et al (2014) Production of hybrids between the 2EBN bridge species *Solanum verrucosum* and 1EBN diploid potato species. Am J Potato Res 91:610–617

Zamir D (2001) Improving plant breeding with exotic genetic libraries. Nat Rev Genet 2:983–989

Zhang Y, Zhao Z, Xue Y (2009) Roles of proteolysis in plant self-incompatibility. Annu Rev Plant Biol 60:21–42

Zhu J, Pearce S, Burke A et al (2014) Copy number and haplotype variation at the VRN-A1 and central FR-A2 loci are associated with frost tolerance in hexaploid wheat. Theor Appl Genet 127:1183–1197

Permissions

All chapters in this book were first published in TPCIANSCH, by Springer; hereby published with permission under the Creative Commons Attribution License or equivalent. Every chapter published in this book has been scrutinized by our experts. Their significance has been extensively debated. The topics covered herein carry significant information for a comprehensive understanding. They may even be implemented as practical applications or may be referred to as a beginning point for further studies.

The contributors of this book come from diverse backgrounds, making this book a truly international effort. We would like to thank all the contributing authors for lending their expertise to make the book truly unique. They have played a crucial role in the development of this book. Without their invaluable contributions this book wouldn't have been possible. They have made vital efforts to compile up to date information on the varied aspects of this subject to make this book a valuable addition to the collection of many professionals and students.

This book was conceptualized with the vision of imparting up-to-date and integrated information in this field. To ensure the same, a matchless editorial board was set up. Every individual on the board went through rigorous rounds of assessment to prove their worth. After which they invested a large part of their time researching and compiling the most relevant data for our readers.

The editorial board has been involved in producing this book since its inception. They have spent rigorous hours researching and exploring the diverse topics which have resulted in the successful publishing of this book. They have passed on their knowledge of decades through this book. To expedite this challenging task, the publisher supported the team at every step. A small team of assistant editors was also appointed to further simplify the editing procedure and attain best results for the readers.

Apart from the editorial board, the designing team has also invested a significant amount of their time in understanding the subject and creating the most relevant covers. They scrutinized every image to scout for the most suitable representation of the subject and create an appropriate cover for the book.

The publishing team has been an ardent support to the editorial, designing and production team. Their endless efforts to recruit the best for this project, has resulted in the accomplishment of this book. They are a veteran in the field of academics and their pool of knowledge is as vast as their experience in printing. Their expertise and guidance has proved useful at every step. Their uncompromising quality standards have made this book an exceptional effort. Their encouragement from time to time has been an inspiration for everyone.

The publisher and the editorial board hope that this book will prove to be a valuable piece of knowledge for students, practitioners and scholars across the globe.

Index

Printed in the USA
CPSIA information can be obtained
at www.ICGtesting.com
LVHW082205201023
761596LV00009B/1574